Discrete Mathematics

Graph Algorithms, Algebraic Structures, Coding Theory, and Cryptography

*தொட்டனைத் தூறும் மணற்கேணி மாந்தர்க்குக்
கற்றனைத் தூறும் அறிவு.*

திருவள்ளுவர்

Translation:

*In sandy soil, when deep you delve, you reach the springs below;
The more you learn, the freer streams of wisdom flow.*

Tamil Poet Tiruvalluvar(500 BC)

Discrete Mathematics

Graph Algorithms, Algebraic Structures, Coding Theory, and Cryptography

R. Balakrishnan
Bharathidasan University, Tiruchirappalli, Tamil Nadu, INDIA

Sriraman Sridharan
Laboratoire LAMPS, Département de Mathématiques et d'Informatique, Université de Perpignan Via Domitia, Perpignan, FRANCE

CRC Press is an imprint of the
Taylor & Francis Group, an **informa** business

CRC Press
Taylor & Francis Group
6000 Broken Sound Parkway NW, Suite 300
Boca Raton, FL 33487-2742

Printed on acid-free paper

International Standard Book Number-13: 978-0-8153-4739-2 (Hardback)

Library of Congress Cataloging-in-Publication Data

Names: Sridharan, Sriraman, author. | Balakrishnan, R. (Rangaswami), author.
Title: Discrete mathematics : graph algorithms, algebraic structures, coding theory, and cryptography / Sriraman Sridharan, R. Balakrishnan.
Description: Boca Raton : CRC Press, Taylor & Francis Group, 2019. | Includes bibliographical references and index.
Identifiers: LCCN 2019011934| ISBN 9780815347392 (hardback : alk. paper) | ISBN 9780429486326 (ebook)
Subjects: LCSH: Mathematics--Textbooks. | Computer science--Textbooks.
Classification: LCC QA39.3 .S7275 2019 | DDC 511/.1--dc23
LC record available at https://lccn.loc.gov/2019011934

Visit the Taylor & Francis Web site at
http://www.taylorandfrancis.com

and the CRC Press Web site at
http://www.crcpress.com

Printed and bound in Great Britain by
TJ International Ltd, Padstow, Cornwall

The first author, R. Balakrishnan, dedicates this book affectionately to his granddaughters:

Amirtha and Aparna

The second author, Sriraman Sridharan, dedicates this book in memory of his thesis supervisor:

Professor Claude Berge

Contents

List of Figures

List of Tables

Preface

Discrete Mathematics and Programming courses are all-pervasive in Computer Science, Mathematics, and Engineering curricula. This book is intended for undergraduate/graduate students of Computer Science, Mathematics, and Engineering. The book is in continuation of our first book, *Foundations of Discrete Mathematics with Algorithms and Programming* (CRC Press, 2018). The student with a certain mathematical maturity and who has programmed in a high-level language like C (although C is *not* such a high-level language) or Pascal can use this book for self-study and reference. A glance at the Table of Contents will reveal that the book deals with fundamental graph algorithms, elements of matrix theory, groups, rings, ideals, vector spaces, finite fields, coding theory, and cryptography. A number of examples have been given to enhance the understanding of concepts. The programming languages used are Pascal and C. The aim of the book is to bring together advanced Discrete Mathematics with Algorithms and Programming for the student community.

Scope of the Book

In Chapter 1 on Graph Algorithms I, we study various algorithms on graphs. Unfortunately, the graphs are not manipulated by their geometrical representation inside a computer. We start with two important representations of graphs, the adjacency matrix representation, and the adjacency list representation. Of course, we have to choose a representation so that the operations of our algorithm can be performed in order to minimize the number of elementary operations.

After seeing how the graphs are represented inside a computer, two minimum spanning tree algorithms in a connected simple graph with weights associated with each edge, one due to Prim and the other invented by Kruskal, are presented.

Then, shortest path problems in a weighted directed graph are studied. First, the single-source shortest path problem due to Dijkstra is presented. Next, the single-source shortest path algorithm for *negative* edge weights is given. Then, the all-pairs shortest path problem due to Floyd and the transitive closure algorithm by Warshall are given. Floyd's algorithm is applied to find the eccentricities of vertices, radius, and diameter of a graph.

Finally, we study a well-known graph traversal technique, called depth-first search. As applications of the graph traversal method, we study the algorithms to find connected components, biconnected components, strongly connected components, topological sort, and PERT (program evaluation and research technique). In the last subsection, we study the famous NP-complete problem, the traveling salesman problem (TSP) and present a brute force algorithm and two approximate algorithms.

In Chapter 2: Graph Algorithms II, we introduce another systematic way of searching a graph, known as breadth-first search. Testing if a given graph is geodetic and finding a bipartition of a bipartite graph are given as applications. Next, matching theory is studied in detail. Berge's characterization of a maximum matching using an alternating chain and the König-Hall theorem for bipartite graphs are proved. We consider matrices and bipartite graphs: Birkhoff-von-Neumann theorem concerning doubly stochastic matrices is proved. Then, the bipartite matching algorithm using the tree-growing procedure (Hungarian method) is studied. The Kuhn-Munkres algorithm concerning maximum weighted bipartite matching is presented. Flows in transportation networks are studied.

Chapter 3: Algebraic Structures I deals with the basic properties of the fundamental algebraic structures, namely, groups, rings, and fields. The sections on matrices deal with the inverse of a non-singular matrix, Hermitian and skew-Hermitian matrices, as well as orthogonal and unitary matrices. The sections on groups deal with Abelian and non-Abelian groups, subgroups, homomorphisms, and the basic isomorphism theorem for groups. We then pass on to discuss rings, subrings, integral domains, ideals, fields, and their characteristics. A number of illustrative examples are presented.

In Chapter 4 on Algebraic Structures II, we deal with vector spaces and their applications to solutions of systems of linear homogeneous and nonhomogeneous equations. To be specific, bases of a vector space, the dimension of a vector space, and linear independence of vectors are discussed. As applications of these concepts, solutions of linear equations over the real field are dealt with. Moreover, the LUP decomposition of a system of homogeneous/nonhomogeneous linear equations is discussed in detail. This is followed by illustrative examples. Finally, a detailed discussion on finite fields is presented.

Coding theory has its origin in communication engineering. It has been greatly influenced by mathematics. Chapter 5, Introduction to Coding Theory, provides an introduction to linear codes. In particular, we discuss generator matrices and parity-check matrices, Hamming codes, sphere packings and syndrome coding, cyclic codes, and dual codes. The algebraic concepts that are used in this chapter have all been discussed in Chapters 3 and 4.

Chapter 6 deals with cryptography. Cryptography is the science of transmitting messages in a secured way. Naturally, it has become a major tool in these days of e-commerce, defense, etc. To start with, we discuss some of the classical cryptosystems like the Caesar cryptosystem, the affine cryptosystem, cryptosystems using matrices, Vigenere ciphers, and the one-time pad. The most important and widely used cryptosystem is the RSA. We discuss this system as well as the ElGamal cryptosystem. RSA is built on very large prime numbers. So, this gives rise to the following natural question: Given a large positive integer, how do we test if the given number is prime or not? We briefly discuss the Miller-Rabin primality testing algorithm. This is a randomized probabilistic algorithm to test if a given number is prime or not. However, the problem of finding a deterministic polynomial-time algorithm to test if a given number is prime or not remained unsolved until the Agrawal-Kayal-Saxena (AKS) primality testing algorithm was proposed in 2002. We present this algorithm, its proof, and some illustrative examples.

Use of the Book

The instructor has a great deal of flexibility in choosing the material from the book. For example, the chapters on graph algorithms I and II may be suitable for a course on "graphs and algorithms." The chapters on algebraic structures I and II and the chapters on coding theory and cryptography may form a course on "Applied Algebra." Many illustrative examples have been given to help the understanding of the concepts. Algorithms are expressed as informal pseudo-codes and programs in Pascal/C. Of course, these can be easily translated into any language like C++, JAVA, etc. Exercises at the end of each chapter/section test the understanding of the concepts developed in the text.

We feel that the presentation of these chapters would go a long way in providing a solid foundation in Discrete Mathematics to the students of Mathematics, Computer Science, and Engineering.

Acknowledgment

Parts of this book have been taught in Indian and French universities. The authors thank N. Sridharan for going through some of the chapters and offering constructive suggestions, and Jaikumar Radhakrishnan and Arti Pandey for their inputs on Chapter 6 to cryptography. They also thank A. Anuradha, R. Dhanalakshmi, N. Geetha, and G. Janani Jayalakshmi for their help in typesetting. We take this opportunity to thank our institutions, Bharathidasan University, Tamil Nadu, India and Université de Perpignan Via Domitia, France, for their academic support. Our thanks are also due to the faculties in our departments whose encouragement proved vital to attain our goal. We also thank the four anonymous reviewers for suggesting some changes on the initial version of the book. Last but not least, we thank Aastha Sharma and Shikha Garg of CRC Press for their kind understanding of our problems and for their patience until the completion of our manuscript.

The second author (S.S.) expresses his deep gratitude to (Late) Professor K. R. Parthasarathy of the Indian Institute of Technology, Chennai, India, for introducing him to Graph Theory and guiding his Ph.D. thesis. He is also indebted to (Late) Professor Claude Berge, one of the greatest pioneers in Graph Theory and Combinatorics, who invited him to CAMS (Centre d'Analyse et de Mathématique Sociale) and guided his doctoral work in Paris. Claude Berge had been a source of immense inspiration to him. Special thanks are also due to Professors R. Balasubramanian (A. M. Jain College, TN, India), Philippe Chrétienne (Université de Pierre et Marie Curie), Robert Cori (Université de Bordeaux I), Alain Fougère (UPVD), Michel Las Vergnas (CNRS), UFR secretaries Mme. Fabienne Pontramont (UPVD), Mme. Dominique Bevilis (UPVD), Mircea Sofonea (Directeur de Laboratoire LAMPS, UPVD), Michel Ventou (UPVD), Annick Truffert (Dean of the Faculty of Sciences, UPVD). Many thanks are also due to my students of UPVD for their feedback. We are responsible for all the remaining errors, but still, we feel that the initial readers of this book could smoke out some more bugs. Though it is not in the Hindu custom to explicitly thank family members, he would like to break this tradition and thank his wife, Dr. Usha Sridharan, and his daughters, Ramapriya and Sripriya, for putting up with unusually prolonged absence, as well as Dheeraj.

The entire book was composed using the TeX and LaTeX systems developed by D. E. Knuth and L. Lamport.

The authors welcome corrections, comments, and criticisms from readers, which would be gratefully acknowledged. They can be sent by email to: rbsri2018@gmail.com.

R. Balakrishnan
Sriraman Sridharan
Tiruchirappalli, Tamil Nadu, India
Perpignan, France
August 2018

Authors

R. Balakrishnan studied at the Annamalai University, India for his Honours degree. He got his Ph.D. degree from the University of Maryland, USA. His fields of interest are combinatorial designs, graph colorings, and spectral graph theory. He is a coauthor for the books *A Textbook of Graph Theory*, Springer, Second Edition (2012) and *Foundations of Discrete Mathematics with Algorithms and Programming*, CRC Press (2018).

Sriraman Sridharan is a member of the LAMPS Research Laboratory and teaches in the Department of Computer Science of the Université de Perpignan Via Domitia, France. He has been with The Université de Bordeaux and The Université de Pierre et Marie Curie, Paris. He is a coauthor of the book *Foundations of Discrete Mathematics with Algorithms and Programming*, CRC Press (2018). He earned his Ph.D. in Paris under the guidance of Professor Claude Berge.

Chapter 1

Graph Algorithms I

> The aim of physical sciences is not the provision of pictures, but the discovery of laws governing the phenomena and the application of these laws to discover new phenomena. If a picture exists, so much the better. Whether a picture exists or not is only a matter of secondary importance.
>
> **P. Dirac**

In this chapter, we study various algorithms on graphs. Unfortunately, the graphs are not manipulated by their geometrical representation inside a computer. We start with two important representations of graphs: the adjacency matrix representation and the adjacency list representation. Of course, we have to choose a representation so that the operations of our algorithm can be performed in order to minimize the number of elementary operations.

After seeing how the graphs are represented inside a computer, two minimum spanning tree algorithms in a connected simple graph with weights associated to each edge, one due to Prim and the other invented by Kruskal are presented. Then, shortest path problems in a weighted directed graph are studied. First, the single-source shortest path problem due to Dijkstra is presented. As an application, an algorithm is given to test the bipartiteness of a graph. Next, a single-source shortest path algorithm for *negative* edge weights is given. Then, all-pairs shortest path problem due to Floyd and the transitive closure algorithm by Warshall are given. Floyd's algorithm is applied to find eccentricities of vertices, radius and diameter of a graph.

Finally, we study a well-known graph traversal technique, called *depth-first search.* As applications of the graph traversal method, we study the algorithms to find connected components, biconnected components, strongly connected components, topological sort and program evaluation and research technique (PERT). In the last subsection, we study the famous NP-complete problem, traveling salesman problem (TSP) and present the brute-force algorithm and two approximate algorithms. For basic properties of graphs and digraphs, the reader may see [6].

1.1 Representation of Graphs

Consider the graph of Figure 1.1 represented geometrically [3].

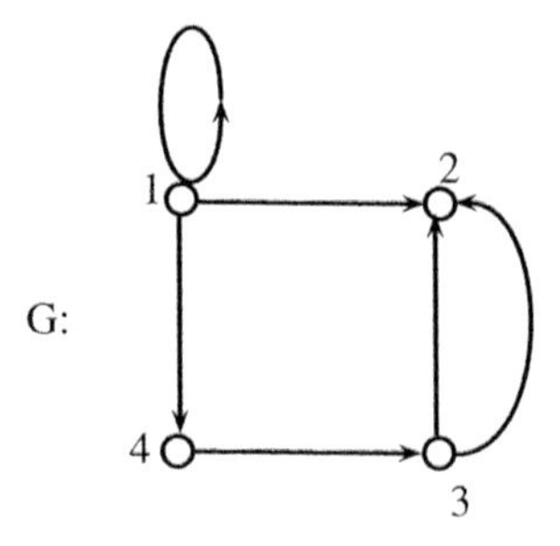

FIGURE 1.1: Geometric representation of a graph.

This graph of Figure 1.1 on 4 vertices is represented by the following 4×4 matrix M: (For a formal treatment on matrices, see Chapter 3.).

$$M = \begin{array}{c} \\ 1 \\ 2 \\ 3 \\ 4 \end{array} \begin{array}{c} \begin{array}{cccc} 1 & 2 & 3 & 4 \end{array} \\ \begin{pmatrix} 1 & 1 & 0 & 1 \\ 0 & 0 & 0 & 0 \\ 0 & 2 & 0 & 0 \\ 0 & 0 & 1 & 0 \end{pmatrix} \end{array}$$

Here, 4 is the number of vertices of the graph. The (i, j) entry of the above matrix M is simply the number of *arcs* with its initial vertex at i and the terminal vertex at j. This matrix is called the adjacency matrix of the graph of figure.

More generally, for a n vertex graph G with vertex set $X = \{1, 2, \ldots, n\}$, the *adjacency matrix* of G is the $n \times n$ matrix $M = (m_{ij})$ where

$$m_{ij} = \text{number of arcs of the form } (i, j).$$

Memory space for the adjacency matrix: Since an $n \times n$ matrix has exactly n^2 entries, the memory space necessary for the adjacency matrix representation of a graph is of order $O(n^2)$. The time complexity of initializing a graph by its adjacency graph is $O(n^2)$. This may preclude algorithms on graphs whose complexities are of order strictly less than n^2.

Properties of the adjacency matrix: Let M denote the adjacency matrix of a graph with vertex set $X = \{1, 2, \ldots, n\}$. Then, by the definition of the adjacency matrix, we have the following properties:

1. The sum of the entries of the ith row of M is equal to the out-degree of the vertex i.

2. The sum of the entries of the jth column of M is equal to the in-degree of the vertex j.

3. The sum of all the entries of the matrix M is the number of arcs of the graph.

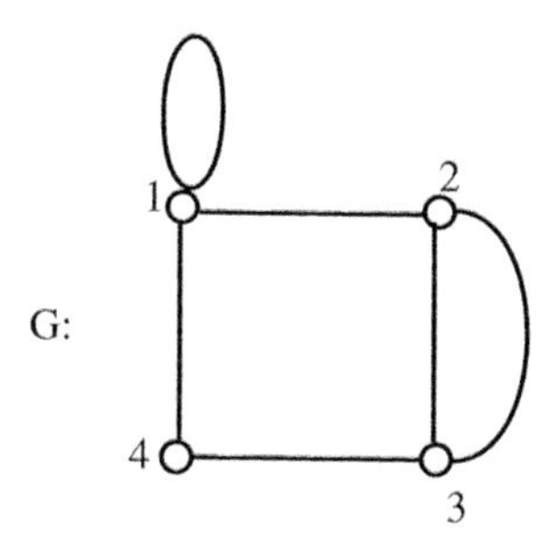

FIGURE 1.2: A multigraph.

Adjacency matrix representation of a multigraph:
Consider the following multigraph of Figure 1.2 which is obtained by ignoring the orientations of the arcs of the graph of Figure 1.1.

Its adjacency matrix is the 4×4 matrix M (see below) with its (i, j) entry, the number of *edges* from the vertex i to the vertex j.

$$M = \begin{array}{c} \\ 1 \\ 2 \\ 3 \\ 4 \end{array} \begin{array}{c} \begin{array}{cccc} 1 & 2 & 3 & 4 \end{array} \\ \begin{pmatrix} 1 & 1 & 0 & 1 \\ 1 & 0 & 2 & 0 \\ 0 & 2 & 0 & 1 \\ 1 & 0 & 1 & 0 \end{pmatrix} \end{array}$$

More generally, the adjacency matrix of a multigraph with vertex set $X = \{1, 2, \ldots, n\}$ is an $n \times n$ matrix $M = (m_{ij})$ where

$$m_{ij} = \text{The number of edges between the vertices } i \text{ and } j.$$

Note that the adjacency matrix of a multigraph is a symmetric matrix. The following properties are easily proved for the adjacency matrix M of a *simple* G.

1. The sum of the entries of the ith row of M is the degree of the vertex i.

2. The sum of all the entries of the matrix M is twice the number of edges.

Incidence matrix representation of a graph:
Consider a directed graph $G = (X, U)$ with vertex set $X = \{1, 2, \ldots, n\}$ of n elements and the arc sequence $U = \{u_1, u_2, \ldots, u_m\}$ of m elements. Then, the incidence matrix M of the graph G is the $n \times m$ matrix $(m_{ij})_{n \times m}$ where

$$m_{ij} = \begin{cases} 1 & \text{if } i \text{ is the initial vertex of } u_j \\ -1 & \text{if } i \text{ is the final vertex of } u_j \\ 2 & \text{if } u_j \text{ is a loop at vertex } i \\ 0 & \text{otherwise} \end{cases}$$

Note that the first subscript i in m_{ij} indicates the vertex i and the second subscript j indicates the arc u_j.

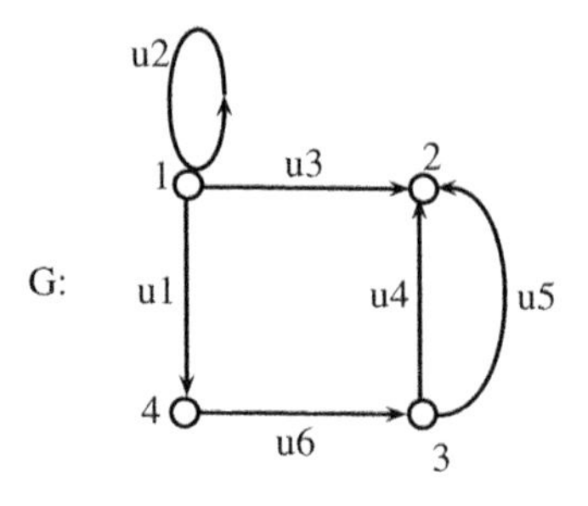

FIGURE 1.3: Incidence matrix of a graph.

The following example clarifies the definition.

Example 1.1: Incidence Matrix

Consider the directed graph G of Figure 1.3. This graph has four vertices and six arcs. The incidence matrix of the graph G is

$$M = \begin{array}{c} \\ 1 \\ 2 \\ 3 \\ 4 \end{array} \begin{array}{c} \begin{array}{cccccc} u_1 & u_2 & u_3 & u_4 & u_5 & u_6 \end{array} \\ \begin{pmatrix} 1 & 2 & 1 & 0 & 0 & 0 \\ 0 & 0 & -1 & -1 & -1 & 0 \\ 0 & 0 & 0 & 1 & 1 & -1 \\ -1 & 0 & 0 & 0 & 0 & 1 \end{pmatrix} \end{array}$$

The following are properties of the incidence matrix of a directed graph:

1. The sum of entries of every column except the ones which represent loops is 0.
2. The sum of entries of every column representing a loop is 2.
3. The sum of entries of the ith row (not containing the entry 2) is $d^+(i) - d^-(i)$.

Incidence matrix representation of a multigraph:
Consider a multigraph $G = (X, E)$ with $X - \{1, 2, \ldots, n\}$ and with edge sequence $E = (e_1, e_2, \ldots, e_m)$. The incidence matrix M is an $n \times m$ matrix $(m_{ij})_{n \times m}$ where

$$m_{ij} = \begin{cases} 1 & \text{if vertex } i \text{ is incident with } e_j \\ 2 & \text{if } e_j \text{ is a loop at vertex } i \\ 0 & \text{otherwise} \end{cases}$$

Example 1.2

Consider the graph G of Figure 1.3 by ignoring the orientations of the arcs and taking $u_j = e_j$. Then, the incidence matrix M of G is

$$M = \begin{array}{c} \\ 1 \\ 2 \\ 3 \\ 4 \end{array} \begin{array}{c} \begin{array}{cccccc} e_1 & e_2 & e_3 & e_4 & e_5 & e_6 \end{array} \\ \begin{pmatrix} 1 & 2 & 1 & 0 & 0 & 0 \\ 0 & 0 & 1 & 1 & 1 & 0 \\ 0 & 0 & 0 & 1 & 1 & 1 \\ 1 & 0 & 0 & 0 & 0 & 1 \end{pmatrix} \end{array}$$

The following are properties of the incidence matrix of a multigraph:

1. The sum of every column entry of an incidence matrix is 2.
2. The sum of the entries of the i-th row of an incidence matrix is the degree of the vertex i which is $d(i)$.

Space complexity to represent a graph in the form of incidence matrix: There are mn entries in the incidence matrix and hence the space complexity is $O(mn)$.

A formal study of matrices is done in Chapter 3.

Adjacency list representation of a digraph:

In the adjacency matrix representation of a graph G, an (i, j) entry *zero* of the matrix represents the absence of arc (i, j) in the graph G. The *adjacency list* representation can be considered as a "condensed form" of the adjacency matrix representation *omitting the zero entries.* Consider the graph of Figure 1.4.

This graph of 4 vertices is represented by 4 adjacency lists: $succ(1)$, $succ(2)$, $succ(3)$, $succ(4)$ where $succ(1) = (1, 2, 4)$, $succ(2) = ()$, the null list, $succ(3) = (2, 2)$, $succ(4) = (3)$. More generally, an n vertex graph G is represented by n lists $succ(1)$, $succ(2), \ldots, succ(n)$ where a vertex j appears k times in the list

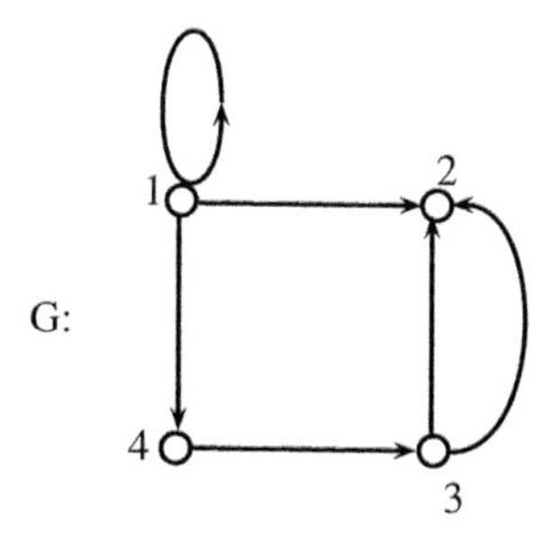

FIGURE 1.4: A graph and its adjacency lists.

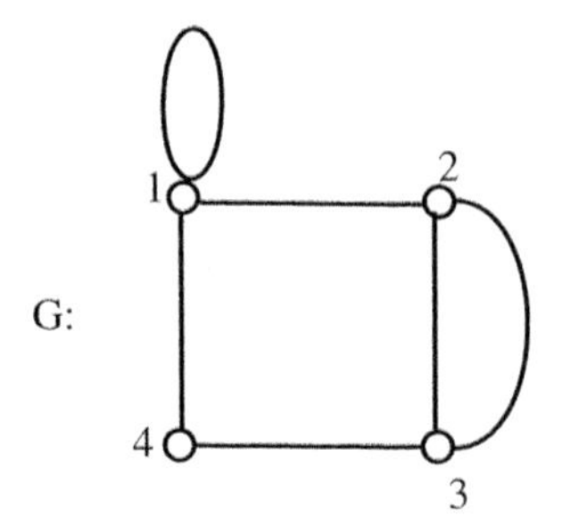

FIGURE 1.5: A multigraph and its adjacency lists.

$succ(i)$ if and only if there are k arcs with initial vertex i and final vertex j, that is, the multiplicity of the arc (i, j) is k.

Adjacency list representation of a multigraph:

> When the sage points his finger to the Moon, the novice looks at its finger.
>
> Tamil Proverb

A multigraph of n vertices is represented by n adjacency lists $adj(1), adj(2), \ldots,$ $adj(n)$ where the list $adj(i)$ consists of all vertices j which are joined to the vertex i. Note that the vertex j appears k times in the list $adj(i)$ if the vertices i and j are joined by k multiple edges. Further, if ij is an edge, then the vertex j is in the list $adj(i)$ and the vertex i is in the list $adj(j)$ (see Figure 1.5).

This graph of 4 vertices is represented by 4 adjacency lists: $succ(1), succ(2),$ $succ(3), succ(4)$ where $succ(1) = (1, 2, 4)$, $succ(2) = (1, 3, 3)$, $succ(3) =$ $(2, 2, 4)$, $succ(4) = (1, 3)$.

Figure 1.6 gives a linked lists representation of graph of Figure 1.5.

Space needed for adjacency linked list representation of a digraph: Consider a graph G with n vertices and m arcs. It is represented by n adjacency lists $succ(1), succ(2), \ldots, succ(n)$. The number of elements in the list $succ(i)$ is clearly the out-degree of the vertex i. In the linked list representation of $succ(i)$, each vertex in the list $succ(i)$ is stored in a node consisting of two fields: a vertex field and a pointer field. Counting each node as two units of space, one for each field, we need $2d^+(i) + 1$ space to represent $succ(i)$ where $d^+(i)$ is the out-degree of the vertex i. Note that the pointer $succ(i)$ needs one unit of space. Hence the total space needed to represent the graph is

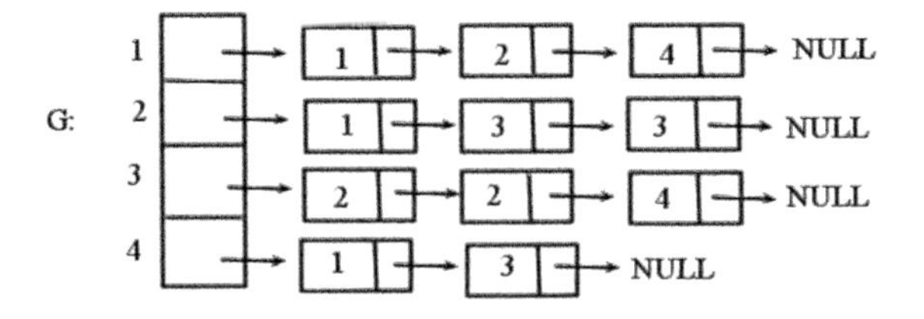

FIGURE 1.6: Linked lists representation of graph of Figure 1.5.

$$\sum_{i=1}^{n}(2d^{+}(i)+1) = n + \sum_{i=1}^{n} 2d^{+}(i) = n + 2m = O(n+m).$$

(since the sum of the out-degrees of the vertices of a graph is equal to the number of arcs).

Hence, we see that the adjacency list representation uses only linear space $O(m+n)$, whereas the adjacency matrix needs a quadratic space $O(n^2)$.

Which representation to choose: adjacency matrix or adjacency linked list?: The answer to this question depends on the operations performed by the algorithm at hand. If a graph is "sparse," that is, the graph does not have too many arcs, then the adjacency list representation may be suitable. Suppose an algorithm on graph frequently needs to test the existence of an arc from a vertex to another vertex, then the adjacency matrix will be appropriate since accessing an arbitrary entry of a matrix can be done in constant time. With the adjacency matrix, since the initialization itself takes $O(n^2)$ time, the importance of algorithms may be diminished with complexity $O(n \log n)$ or $O(n^{3/2})$.

Let us write a program in C which reads a directed graph G represented by a linked list. The program finally writes the successors of each vertex (see Figure 1.6).

```
#include <stdio.h>
#define max_n = 20 /* maximum number of vertices */
#define max_m = 100 /* maximum number of  arcs */
struct node
      {int v; struct node *next;};/* v for vertex */
int n, m; /*n = number of vertices. m = number of arcs */
struct node *succ[max_n]; /* graph is represented as an
                           array of pointers */
/* We work with a global graph.Otherwise the graph should be
declared as a variable parameter*/
void adj_list ( )
{
  int i, a, b; /* i for the loop.(a,b) is an arc.*/
  struct node *t; /* t, a temporary pointer */
  printf("Enter two integers for n and m\n");
  scanf(%d %d\n", &n, &m); /* read n and m */
  /* initialize the graph with n vertices and 0 arcs */
  for (i = 1; i <= n; i++) succ[i] = NULL;
  /* read the m arcs */
  for (i = 1; i <= m; i++)
  {
    printf("Enter the arc number %d ", i);
    scanf("%d %d \n", &a, &b);/* un arc is an ordered pair of
                             vertices. b will be in succ(a) */
    /* create a node referenced by t*/
    t = (struct node *) malloc(sizeof *t);
```

```
    t->v = b; /* assign b to v field of t-> */
    /* attach t-> at the head of list succ[a] */
    t->next = succ[a];
    succ[a] = t;
  }/*for*/
}/*adj_list*/

void print_list ( )
{
   /* print_list writes the list of successors of
   each vertex of the graph G*/
   int i; /* i for the loop */
   struct node *t; /* t, a temporary pointer */
   for (i = 1; i <= n; i++)
   {
     /* write the list succ[i]*/
     t = succ[i];
     if (t == NULL)
     printf(" No successors of %d\n ", i);
     else
     {
       printf(" The successors of %d are :", i);
       /* scan the list succ[i] and write the v fields of nodes */
       while (t != NULL)
       {
         printf("%d ", t->v);
         t = t->next; /* move t to next node */
       }/*while*/
       printf("\n");
     }/*else*/
   }/*for*/
}/* print_list*/

int main( )
{
  adj_list ( ); /* call */
  print_list ( ); /* call */
}
```

Let us write a program in C to represent a multigraph G with adjacency matrix. The Program finally prints the adjacency matrix.

```
#include <stdio.h>
#define max_n = 20 /* maximum number of vertices*/
int i, j, n, m;/* n, the number of vertices. m, the number of edges
                i, j for the loops*/
```

```
int adj[max_n][max_n];/* The graph is represented by the
                         matrix adj. We don't use row 0 and column 0*/
/* We work with a global graph.Otherwise the graph should be
declared as a variable parameter*/
void adj_matrix ( )
{
  int a, b; /* ab is an edge. An edge is an unordered pair*/
  printf("Enter two integers for n and m\n");
  scanf("%d %d\n", &n,&m);
  /* initialize the graph with n vertices and 0 edges */
  for ( i =1; i <= n; i++)
    for ( j= 1; j <= n; j++)
      adj[i][j] = 0;
  /* end of initialization */
  /* read the m edges*/
  for (i = 1; i <= m; i++)
  {
    printf("Enter the edge number %d " , i);
    scanf("%d %d \n", &a, &b);
    adj[a][b]++;adj[b][a]++;
  }/*for*/
}

void print_adj_matrix ( )
{
  int i, j;/* i, j for the loops */
  for (i = 1; i <= n; i++)
  {  for( j =1; j <= n; j++)
        printf("%d ", adj[i][j]);
     printf("\n");
  }
}
int main ( )
{
  adj_matrix( );/* call */
  print_adj ( ); /* call */
}
```

1.2 Minimum Spanning Tree Algorithms

A *free tree* or simply a *tree* is a connected graph without an elementary cycle, that is, a tree is a connected *acyclic graph*. Hence, a tree does not contain loops or multiple edges, that is, a tree is a simple graph. In the following figure, a tree with ten vertices is drawn (see Figure 1.7).

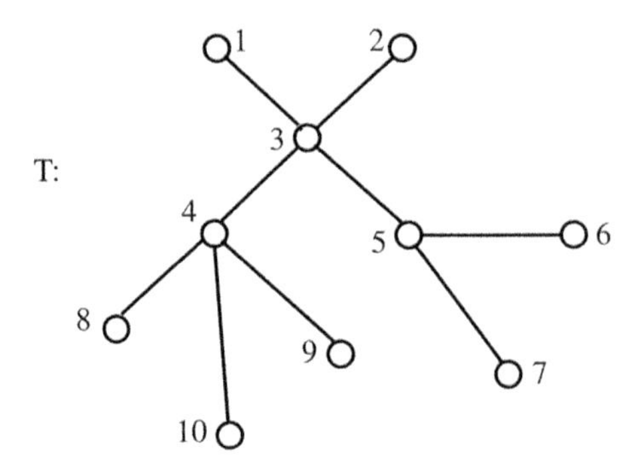

FIGURE 1.7: A tree.

In the tree of Figure 1.7 with ten vertices, we observe the following properties:

1. There is a unique elementary walk, that is, a path between any two vertices.
2. The number of edges is exactly one less than the number of vertices.

We now prove the above two properties for any general tree.

Proposition 1.1. *In any tree, there is a unique path joining any two vertices.*

Proof. We shall prove the proposition by contradiction. If not, consider two different paths P_1 and P_2 joining the vertices x and y. If the paths P_1 and P_2 are vertex-disjoint, except of course the initial and final vertices, then clearly the union of the paths $P_1 \cup P_2$ is an elementary cycle in the tree, which is a contradiction. Otherwise, let $z(\neq y)$, z'(possibly x) be the last and next-to-last vertices common to the paths P_1 and P_2, respectively. Then, the union of the two subpaths of the paths P_1 and P_2 from z' to z is an elementary cycle, a contradiction. Hence the result. □

For the second property, we need the following simple lemma.

Lemma 1.1. *Every simple graph with the degree of each vertex at least two contains an elementary cycle.*

Proof. Consider a *longest* possible path P in the graph. (The length of a path is the number of edges in the path.) Let x_0 be the initial vertex of the path $P = (x_0, x_1, \ldots, x_l)$ (the length of the path P is l) and consider the vertices of the graph adjacent to the initial vertex x_0. Since the degree of $x_0 \geq 2$, there is a vertex $y \neq x_1$ such that $x_0 y$ is an edge of the graph.

If $y \neq x_i$ for $2 \leq i \leq l$, then we have a path $(y, x_0, x_1, \ldots, x_l)$ whose length $l+1$ exceeds that of P, a contradiction.

If $y = x_i$ for some i with $2 \leq i \leq l$, then $(x_0, x_1, \ldots, x_i, x_0)$ (since $x_0 y$ is an edge) is an elementary cycle. □

Proposition 1.2. *In any tree with n vertices, the number of edges m is exactly $n-1$.*

Proof. The proof proceeds by induction on the number of vertices of the tree.
Basis: If the number of vertices $n = 1$, then the tree consists of only one vertex and no edges, that is, the number of edges $m = 0$. Hence the proposition is trivially satisfied.

Induction hypothesis: Suppose the proposition is true for all trees with the number of vertices n at least two.

We shall prove the proposition for trees with $n+1$ vertices. Consider a tree T with $n+1$ vertices. We claim that the tree T contains a vertex of degree *one*. If not, the degree of each vertex is at least two and by Lemma 1.1, T contains a cycle, a contradiction. Let x be a vertex of degree one in the tree T. Then, the vertex deleted graph $T-x$ is still a tree. Note that T and $T-x$ differ by exactly one vertex x and exactly one edge, the edge incident with x. Now $T-x$ is a tree on n vertices and the number of edges m of the tree $T-x$, verifies, by induction hypothesis, the following equation,

$$m = n - 1.$$

Adding one on both sides, we get $m+1 = (n+1)-1$, that is, the number of edges of T is exactly one less than the number of vertices of T. Thus, the proof is complete. □

Spanning tree of a connected graph:
Consider a connected graph G. A *spanning* subgraph of G, which is a tree, is called a *spanning tree* of G. In a connected graph, a spanning tree always exists as we shall see below:

Since a tree is acyclic, we intuitively feel that if we remove all the cycles of the graph G without disconnecting G, we will be left with a spanning tree. In fact, this is the case.

Consider any elementary cycle of G and delete exactly one of its edges. This results in a *connected* spanning subgraph G_1 of G, since in an elementary cycle, between any two vertices of the cycle, there are exactly two different internally vertex-disjoint paths. If the resulting graph is a tree, then we have obtained a desired spanning tree. Otherwise, we find an elementary cycle of G_1 and delete one of its edges. This gives us a connected spanning graph G_2. If G_2 is a tree, then G_2 is a desired spanning tree of G. Otherwise, we continue the procedure as before till a spanning tree of G is found. If a graph G contains a spanning tree, then clearly it must be connected. Thus, we have proved the following result.

Theorem 1.1. *A graph G has a spanning tree if and only if it is connected.*

1.2.1 Prim's minimum spanning tree algorithm

Input: The input of this algorithm is a connected simple graph G with weights associated to each edge of the graph (see Figure 1.8). Such graphs are also

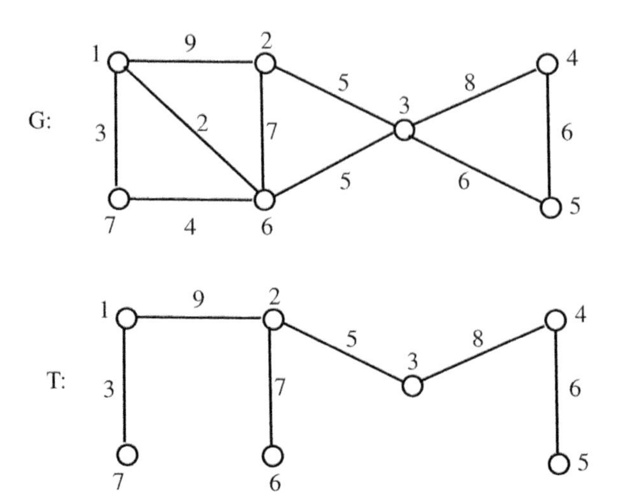

FIGURE 1.8: A weighted graph G and its spanning tree T.

called *networks*. The vertices are interpreted as cities and the edges as highways connecting its end vertices. The weight associated with an edge is interpreted as the cost of constructing a highway connecting its end vertices (cities). The total cost of the network is the sum of the costs on each edge of G. For example, the total cost of the figure G is 55 (see Figure 1.8).

How to represent a weighted graph?:
One way to represent a weighted graph is by *weighted adjacency matrix* $M = (m_{ij})_{n \times n}$ where

$$m_{ij} = \begin{cases} \text{weight of the edge } ij & \text{if } ij \text{ is an edge of the graph} \\ 0 & \text{if } i = j \\ \infty & \text{if } ij \text{ is not an edge of the graph.} \end{cases}$$

In the computer implementation, the ∞ is replaced by, for example, "INT_MAX" in the language C if the weights are integers, "maxint" in the language Pascal, or a very large number which will not be a *"legal"* cost of constructing a highway.

The weighted graph G of Figure 1.8 is represented by its weighted adjacency matrix as follows (the weight of a missing edge is taken as 100):

$$M = \begin{array}{c} \\ 1 \\ 2 \\ 3 \\ 4 \\ 5 \\ 6 \\ 7 \end{array} \begin{array}{c} \begin{array}{ccccccc} 1 & 2 & 3 & 4 & 5 & 6 & 7 \end{array} \\ \begin{pmatrix} 0 & 9 & 100 & 100 & 100 & 2 & 3 \\ 9 & 0 & 5 & 100 & 100 & 7 & 100 \\ 100 & 5 & 0 & 8 & 6 & 6 & 100 \\ 100 & 100 & 8 & 0 & 6 & 100 & 100 \\ 100 & 100 & 6 & 6 & 0 & 100 & 100 \\ 2 & 7 & 5 & 100 & 100 & 0 & 4 \\ 3 & 100 & 100 & 100 & 100 & 4 & 0 \end{pmatrix} \end{array}$$

The other way of representing a weighted graph is by *weighted adjacency lists*. In this representation, we introduce a new field called *weight* in each node, in addition to the two existing fields, namely, the field containing the vertex and the field containing the pointer. In the weight field, the weight of the edge ij will be stored, if the content of the vertex field is j and the node in question belongs to the list $adj(i)$.

Output: A *minimum spanning tree* of G, that is, a spanning tree T of the graph G such that the sum of the costs of all the edges of the tree T is a *minimum* (see Figure 1.8). In Figure 1.8, a spanning tree T of G is depicted and its total cost is 38.

Algorithm 1.1 (Brute-force method). *One way to find a minimum cost spanning tree in the graph G is the following exhaustive search.*

Generate all possible spanning trees of G and find the total cost of each of these spanning trees. Finally, choose one tree for which the total cost is minimum. This procedure will work "quickly" if the number of vertices of the input graph G is sufficiently small. If the number of vertices is *not* sufficiently small, this procedure takes an enormous amount of time, that is, exponential time since by Caley's theorem [1] there are n^{n-2} possible non-identical trees on n given vertices.

Algorithm 1.2 (Prim's algorithm). *This is a polynomial time algorithm. Prim's algorithm, though it is greedy, provides us with an exact solution. We shall now describe what is meant by a greedy algorithm.*

A Greedy algorithm: A greedy algorithm is an algorithm which "hopes" to find a *global* optimum by choosing at each iteration of the algorithm a local optimum. A greedy algorithm sees only "short-term" gains. More generally, a heuristic is an algorithm which "quickly" produces a suboptimal solution but not necessarily an optimal solution. The word "quickly" means "in polynomial time."

For example, greedy is a heuristic. Greedy does not always give an optimal solution, but it gives "somewhat" of an acceptable solution. But in the case of Prim's algorithm, greedy provides us with an exact solution. First of all, we shall illustrate Prim's algorithm with our graph of Figure 1.8.

Observation 1.1. *Trees can be grown by starting with one vertex (initialization) and adding edges and vertices one-by-one (iteration).*

According to Observation 1.1, we start with any vertex, say the vertex 1. The vertex 1 forms a tree T by itself. The algorithm chooses an edge e_1 of the graph G such that one end of the edge e_1 lies in tree T and the other end lies outside of the tree T (this is to avoid cycle in T) with the weight of e_1 *minimum*. Then, we add the edge e_1 to the tree T. In our figure, the edge $e_1 = 16$. Then, we select an edge e_2 such that one end of the edge e_2 lies in the tree T under construction, and the other end lies outside of the tree

TABLE 1.1: Execution of Prim's algorithm on graph of Figure 1.8

Iteration number	$S \neq X$?	s	t	$S \leftarrow S \cup \{t\}$	$T \leftarrow T \cup \{st\}$
0 (Initial)	–	–	–	$\{1\}$	$\emptyset$
1	yes	1	6	$\{1, 6\}$	$\{16\}$
2	yes	1	7	$\{1, 6, 7\}$	$\{16, 17\}$
3	yes	6	3	$\{1, 6, 7, 3\}$	$\{16, 17, 63\}$
4	yes	3	2	$\{1, 6, 7, 3, 2\}$	$\{16, 17, 63, 32\}$
5	yes	3	5	$\{1, 6, 7, 3, 2, 5\}$	$\{16, 17, 63, 32, 35\}$
6	yes	5	4	$\{1, 6, 7, 3, 2, 5, 4\}$	$\{16, 17, 63, 32, 35, 54\}$
Exit the loop	no				

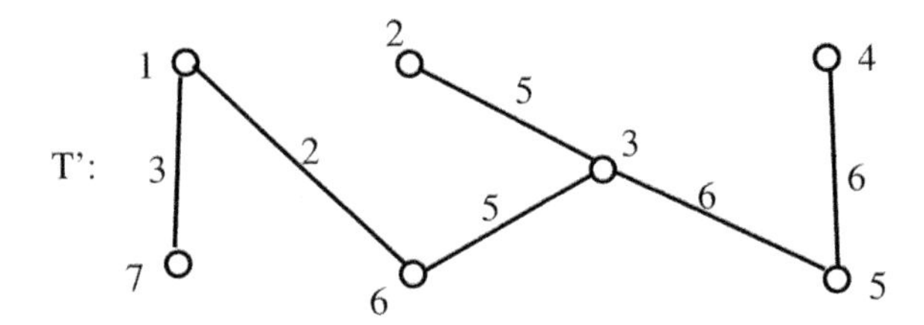

FIGURE 1.9: A minimum spanning tree T' of G.

T with the weight of e_2 minimum. In our case, the edge $e_2 = 17$ and so on. The different steps of the tree-growing procedure are illustrated in Table 1.1. In Table 1.1, the set S denotes the vertices of the tree T under construction. T denotes the set of edges of the tree which is grown by the algorithm, s and t denote vertices of the graph G with $s \in S$ and $t \in X \setminus S$, and X represents the set of vertices of the graph G. The procedure ends as soon as the tree T under construction contains all of the vertices of the graph G. The edges of a minimum spanning tree obtained are found in the 6th column of iteration number 6 in Table 1.1. The minimum spanning tree obtained by Table 1.1 is shown in Figure 1.9 and its cost is the sum of the costs of its edges which is 27. Now let us write the algorithm in pseudo-language.

```
procedure prim(var G : Graph; var T: Set of Edges);
(* The procedure prim takes as input a connected weighted
graph G and returns as output a minimum spanning tree T of G.
Note that the representation of the graph G as well as the
tree T are left unspecified*)
var S : Set of Vertices;(* representation of S unspecified *)
    s, t : Vertex;
begin(* prim *)
```

```
    (* initialization of S and T*)
    (* S denotes the set of vertices of tree T grown by prim in a
    step by step manner. T denotes the set of edges of tree T *)
    S := {1}; T := empty_set;
    (* iteration *)
    while S not_equal_to X do (* X = {1,2,...,n} denotes
    the set of vertices of G *)
      begin(* while *)
        (1) select an edge st such that s in S and t in X - S
            with the weight of st minimum;
        (2) add the vertex t to the set S;
        (3) add the edge st to the tree T;
      end; (* while *)
  end;(* prim *)
```

Complexity of Prim's algorithm:
Clearly, the initialization takes a constant amount of time O(1), which involves two assignments. The while loop is executed exactly $n-1$ times, because of the following two reasons:

1. At each iteration, the number of vertices in the set S is incremented by one unity, thanks to the instruction (2) in the while loop.

2. The loop terminates as soon as $S = X$ and $|X| = n$.

We shall now consider the instruction number (1) of the while loop of Prim. This statement can be *implemented* in O(n) time. Hence, each execution of the while loop demands O(n) time (Recall that when we compute complexity, the lower-order terms and multiplicative constants can be neglected [6]).

Therefore, complexity of Prim's algorithm is $(n-1)O(n) = O(n^2)$, which is a polynomial of degree 2.

Why does Prim's greedy algorithm give the exact optimum?:
Prim's algorithm chooses at each iteration what appears "locally" as the best thing to do. At the end of the iteration, it obtains the best overall! The following result explains the magic.

Minimum spanning tree property:

Property 1.1. *Let G be a connected simple graph with weight attached to each edge of the graph G. Consider a* proper *subset S of vertices of G and an edge e having one end in S and the other end outside of S with the weight of the edge e* minimum. *Then, there is a spanning tree of* minimum *weight in G* containing *the edge e.*

Proof. Since G is a connected graph, by Theorem 1.1, it contains a spanning tree. Consider a spanning tree T of the graph G. If the tree T contains the edge $e = xy$, then we are done. Otherwise, the tree T surely will contain an edge $e' = x'y'$ having one end x' in S and the other end y' outside of S, for, if not, the tree T will not be connected. (In fact, more generally, a graph is connected, if and only if for any partition of the vertex set into two proper subsets, there is an edge having one end in one subset of the partition and the other end in the other subset of the partition.) Since the weight of the edge e is a minimum among all edges joining S and and its complement, we have the weight of $e' \geq$ the weight of e.

Now consider the spanning subgraph obtained by adding the edge e to the tree T, that is, consider the graph $G' = T + e$. Then, G' contains *exactly* one elementary cycle. This is because of the following argument: By Proposition 1.1, in the spanning tree T, there is a unique path P joining the end vertices x and y of the edge e. But then, $G' = T + e$ contains the unique elementary cycle $P + e$ (A path plus the edge joining the initial and final vertices of the path is an elementary cycle) (see Figure 1.10).

Now form the graph $T' = G' - e'$. T' is *still* connected, because the end vertices x' and y' of the deleted edge e' are connected by the path $P + e - e'$ in the graph T'. Moreover, T' is a *tree*, because the *only* cycle $P + e$ of G' is destroyed by the removal of the edge e' of this cycle. Not that the trees T and T' differ by only *two edges* e and e'. Since the weight of the edge $e \leq$ the weight of the edge e', we have, the weight of the tree $T' \leq$ the weight of the tree T.

Since T is a minimum weight spanning tree, we must have equality of the weights of T and T'. Thus, we have constructed a minimum spanning tree T' containing the edge e. □

We shall now write a computer program in C to implement Prim's algorithm. The given weighted graph is represented by its cost matrix M where $M[i][j]$ is the cost of the edge ij. If there is no edge between the vertices i and j, we set $M[i][j] = \infty$. In C, ∞ can be replaced by the constant INT_MAX if costs of the edges are "small" integers. In fact, in a computer implementation of Prim's algorithm, ∞ is replaced by a cost which cannot be a "legal" cost

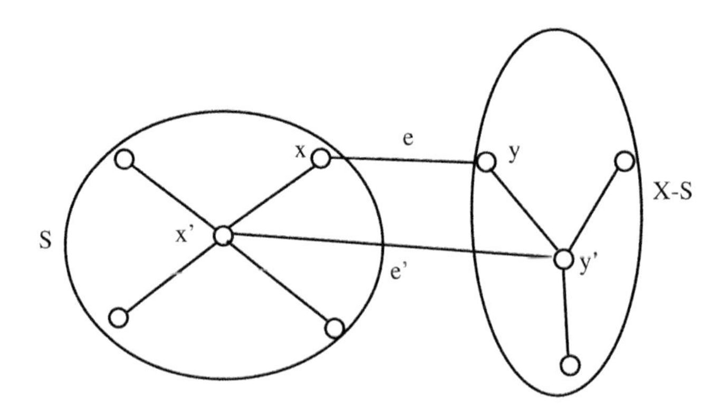

FIGURE 1.10: A step in the proof of Prim's algorithm.

of any edge. We define two arrays dad and cost_mins. $dad[i]$ gives the vertex in S that is currently "closest" (with respect to the cost of edges) to vertex i in $X \setminus S$. cost_min[i]= the cost of the edge $(i, dad[i])$.

```
Implementation of Prim's algorithm in C:
#include <stdio.h>
#include <stdlib.h>
#include <limits.h>
/* INT_MAX is defined in this library */
#define n_max 10//maximum vertices

int M[n_max][n_max],n,m;//M, the cost matrix
//n,m the number of vertices, edges
void graph ()

     {/*keyboarding the  M*/
     int i,j,x,y,c;
     /*i,j for the loops
     x--y is an edge, its cost c */
     printf ("enter the values of n,m \n ");
     scanf ("%d %d",&n,&m ) ;

     /* initialization of M*/

        for (i=1;i<=n;i++)
            for ( j=1;j<=n;j++)
              M[i][j]=INT_MAX;

            /*end of initialization */

            /* enter the costs of edges */
            for (i=1;i<=m;i++)
               {
               printf ("enter the two ends of edge %d \n",i );
               scanf ("%d%d", &x,&y );
               printf (" enter the edge cost c : \n");
               scanf ("%d",&c);
               M[x][y]=c; M[y][x]=c;
               }
      }
        void print_graph ()
             {
             int i,j;
             for (i=1;i<=n;i++)
                       {
```

```
                    for (j=1;j<=n;j++)
                    printf ("%5d ",M[i][j]);
                    printf ("\n\n\n");
                    }
        }

//begin Prim
void prim()
{ int dad[n_max], cost_min[n_max];
  //array dad represents the tree under construction
  int  i,j,k,min;
  //initialization
  for(i=1;i<=n;i++)

                      dad[i]=1;//tree consists of only one
                      //vertex 1
                      cost_min[i]=M[i][1];
                      //if there is no edge (1,i), M[1][i]=
                      //INT_MAX
                      }

 for(i=1;i<=n;i++)
  {//find a vertex k outside the tree to be added to the
   //tree
        k=2;
        min=cost_min[2];

 for(j=3;j<=n;j++)
 if(cost_min[j]<min)
        {
           k=j;
           min=cost_min[j];
         }

 //print edge
printf("%d%d\n",k,dad[k]);
cost_min[k]=INT_MAX;//k is added to tree
//update arrays cost_min, dad
for(j=2;j<=n;j++)

if((M[k][j]<cost_min[j])&&(cost_min[j]<INT_MAX))
{
      cost_min[j]=M[k][j];
        dad[j]=k;
        }
```

```
                }


            }
            //end of prim

    int main ()
                {
                                graph () ;
                                print_graph () ;
                                 prim();
                                return 0;

                }
```

1.2.2 Kruskal's minimum spanning tree algorithm

This is again a greedy algorithm which furnishes us an *exact* optimum. An acyclic simple graph is often referred to as a *forest*, that is, each connected component of a forest is a tree. This algorithm proceeds somewhat in the "opposite direction" of Prim's algorithm. We shall now describe Kruskal's algorithm.

Input: A weighted connected simple graph, that is, a simple connected graph with a weight attached to each edge. As in Prim's algorithm, the vertices represent different cities and the weight associated with an edge $e = xy$ may represent the cost of constructing a highway or railway line between the two cities x and y.

Output: A minimum weight spanning tree of the graph.

Algorithm: We shall first illustrate the algorithm on the graph of Figure 1.8.

First of all, the edges of the given graph are sorted in order of increasing cost. Initially, we start with a forest F consisting of all the vertices of the graph but *no* edges. Let us denote by PX the set vertices of the connected components of the forest F which evolves during the execution of the algorithm. PX defines a partition of the vertex set X of the graph G.

Then, we consider each edge $e = xy$ in turn among the sorted list of edges. If the end vertices x and y of the edge e under consideration are in *distinct components* of the forest F, then we add the edge e to the forest. Note that this operation reduces the number of components of the forest F by one. If the vertices x and y are in the *same component* of the forest F, we discard the edge e. The algorithm proceeds like this, until the forest F consists of only one connected component, that is, until F becomes a tree. This terminal condition can be tested by the equality: $|PX| = 1$, that is, the number of subsets in the partition is 1.

TABLE 1.2: Sorting the edges of graph Figure 1.8 in order of increasing cost

Edges	Costs	Edges	Costs
16	2	35	6
17	3	45	6
67	4	26	7
23	5	34	8
36	5	12	9

TABLE 1.3: Execution of Kruskal's algorithm on the graph G of Figure 1.8

Iteration number	Edges	Add or reject	PX	F	$\|PX\| = 1?$
0(Initial)	–	–	$\{1\}, \{2\}, \{3\}$ $\{4\}, \ldots, \{7\}$	$\emptyset$	–
1	16	add	$\{1,6\}, \{2\}, \{3\}$ $\{4\}, \{5\}, \{7\}$	$\{16\}$	no
2	17	add	$\{1,6,7\}, \{2\}$ $\{3\}, \{4\}, \{5\}$	$\{16, 17\}$	no
3	67	reject			no
4	23	add	$\{1,6,7\}, \{2,3\}$ $\{4\}, \{5\}$	$\{16, 17, 23\}$	no
5	36	add	$\{1,6,7,2,3\}$ $\{4\}, \{,5\}$	$\{16, 17, 23, 36\}$	no
6	35	add	$\{1,6,7,2,3,5\}, \{4\}$	$\{16, 17, 23, 36, 35\}$	no
7	45	add	$\{1,6,7,2,3,5,4\}$	$\{16, 17, 23, 36, 35, 45\}$	no
Exit loop					yes

The different steps in the execution of the algorithm on the graph G of Figure 1.8 are shown in the following table.

Let us first sort the edges in order of increasing cost (Table 1.2). The edges of a minimum spanning tree are available in the fifth column of iteration number 7 in Table 1.3. The reader can see that the spanning tree obtained by Kruskal's algorithm is the *same* as the one obtained in Prim's algorithm (see Figure 1.11).

We shall now write Kruskal's algorithm in pseudo-code.

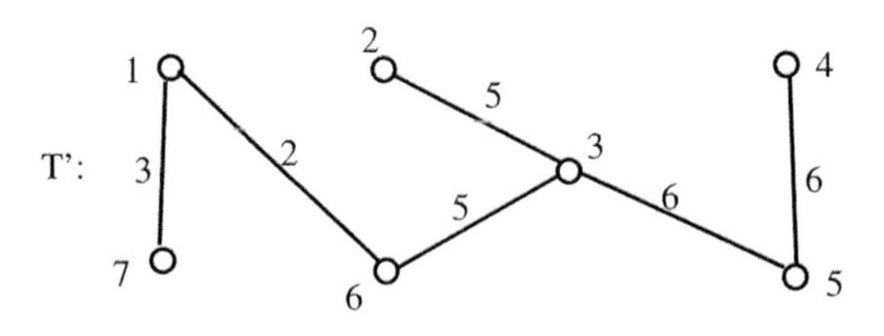

FIGURE 1.11: Kruskal's minimum spanning tree T' of G.

```
Kruskal's Algorithm
procedure Kruskal( var G: Graph; var F: Set of Edges);
(* Kruskal' takes as input a connected weighted graph G and
outputs a minimum spanning tree F*)
var PX : Partition of the vertex set X;
(* PX denotes the partition of the vertex set X, induced by
the connected components of the forest F*)
     e : Edge; L : List of Edges;
   x,y : Vertex;(* x, y are the ends of edge e*)
begin (* Kruskal *)
  (* initialization of PX and F *)
  PX := empty_set; F := empty_set;
  Form a list L of  the edges of G in increasing order of cost;
     (* Forest consists of isolated vertices and no edges *)
  for each vertex x of the graph G do
    add {x} to PX;
   (* Iteration *)
  while | PX | > 1 do
  begin (* while *)
    (1) choose an edge e = x y, an edge of minimum cost from the
    list L;
    (2) delete the edge e from the list L;
    (3) if x and y are in two different sets S1 and S2 in PX
        then begin
    (4) replace the two sets S1 and S2 by their union S1
        U S2;
               add the edge x y to F;
         end;
  end;(* while *)
end;
```

Complexity of Kruskal's algorithm:
The "for" loop in the initialization part takes $O(n)$ time where n is the number of vertices of the graph G. In fact, the edges are not sorted in the initialization part into a list but are kept in a *heap data structure* [3]. A heap is an array $a[1], a[2], \ldots, a[n]$ of numbers such that $a[i] \leq a[2i]$ for $1 \leq i \leq n/2$ and $a[i] \leq a[2i+1]$ for $1 \leq i < n/2$. This array structure can be conveniently viewed as a quasi-complete binary tree [3] in which the value at a node is less than or equal to the values at its children node. This organization of the edges require $O(m)$ time.

The major cost of the algorithm is dominated by the while loop. In the while loop, each edge is examined one-by-one in order of increasing cost. Either the edge examined is added to the forest F under construction or it is rejected. Hence, the while loop is executed at most m times where m is the number of edges of the graph G. The instruction (1) choosing an edge of minimum cost can be performed in $O(\log m)$ time. In fact, $\log m$ is the initial height of the heap viewed as a binary tree. The total time needed to execute the instructions (3) and (4) is $O(m)$ for all "practical" values of m, if we use the fast disjoint-set union algorithm [3]. The remaining instructions in the while loop can be performed in $O(1)$ time. Hence, the complexity of the Kruskal algorithm is $O(m \log m)$.

1.2.3 Rooted ordered trees and traversal of trees

A rooted tree T can be defined recursively as follows:

Definition 1.1 (Rooted tree). *A single vertex r alone is a tree and r is its root. (basis of recursion)*

If $T_1, T_2, \ldots, T_k$ are trees with roots $s_1, s_2, \ldots, s_k$, respectively, then we may construct a new tree $T = (r; T_1, T_2, \ldots, T_k)$ with root r as follows (see Figure 1.12):

$T_1, T_2, \ldots, T_k$ are called the subtrees of the root r and $s_1, s_2, \ldots, s_k$ are called the sons/daughters *of the root r taken in this order. (recursion) T_1 is the leftmost subtree of r, and T_k is the rightmost subtree. Note that the order of the sons is taken into account. The edges are supposed to be directed downwards (even though it is not represented in Figure 1.12. If there is a downward path from a vertex x to a vertex y, then y is called a* descendant *of x, and x is called an* ancestor *of y. A direct descendant is a* son/daughter *and a direct ancestor is called the* dad or mom. *A vertex with no son/daughter is called a* leaf.

Traversal of rooted ordered trees:

We define three tree traversals: (1) Preorder, (2) Inorder, (3) Postorder.

These traversals are again defined recursively. In fact, they are simply mappings of the two-dimensional tree into a one-dimensional list.

If the tree consists of only one vertex, the root r, $T = (r)$ then the list (r) consisting of only one element is the preorder, inorder and postorder traversal of T. That is, preorder(T)=inorder(T)=postorder(T)=(r).(basis).

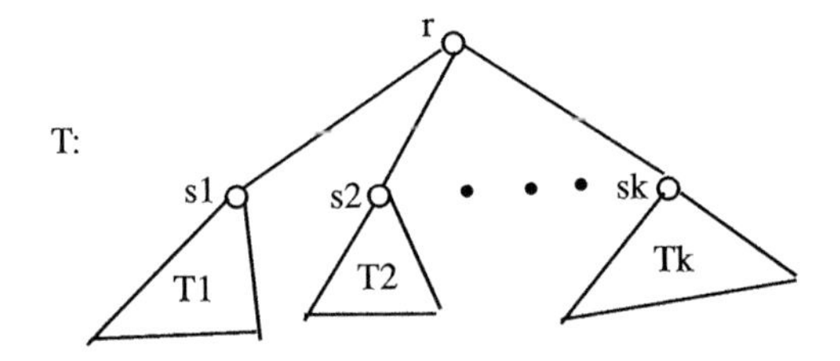

FIGURE 1.12: Recursive definition of a rooted tree.

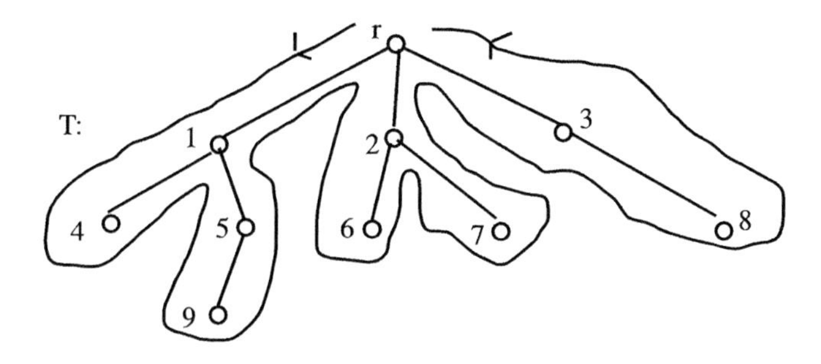

FIGURE 1.13: Illustration of tree traversals.

If $T = (r; T_1, T_2, \ldots, T_k)$, then we define the preorder, inorder, and postorder as the lists of vertices

$$preorder(T) = (r, preorder(T_1), preorder(T_2)), \ldots, preorder(T_k))(\text{recursion}).$$
$$inorder(T) = (inorder(T_1), r, inorder(T_2) \ldots, inorder(T_k))(\text{recursion}).$$
$$postorder(T) = (postorder(T_1), postorder(T_2), \ldots, postorder(T_k), r)(\text{recursion}).$$

Note that the root r comes first in preorder, last in postorder. In the inorder, the root r comes after the listing of vertices of the first subtree of r[3].

A useful trick to find these orders:
We shall illustrate this trick by an example. Consider the tree of Figure 1.13. Think of a walk around the tree in the counter-clockwise direction, starting from the root and staying as close to the tree as possible.

For preorder, we write the vertex in the list the very first time we meet it during the walk.

For postorder, we write the vertex in the list the last time we meet it during the walk, that is, as we move upwards.

For inorder, we write the leaf (a leaf is vertex of the tree with no son/daughter) the first time we meet it during the walk and other vertices during the second time during the walk (in the bay).

Note that each vertex is written exacly once in the list. In the graph of Figure 1.13, preorder(T) $= (r, 1, 4, 5, 9, 2, 6, 7, 3, 8)$, postorder($T$) $= (4, 9, 5, 1, 6, 7, 2, 8, 3, r)$, inorder($T$) $= (4, 1, 9, 5, r, 6, 2, 7, 8, 3)$. The sons/daughters of the root r are $1, 2, 3$ taken in this order: the leftmost subtree of r is the vertex 1 and all its descendants in the tree T.

1.3 Shortest Path Algorithms

In this section, we may consider without loss of generality directed graphs with no multiple arcs, that is, 1-graphs, without loops. We shall first study the *single-source shortest path algorithm* invented by Dijkstra. This is again a *greedy algorithm*. We use "cheapest cost path" as synonym for the "shortest path." "Path" always means in this section a "directed path."

1.3.1 Single-source shortest path algorithm

Input: A directed graph $G = (X, U)$ where $X = \{1, 2, \ldots, n\}$. To each arc is associated *a non-negative* weight. We interpret the vertices of the graph G as different cities and the weight of an arc $u = (x, y)$ as the *cost* of traveling from the city x to the city y along the arc u. The weight function on the set of arcs is denoted by w, that is, the weight of the arc $u = (x, y)$ is $w(x, y)$. The vertex 1 is designated as the *source* or *origin*. This vertex 1 may be interpreted as the capital city.

If there is no arc from a vertex s to a vertex t, we set the weight of the arc (s, t), $w(s, t) = \infty$. In the computer implementation of the algorithm, ∞ will be replaced by a large number which cannot be a *legal* cost of traveling between two cities. For example, if the weights are non-negative integers, we may replace ∞ by INT_MAX in the language C or maxint in the language Pascal.

Output: The *cheapest* cost of traveling from the vertex 1 to all other vertices of the graph is along a directed path. The cost of a directed path is the *sum* of the costs of its arcs (see graph G of Figure 1.14).

The cheapest dipath from the vertex 1 to the vertex 5 is $(1, 2, 4, 5)$ and its cost is $5 + 4 + 3 = 13$ and the cheapest path from the source vertex 1 to the vertex 3 is the path $(1, 3)$ which consists of the directed edge 13 and its cost is just 3.

Algorithm: To explain the algorithm, we need a few ideas. A vertex x of a directed graph G, is called a *root* of G if there is a path from the root vertex to all other vertices of G. A root may not always exist.

An arborescence is a graph possessing a root and having no cycles. Note that the root of an arborescence is *unique*. The following figure is a spanning arborescence of the graph of Figure 1.14 (see Figure 1.15). In fact, it is a "minimum weighted" spanning arborescence.)

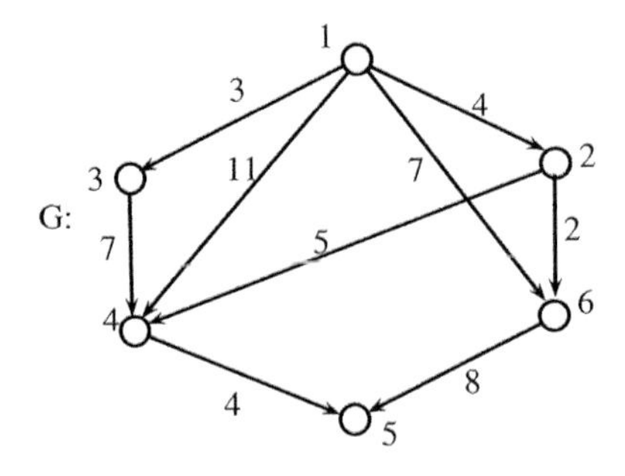

FIGURE 1.14: A directed weighted graph G.

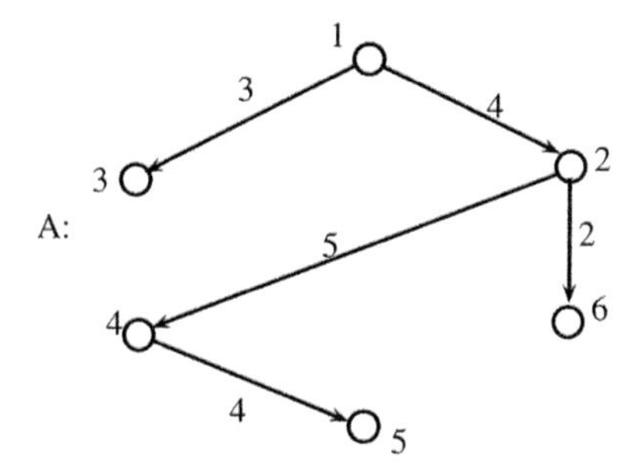

FIGURE 1.15: A minimum spanning arborescence A of Figure 1.14.

In an arborescence, there is a unique path from the root to all the other vertices. In fact, Dijkstra's algorithm constructs a spanning arborescence of the given graph G having its root as the source of the graph G and its unique path from the root to a vertex x as a cheapest path in the graph G. In the algorithm, we use the variable S denoting the set of vertices of a cheapest arborescence under construction. (Note that in Prim's algorithm also the variable S denotes the set of vertices of a minimum spanning tree under construction.)

The variable S satisfies the following two properties, which can be considered as *loop invariant*:

1. For all vertices s in the set S, a cheapest cost of a path (also called *shortest path*) from the source 1 to s is *known* and this path lies *entirely* in the set S, that is, all the vertices and the arcs of a shortest path from 1 to s lies *fully* in S. This condition implies that the set S must contain the source vertex 1. In fact, the assignment $S := \{1\}$ is used as initialization of the set S.

2. For a vertex t *outside* of the set S (if such a vertex exists), a *path relative to the set* S is a path from the source vertex 1 to the vertex t such that all the vertices of this path lie completely in the set S but for the last vertex t. We shall now state the second property of the set S in terms of paths relative to the set S. For each vertex t *outside* of the set S, a *shortest* path from 1 to t *relative to* S lies completely in the set S except for the *final* vertex t.

Interpretation of two properties of S when S equals the whole of the vertex set X of G:
According to the first property of $S = X$, a shortest path from the origin vertex 1 to any vertex x of the graph is known, and we have solved the problem! Note that the second property is *not* applicable, since $X \setminus S = \emptyset$. Hence, the terminal condition of the algorithm is $S = X$.

In the Dijkstra's algorithm, the following elementary observation is used:

A subpath of a shortest path is a shortest path!

Note that the above observation will not be valid if the weights of the arcs are *not* non-negative numbers. Another property on which Dijkstra's algorithm works is the following: This is the point where the algorithm becomes *greedy*!

Consider a shortest path P among the shortest paths relative to the set S. Then, the path P is a shortest path in G itself.

Let us write Dijkstra's algorithm in pseudo-code. The weighted graph is represented by its weighted adjacency matrix $W = (w_{ij})_{n \times n}$ where

$$w_{ij} = \begin{cases} 0 & \text{if } i = j \\ \text{weight of the arc } ij & \text{if } ij \text{ is an arc} \\ \infty & \text{otherwise.} \end{cases}$$

The vertex set is $X = \{1, 2, \ldots, n\}$. We use the array D where

$$D[i] = \begin{cases} \text{The cost of the shortest path from the vertex 1 to i if } i \in S, \\ \text{The cost of the shortest path relative to S if } i \notin S. \end{cases}$$

Note that we obtain the final result in the array D as soon as $S = X$. We use the types, Matrix and T-Array where Matrix is a two-dimensional $n \times n$ array of real numbers, and T-Array is a one-dimensional array of n numbers in the algorithm:

```
procedure Dijkstra( var W : Matrix; var D : T_Array);
(* Dijkstra takes as input a matrix of non-negative weights
and it returns an array D of cheapest cost paths.*)
var S : set of vertices; x, y : vertex;
begin(* Dijkstra*)
  (* Initialization of S and D*)
  S := {1};
  D[1] := 0;(* The cheapest cost from 1 to itself of 0*)
  for i := 2 to n do
    D[i] := w(1, i);(* The shortest path relative to S = {1}
    is the arc (1, i). S verifies the two properties *)
  (* iteration *)
  while S not equal to X do
  begin (* while *)
    (1) choose a vertex y in X - S such that D[y] is minimum
        (* y is called the pivot vertex*)
    (2) add y to the set S;(* The shortest path among the
        shortest paths relative to S is a shortest path in G.
        This is the statement where "greedy" comes into
        play *)
    (3) for each x in X - S do
    (4) D[x] := min( D[x], D[y] + w(y, x));
        (* This "for" loop is the crux of the Dijkstra's
               algorithm.
```

```
          In this loop, we adjust the values of D[x] to take
          into account the vertex y just added to S, to preserve
          the second property of the set S*)
    end;(*while*)
  end;(*Dijkstra*)
```

Remark 1.1. *The vertex y in Dijkstra's algorithm is called the* pivot vertex. *Of course, the pivot changes at each iteration.*

Let us now execute Dijkstra's algorithm on the graph of Figure 1.14. The following Table 1.4 traces the different steps of the algorithm.

The final result is found in columns 5, 6, 7, 8, 9 of iteration number 5 in Table 1.4.

We observe the following properties of Dijkstra's algorithm:

1. The vertices are added to the set S in order of their increasing costs of D.

2. Once a vertex x is added to the set S, the value of $D[x]$ will remain *constant* till the end of the algorithm. Hence the set S can be interpreted as the set of Stabilized vertices.

3. The statement number (4) of the algorithm can be described as follows: Can we reduce the value of $D[x]$ for each x in the set $X \setminus S$ by using the newly added vertex y to the set S, that is, is there a cheaper path *relative to* S from 1 to x by using the vertex y as the next-to-last vertex in the path? We shall call the vertex y as the *pivot vertex*. Of course, the pivot vertex changes at each iteration.

Complexity of Dijkstra's algorithm:
The initialization of the set S takes O(1) time and the initialization of the D array costs $O(n)$ time. Hence the total cost of the initialization step is $O(n)$.

TABLE 1.4: Execution of Dijkstra's algorithm on the graph of Figure 1.14

Iteration number	$S \neq X$	y	$S \leftarrow S \cup \{y\}$	D[2]	D[3]	D[4]	D[5]	D[6]
0(Initial)	–	–	$\{1\}$	4	3	11	∞	7
1	yes	3	$\{1,3\}$	4	3	10	∞	7
2	yes	2	$\{1,3,2\}$	4	3	9	∞	6
3	yes	6	$\{1,3,2,6\}$	4	3	9	14	6
4	yes	4	$\{1,3,2,6,4\}$	4	3	9	13	6
5	yes	5	$\{1,3,2,6,4,5\}$	4	3	9	13	6
Exit loop	no							

The "while" loop is executed exactly $n-1$ times, since the number of vertices in the set S is incremented exactly by one at each iteration and the loop terminates as soon as $|S| = n$. Statement (1) requires $O(n)$ steps (finding a minimum in an array of n integers requires $O(n)$ steps) and similarly, the "for" loop demands $O(n)$ time. Hence, the complexity of the algorithm is $O(n) + (n-1)O(n) = O(n^2)$. (Recall that while computing the complexity of an algorithm, we can ignore the lower-order terms and the multiplicative constants.)

Proof of Dijkstra's algorithm:

Proof. We have to prove the following two properties (the loop invariant) of the set S at the beginning of the "while" loop of the algorithm:

1. For each vertex, s in S, the cost $D[s]$ of a cheapest path from the vertex 1 to the vertex s is known, and this path lies entirely within the set S.

2. For all t outside of the set S, the cost $D[t]$ of the cheapest path *relative* to S is known.

Proof proceeds by the induction on the number of elements of S.
Basis. $|S| = 1$. If $S = \{1\}$, then clearly, the cheapest path from 1 to itself has no arcs and hence its cost is zero. Further, to each vertex t outside of the set S, a path relative to S is simply an arc $(1, t)$ and its cost is $w(1, t)$. Thanks to the initialization of S and the array D, the two properties of S are satisfied.

Induction hypothesis: Suppose the two properties for the set S with $|S| = k < n$, that is, the properties are true after $(k-1)$th iteration. We shall prove the two properties *after* performing the kth iteration.

During the k-th iteration, the pivot vertex y is added to the set S. We claim that $D[y]$ is the cost of the shortest dipath from 1 to y. Otherwise, there is a shorter dipath P to the vertex y. By the choice of the pivot vertex y, this path P must have a vertex $\neq y$ outside of S (see Figure 1.16).

In Figure 1.16, illustrating Dijkstra's algorithm, the wavy lines represent directed paths. The directed path P starts from the source vertex 1 passing

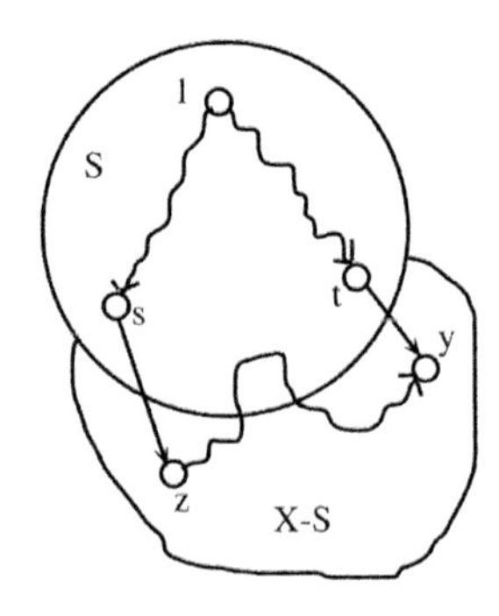

FIGURE 1.16: A step in the proof of Dijkstra's algorithm.

through the vertex s, then goes directly to the vertex z, then again goes possibly inside the set S and finally reaching the vertex y. The directed path from the vertex 1 to the vertex y relative to the set S goes through the vertex t and then directly reaches the vertex y from the vertex t through an arc.

Let z be the first vertex of path P outside of S. Then, the *subpath* $P(1, z)$ of P from 1 to z is a path *relative* to S and its cost $D[z]$ is clearly strictly less than the length of the path P which is strictly less than the number $D[y]$, by induction hypothesis. This is impossible since $D[y]$ was a minimum among all vertices outside of the set S during the $(k+1)$-th iteration (statement number (1) of the procedure Dijkstra). Note that we use here the fact that the costs of the arcs are non-negative numbers. Otherwise, our argument will not be valid. Hence the first property of the set S after the iteration number $k+1$.

The second property of S remains true even after adding the pivot vertex, y, because of statement number (4) of the procedure Dijkstra of Section 1.3.3, which adjusts the cost of each $D[x]$, $x \in X \setminus S$ to take into account the arrival of the pivot vertex into the set S. □

Recovering paths in Dijkstra's algorithm:
Dijkstra's algorithm finds the cost of a cheapest path from the source vertex 1 to all vertices of a weighted digraph. We are now interested in finding paths *realizing* the minimum costs.

An arborescence can be conveniently represented by an array "dad" where dad$[i]$ gives the father of the vertex i. Stated differently, dad$[i] = j$ if and only if (i, j) is an arc of the arborescence. Initially, the paths relative to $S = \{1\}$ are the arcs $(1, i)$. Hence, we add the following loop in the initialization part:

```
for i := 2 to n do
  dad[i] := 1; (* The vertex 1 is the root of the arborescence
  and hence has no dad*)
```

When and where to update the array dad?
Consider statement number (4) of Dijkstra's algorithm. In this assignment, if $D[y] + w(y, x) < D[x]$, that is, if we could use the pivot vertex y as the next-to-last vertex to reach x via a path from the source vertex 1, then we update dad$[x] := y$. Hence, we assign and replace the "for" loop numbered (3) and the statement numbered (4) of the procedure Dijkstra 1.3.1 by the following loop:

```
for each x in X - S do
  if D[y] + w(y, x) < D[x]
  then
  begin
    D[x] := D[y] + w(y, x);
    dad[x] := y;(* update dad of x*)
  end;
```

Once we have recorded the dads of different vertices, a shortest path from the vertex 1 to any vertex i can be printed by "climbing" up the arborescence from the vertex i till the root 1. The following procedure prints the vertices of a shortest path from the vertex 1 to the vertex i. The vertices which are encountered while climbing up the tree will be pushed into the array a and finally will be popped to get the right order of vertices on the path from the vertex 1 to the vertex i.

```
procedure path ( i : vertex);
var a : array[vertex] of vertex;
    k : integer;
begin(* path*)
  k := 1;
  while i < > 1 do
  begin
    a[k] := i; k := k + 1;
    i := dad[i];
  end;
  a[k] := 1;
  (* print the vertices of the path*)
  for i := k down to 1 do
    write(a[i], ' ');
end;(* path *)
```

We shall now write a computer program in C to implement Dijkstra's algorithm. The given weighted graph is represented by its cost matrix M where $M[i][j]$ is the cost of the arc/directed edge (i,j). If there is *no* edge between the vertices i and j, we set $M[i][j] = \infty$. In C, ∞ can be replaced by the constant INT_MAX if costs of the edges are "small" integers. In fact, in a computer implementation of Dijkstra's algorithm, ∞ is replaced by a cost which cannot be a "legal" cost of any directed edge. We define two arrays D (for Distance) and S. $D[i]$ represents the minimum cost of a directed path from the root vertex 1 to the vertex i with respect to the set S. $S[i] = 1$ if $i \in S$ and $S[i] = 0$ if $i \notin S$. $dad[i]$ is the father/mother of the vertex i in the arborescence under construction.

```
Dijkstra's algorithm in C:
#include<stdio.h>
#include<stdlib.h>
#include<limits.h>
#define nmax 10//maximum vertices
int M[nmax][nmax];//cost matrix
int n,m;//number of vertices and arcs
void dijkstra(){

    int D[n+1],S[n+1],dad[n+1],stack[n+1];
    //S[i]=1 if i in S, otherwise S[i]=0.
```

```
//D, the distance array Dijkstra
    int p,i,j,min,k,head=0;//p, the pivot vertex
    // in the pseudo-code we have used y for pivot instead of p
    S[1]=1;//arborescence consists of only one vertex
    for (i=2; i<=n;i++)
    {
        S[i]=0;
        D[i]=M[1][i];
        dad[i]=1;
   }
for (i=2;i<=n;i++)
{
    k=1;//find the least vertex k outside S
    while (S[++k] !=0);
    p=k;min=D[p];
    for (j=k+1;j<=n;j++)
    if(S[j]==0)
    if (D[j]<D[p])
    {p=j;
    min=D[p];
    }
    S[p]=1;//add p to S
    //update arrays D, dad
    for (j=2; j<=n;j++)
    if (S[j]==0)
    if ((D[p]!=INT_MAX) && (M[p][j] !=INT_MAX))
    if (D[p] +M[p][j]< D[j])
    { D[j]= (D[p]+M[p][j]);
      dad[j]=p;
}}

for (i=2;i<=n;i++)
{printf("%5d  ",D[i]);
}

  //print a shortest path from 1 to a vertex i
printf("enter a vertex to which you want to print a minimum
path\n");
scanf("%d",&i);
//climb the arborescence from i to the root 1
//and stack the vertices encountered
while(i!=1)
{
stack[++head]=i;
i=dad[i];}
```

```
stack[++head]=i;//i=1
//print the path from 1 to i
printf("the path is   ");
for(i=head;i>0;i--)
printf("%d ",stack[i]);

}

void graph(){

     int i,j,x,y,c;
     printf("enter n",\n);
     scanf("%d",&n);
     printf("\n");
     printf("enter  m",\n);
     scanf("%d",&m);

     for(i=1;i<=n;i++)
         for(j=1;j<=n;j++)
         {
            M[i][j]=INT_MAX;
            if(i==j)
            M[i][j]=0;
            }

              for(i=1;i<=m;i++)
              {
               printf("enter the arc %d",i) ;
               scanf("%d %d",&x,&y) ;
                printf("enter the cost %d",i) ;
               scanf("%d",&c) ;
               M[x][y]=c;
                              }}
  void print_graph()
  {
      int i,j;

       for(i=1;i<=n;i++)
              {
              for(j=1;j<=n;j++)
              printf("%5d  ",M[i][j]);
                  printf("\n\n\n");
                  }

                            }
```

```
int main()
{
      graph();
      print_graph();
      dijkstra();
      return 0;
      }
```

1.4 Dijkstra's Algorithm for Negative Weighted Arcs

Dijkstra's algorithm does not work if some of the arcs are associated with negative weights. We shall consider Figure 1.17.

Clearly, in Figure 1.17, the cheapest cost of a path from the source 1 to the vertex 3 is $3 - 2 = 1$. But, we shall see that Dijkstra's algorithm as presented above will give the answer 2, which is false. This is an example where the greedy method does not always work!

Let us execute the algorithm on the graph of Figure 1.17. The following table traces the different steps during execution. (Table 1.5).

The result of the execution can be read from columns 5 and 6 of the iteration number 2. Column 6 of the iteration number 2 says that the cost of a shortest path from the vertex 1 to the vertex 2 is 2, which is *not* correct.

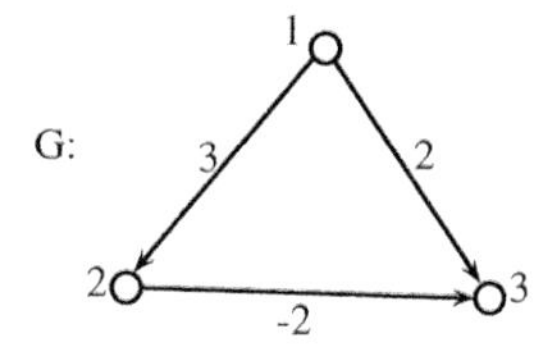

FIGURE 1.17: Illustration of Dijkstra's algorithm for negative weighted arcs.

TABLE 1.5: Execution of Dijkstra's algorithm on Figure 1.17

Iteration number	$S \neq X$	y	$S \leftarrow S \cup \{y\}$	D[2]	D[3]
0(Initial)	–	–	$\{1\}$	3	2
1	yes	3	$\{1,3\}$	3	2
2	yes	2	$\{1,3,2\}$	3	2
Exit loop	no				

Modification of Dijkstra's algorithm for negative weight arcs: Consider a directed graph with weight (possibly negative) associated to each arc; however, the graph has no negative circuits, that is, the graph does not possess a circuit with the sum of the weights of its arcs < 0.

Note the following fact:

Fact 1.1. *In the case of graphs with negative-weight arcs, the "greedy" statement that the shortest path among the shortest paths relative to S is a shortest path in G is no longer valid!*

The following algorithm computes the cost of a shortest path from the source vertex 1 to all other vertices of the graph. As in Dijkstra's algorithm for graphs with non-negative weights, we use the variables S, the array D, and the variable vertices x, y and the integer variable i.

```
procedure Dijkstra( var W : Matrix; var D : T_Array);
(* Dijkstra takes as input a matrix whose entries can be < 0
  but no negative cycles and it returns an array D
  of cheapest cost paths.*)
var S : set of vertices; x, y : vertex;
begin(* Dijkstra*)
  (* Initialization of S and D*)
  S := {1};
  D[1] := 0;(* The cheapest cost from 1 to itself of 0*)
  for i := 2 to n do
    D[i] := w(1, i);(* The shortest path relative to S = {1}
    is the arc (1, i). *)
  (* iteration *)
  while S not equal to X do
  begin (* while *)
    (1) choose a vertex y in X - S such that D[y] is minimum;
    (2) add y to the set S;
    (3) for each x in X with D[x] > D[y] + w(y, x) do
        begin(* for *)
    (4)   D[x] :=  D[y] + w(y, x);
    (5)   S := S - {x};(* x not stabilized. so delete x
          from S *)
        end;(* for *)
  end;(*while*)
end;(* Dijkstra *)
```

Let us execute the negative-weight version of Dijkstra's algorithm on the graph of Figure 1.17. The following table illustrates the different steps during execution (Table 1.6).

The final result of the execution is found in columns 5 and 6 of the last iteration number 3, which is the correct answer.

TABLE 1.6: Execution of improved Dijkstra's algorithm on Figure 1.17

Iteration number	$S \neq X$	y	$S \leftarrow S \cup \{y\}$	D[2]	D[3]	$S \leftarrow S \setminus \{x\}$
0(Initial)	–	–	$\{1\}$	3	2	
1	yes	3	$\{1,3\}$	3	2	$\{1,3\}$
2	yes	2	$\{1,3,2\}$	3	1	$\{1,2\}$
3	yes	3	$\{1,2,3\}$	3	1	$\{1,2,3\}$
Exit loop	no					

1.5 All-Pairs Shortest Path Algorithm

In this section, we may consider without loss of generality directed graphs with no multiple arcs, that is, 1-graphs, without loops. We study *all-pairs shortest path algorithm* due to Floyd. Floyd's algorithm finds the cost of a cheapest path between *any two* vertices of a directed graph with a non-negative weight associated to each arc. We use "cheapest cost path" as synonym for the "shortest path." "Path" always means in this section "directed path."

Input: A directed graph with a non-negative weight associated to each arc. As in Dijkstra's algorithm, the vertices represent the different cities and the weight of an arc (i,j) represents the cost of traveling from the city i to the city j along the arc (i,j). As in Dijkstra's algorithm, the given graph is represented by a weighted matrix $W = (w_{ij})_{n\times n}$ as follows:

$$w_{ij} = \begin{cases} 0 & \text{if } i = j \\ \infty & \text{if } ij \text{ is not an arc} \\ \text{weight of the arc } ij & \text{otherwise} \end{cases}$$

In a computer program implementing Floyd's algorithm, the symbol ∞ will be replaced by a large number which cannot be a legal cost of traveling from city i to city j. For example, if the weights of arcs are integers, we may replace the symbol ∞ by INT_MAX in the language C or maxint in the language Pascal. We assume that the vertex set $X = \{1, 2\ldots, n\}$.

Output: The cheapest cost of a directed path from city i to city j for all $1 \leq i,j \leq n$. The cost of a directed path is the sum of the cost of its arcs.

Algorithm 1.1: The first algorithm which comes to our mind is to invoke Dijkstra's algorithm n times for an n vertex directed graph by varying the source vertex among all vertices of the graph.

The complexity of calling n times Dijkstra's algorithm is $nO(n^2) = O(n^3)$, since the complexity of Dijkstra's algorithm is $O(n^2)$.

We will now see one direct algorithm to compute the shortest directed path between any two vertices of a directed weighted graph.

Algorithm 1.3 (Floyd's Algorithm). *For a directed path from the vertex i to the vertex j, the set of vertices of the path* other than *the initial vertex of the path i and the final vertex of the path j are called its intermediate vertices.*

The algorithm constructs a sequence of $n+1$ matrices $M_0, M_1, \ldots, M_n$ where in the matrix $M_k = (m_{ij}^{(k)})_{n\times n}$ where $m_{ij}^{(k)}$ = The cost of a shortest directed path from the vertex i to the vertex j not *passing through any intermediate vertices $k+1, k+2, \ldots, n$, that is, the cost of a cheapest dipath from i to j where the set of its intermediate vertices is a* subset *of the set $\{1, 2, \ldots, k\}$.*

The subscript k in the matrix M_k represents the matrix obtained after k *iterations* and the superscript k in the entry $m_{ij}^{(k)}$ is the (i, j) entry of the matrix M_k after k iterations. These subscripts and superscripts will not be present while expressing Floyd's algorithm, and we will see that we can work with only one copy of the matrix M.

With this definition of M_k for $0 \leq k \leq n$, let us interpret the matrices M_0 and M_n.

Interpretation of M_0:
$M_0 = (m_{ij}^{(0)})_{n\times n}$, where $m_{ij}^{(0)}$ is the weight of a shortest dipath from the vertex i to the vertex j not passing through any intermediate vertices $1, 2, \ldots, n$. Since the vertex set of the graph is $X = \{1, 2, \ldots, n\}$, this means that $m_{ij}^{(0)}$ is the weight of the shortest dipath *not* using *any* intermediate vertices at all, that is, $m_{ij}^{(0)}$ is the weight of (i, j), if (i, j) is an arc. Hence, $M_0 = W$, the weighted adjacency matrix. This assignment is used as *initialization* in Floyd's algorithm.

Interpretation of M_n:
By definition, $M_n = (m_{ij}^{(n)})$, where $m_{ij}^{(n)}$ is the weight of a shortest path from the vertex i to the vertex j *not* passing through any intermediate vertices of the set $\{n+1, n+2, \ldots\}$. But this set is the empty set $\emptyset$, because the vertex set of the graph is $X = \{1, 2, \ldots, n\}$. Hence, $m_{ij}^{(n)}$ is simply the cost of a shortest dipath whose internal vertices are from the set of vertices $\{1, 2, \ldots, n\}$ of the graph G. This means that $m_{ij}^{(n)}$ is the cost of a shortest dipath in the graph G.

Therefore, the matrix M_n is the desired result of the algorithm.

We shall now see how to find the matrix $M_k = (m_{ij}^{(k)})$, given that the matrix $M_{k-1} = (m_{ij}^{(k-1)})$, that is, how to move from the $(k-1)$-th iteration to the k-th iteration?

To do this, we again exploit the definition of the matrices M_{k-1} and M_k. Consider a shortest cost path $P_{(k-1)}(i, k)$ from i to k *not* passing through any intermediate vertices $k, \ldots, n$ and a shortest cost path $P_{(k-1)}(k, j)$ from the vertex k to the vertex j *not* passing through any intermediate vertices $k, \ldots, n$. For the path $P_{(k-1)}(i, k)$, k is the final vertex and for the path $P_{(k-1)}(k, j)$ the vertex k is the initial vertex. The concatenation or juxtaposition of the dipaths $P_{(k-1)}(i, k)$ and $P_{(k-1)}(k, j)$ taken in this order which gives us either a directed elementary path from i to j with k as an *intermediate vertex* (see Figure 1.18)

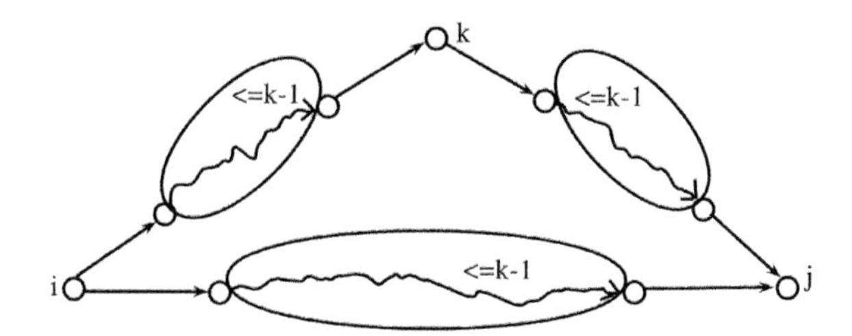

FIGURE 1.18: A step illustrating Floyd's algorithm.

or there is no directed elementary path (but there will be a simple directed path but not elementary, since there is a directed circuit passing through k in the concatenation) from vertex i to vertex j passing through the vertex k.

If the dipaths $P_{(k-1)}(i,k)$ and $P_{(k-1)}(k,j)$ are internally vertex-disjoint, then their concatenation will still give us a directed path from the vertex i to the vertex j for which the set of intermediate vertices is a subset of $\{1, 2, \ldots, k\}$, that is, this concatenated path includes the vertex k as an intermediate vertex and other intermediate vertices p satisfy the inequalities $1 \leq p \leq k-1$. Hence we derive the following formula to compute the entries of the matrix $M_k = (m_{ij}^{(k)})$ in terms of the entries of the matrix $M_{k-1} = (m_{ij}^{(k-1)})$.

$$m_{ij}^{(k)} = \min\left(m_{ij}^{(k-1)}, m_{ik}^{(k-1)} + m_{kj}^{(k-1)}\right), \quad \text{for } 1 \leq i, j \leq n.$$

In words, the above formula can be stated as follows:

The cost of a shortest path from the vertex i to the vertex j not passing through any intermediate vertices higher than k is the *minimum* of the cost of a shortest path from i to j not passing through any intermediate vertex higher than $k-1$, and the cost of a shortest path from vertex i to k not passing through any intermediate vertices higher than $k-1$, plus the cost of a shortest path from k to j not passing through any intermediate vertices higher than $k-1$.

By setting $i = k$ in the above formula, we have $m_{kj}^{(k)} = m_{kj}^{(k-1)}$ since $m_{kk}^{(k-1)} = 0$. Similarly, by setting $j = k$ in the formula, we get $m_{ik}^{(k)} = m_{ik}^{(k-1)}$. This means that during the kth iteration of Floyd's algorithm, that is, during the computation of the entries of the matrix M_k, the entries of the kth row and the entries of the kth column of the matrix M_{k-1} will remain unchanged. Hence, we can carry out our computation in Floyd's algorithm with *only one* copy of the matrix M.

The vertex k during the kth iteration is called the *pivot vertex*. This vertex k is equivalent to the pivot vertex y in Dijkstra's algorithm. We now write the algorithm:

```
procedure Floyd(var W, M : Matrix);
(* Floyd takes as input parameter a matrix W which is the
weighted adjacency matrix of a directed weighted graph. The
```

```
procedure returns as output parameter the matrix M of the cost
of a shortest directed path of the graph G. *)
var i, j, k : integer; (* i, j, k for the loops *)
begin (* Floyd *)
    (* Initialization of M = W *)
    for i := 1 to n do
        for j := 1 to n do
            M[i, j] := W[i, j];
    [* iteration *)
    for k := 1 to n do
    (* compute the matrix M sub k *)
        for i := 1 to n do
            for j := 1 to n do
                if M[i, k] + M[k, j] < M[i, j]
                then M[i, j] := M[i, k] + M[k, j];
end;(* Floyd *)
```

The complexity of Floyd's algorithm:
The initialization part has two nested loops, and each loop is executed exactly n times. Hence, the number of steps needed for the initialization part is $O(n^2)$. The iteration part is constituted by three nested loops and the conditional statement inside obviously requires $O(1)$ step. Therefore, the iteration part requires $O(n^3)$ steps.

Hence, the time taken by Floyd's algorithm is $O(n^2) + O(n^3) = O(n^3)$.

Proof of Floyd's algorithm:

Proof. The proof is by induction on k. *Base*: $k = 0$. By the definition of the matrix M_0, we have the equality $M_0 = W$, where W is the weighted adjacency matrix of the given graph. Hence, we have correctly initialized the matrix M.

Induction Hypothesis: Suppose the entries of the matrix $M_{k-1} = (m_{ij}^{(k-1)})$ satisfy the following property:

$m_{ij}^{(k-1)}$ = The cost of a cheapest path from the vertex i to the vertex j not passing through any intermediate vertex $> k-1$, for $1 \leq i, j \leq n$.

We shall show that the entries of the matrix $M_k = (m_{ij}^{(k)})$ obtained after k iterations are such that $m_{ij}^{(k)}$ is equal to the cost of a cheapest path from the vertex i to the vertex j not passing through any intermediate vertices $> k$.

The induction step follows because of the fact that the algorithm simply implements the formula

$$m_{ij}^{(k)} = \min\left(m_{ij}^{(k-1)}, m_{ik}^{(k-1)} + m_{kj}^{(k-1)}\right).$$

Thus, the proof is complete. □

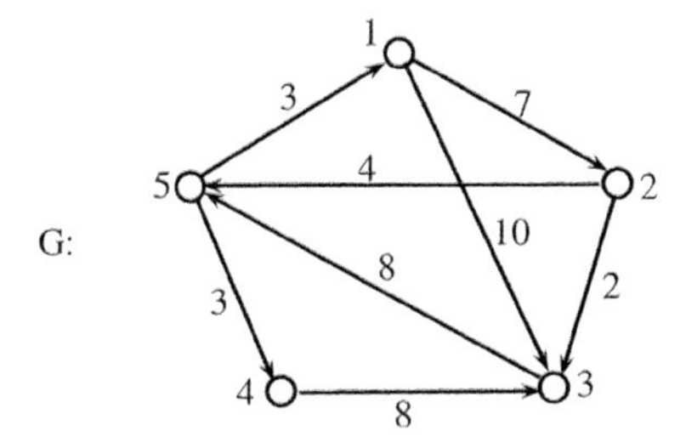

FIGURE 1.19: A digraph on which Floyd's algorithm is executed.

Let us execute Floyd's algorithm on a given graph.

Example 1.3

Execution of Floyd's algorithm:
Consider the graph of Figure 1.19.

The graph is represented by the weighted adjacency matrix W where

$$W = \begin{array}{c} \\ 1 \\ 2 \\ 3 \\ 4 \\ 5 \end{array} \begin{array}{c} \begin{array}{ccccc} 1 & 2 & 3 & 4 & 5 \end{array} \\ \begin{pmatrix} 0 & 7 & 10 & \infty & \infty \\ \infty & 0 & 2 & \infty & 4 \\ \infty & \infty & 0 & \infty & 8 \\ \infty & \infty & 8 & 0 & \infty \\ 3 & \infty & \infty & 3 & 0 \end{pmatrix} \end{array}.$$

The indices in the entry represent the last iteration number that changed the entry. These indices are useful in finding the paths realizing the minimum costs. Initially, $M_0 = W$. Can we use the vertex 1 as an intermediate vertex to reduce the entries of the matrix M_1?
Iteration 1:

$$M_1 = \begin{bmatrix} 0 & 7 & 3 & \infty & \infty \\ \infty & 0 & 2 & \infty & 4 \\ \infty & \infty & 0 & \infty & 8 \\ \infty & \infty & 8 & 0 & \infty \\ 3 & 10_1 & 13_1 & 3 & 0 \end{bmatrix}.$$

Can we use some vertices of the set $\{1, 2\}$ as intermediate vertices to reduce the entries of the matrix M_1?
Iteration 2:

$$M_2 = \begin{bmatrix} 0 & 7 & 9_2 & \infty & 11_2 \\ \infty & 0 & 2 & \infty & 4 \\ \infty & \infty & 0 & \infty & 8 \\ \infty & \infty & 8 & 0 & \infty \\ 3 & 10_1 & 12_2 & 3 & 0 \end{bmatrix}.$$

Can we use some vertices of the set $\{1, 2, 3\}$ as intermediate vertices to reduce the entries of the matrix M_2?
Iteration 3:

$$M_3 = \begin{bmatrix} 0 & 7 & 9_2 & \infty & 11_2 \\ \infty & 0 & 2 & \infty & 4 \\ \infty & \infty & 0 & \infty & 8 \\ \infty & \infty & 8 & 0 & 16_3 \\ 3 & 10_1 & 12_2 & 3 & 0 \end{bmatrix}.$$

Can we use some vertices of the set $\{1, 2, 3, 4\}$ as intermediate vertices to reduce the entries of the matrix M_3?
Iteration 4:

$$M_4 = \begin{bmatrix} 0 & 7 & 9_2 & \infty & 11_2 \\ \infty & 0 & 2 & \infty & 4 \\ \infty & \infty & 0 & \infty & 8 \\ \infty & \infty & 8 & 0 & 16_3 \\ 3 & 10_1 & 11_4 & 3 & 0 \end{bmatrix}.$$

Can we use some vertices of the set $\{1, 2, 3, 4, 5\}$ as intermediate vertices to reduce the entries of the matrix M_4?
Iteration 5:

$$M_5 = \begin{bmatrix} 0 & 7 & 9_2 & 14_5 & 11_2 \\ 7_5 & 0 & 2 & 7_5 & 4 \\ 11_5 & 18_5 & 0 & 11_5 & 8 \\ 19_5 & 26_5 & 8 & 0 & 16_3 \\ 3 & 10_1 & 11_4 & 3 & 0 \end{bmatrix}.$$

The final result is made available by the entries of the matrix M_5.

Recovering paths in Floyd's algorithm:
Let us again refer to Example 1.3. As we have already remarked, the subscripts in the entries of the final matrix represent the *last* iteration number which has modified the corresponding entry. We shall denote the entries corresponding to the subscripts of the final matrix M_5 by INTER (for intermediate vertex) where

$$\text{INTER}[i, j] = \begin{cases} 0 \text{ if there is no subscript in the entry (i,j) of } M_5 \\ \text{subscript of the entry (i,j) of } M_5 \end{cases}$$

With this definition, the matrix INTER of the graph Figure 1.19 is given by

$$\text{INTER} = \begin{bmatrix} 0 & 0 & 2 & 5 & 2 \\ 5 & 0 & 0 & 5 & 0 \\ 5 & 5 & 0 & 5 & 0 \\ 5 & 5 & 0 & 0 & 3 \\ 0 & 1 & 4 & 0 & 0 \end{bmatrix}$$

This matrix INTER will be used to recover paths realizing the cheapest costs. Note that if an (i,j) entry of the matrix INTER is 0, then the path realizing the minimum cost from the vertex i to the vertex j is just the arc (i,j), with no intermediate vertices.

For example, the cost of a cheapest path from the vertex 4 to the vertex 2 is the (4,2) entry of the matrix M_5, which is 26. What are the intermediate vertices of a path from 4 to 2 realizing this cost 26? To find the intermediate vertices, we read first the (4,2) entry of the matrix INTER which is 5. This means that 5 is an intermediate vertex between 4 and 2. Now we read the entry (4,5) of the matrix INTER to find an intermediate vertex between 4 and 5 which is 3. Hence, 3 is an intermediate vertex between 4 and 5. Next, we read the entry (4,3) which 0. This means that there is no intermediate vertex between 4 and 3. We read the entry (3,5) which is again 0. We read the entry (5,2) which is 1. This means that 1 is an intermediate vertex between 5 and 2. Now the entries (5,1) and (1,2) are 0 meaning that no intermediate vertices are found between 5,1 and 1,2.

Hence, the intermediate vertices of a cheapest path from 4 to 2 are 3, 5, 1 and the cheapest cost path is (4, 3, 5, 1, 2).

Let us now write an algorithm to find the matrix INTER. To do this, we have only to initialize the matrix INTER and update the entries of INTER at the appropriate point in Floyd's algorithm.

```
procedure Floyd( var W, M, INTER : Matrix);
 (* Floyd takes as input parameter a matrix W which is the
 weighted adjacency matrix of a directed weighted graph G. The
 procedure returns as output parameters the matrix M giving the
 cost of a shortest directed path of the graph G and the matrix
 INTER with its (i, j) entry, the last iteration number k which
 has changed the (i, j) entry of M *)
  var i, j, k : integer; (* i, j, k for the loops *)
  begin (* Floyd *)
      (* Initialization of M = W and INTER = 0 *)
      for i := 1 to n do
          for j := 1 to n do
          begin
              M[i, j] := W[i, j];
              INTER[i, j] := 0;
          end;
      (* iteration *)
      for k := 1 to n do
      (* compute the matrix M sub k *)
          for i := 1 to n do
              for j := 1 to n do
                  if M[i, k] + M[k, j] < M[i, j]
                  then
```

```
                    begin(* we use k to reduce the cost*)
                        M[i, j] := M[i, k] + M[k, j];
                        INTER[i, j] := k;(* update INTER[i, j]*)
                    end;
    end;(* Floyd *)
```

We shall now write a *recursive* algorithm to print the intermediate vertices of a shortest path from the vertex i to the vertex j using the matrix INTER.

```
procedure interpath(i, j : integer);
    (* print_path writes the intermediate vertices of a shortest path
    from the vertex i to the vertex j using the matrix INTER *)
    var k : integer;
begin
    k := INTER[i, j];
    if k = 0 then
        return;
    interpath(i, k);(* recursive call writing intermediate vertices
                       between i and k *)
    write(k, ' ');
    interpath(k, j);(* recursive call writing intermediate vertices
                       between k and j *)
end;
```

The following example illustrates the call of procedure "interpath(4, 2)" on the graph Figure 1.19 with the help of the matrix INTER.

Example 1.4

Hand simulation of the procedure call "interpath(4, 2)": (see the tree of Figure 1.20).

The call interpath(4, 2) results in two calls: interpath(4, 5) and interpath(5, 2).

The call interpath(4, 5) leads to two calls interpath(4, 3) and interpath(3, 5). The calls interpath(4, 3) and interpath(3, 5) lead to *no* calls, since INTER[4, 3] = INTER[3, 5] = 0. This results in printing of the vertices : 3, 5

Now interpath(5, 2) calls interpath(5, 1) and interpath(1, 2). The calls interpath(5, 1) and interpath(1, 2) lead to *no* calls, since INTER[5,1] = INTER[1, 2] = 0. This leads to printing of the vertex: 1.

Hence, the intermediate vertices of a shortest path from the vertex 4 to the vertex 2 are 3, 5, 1.

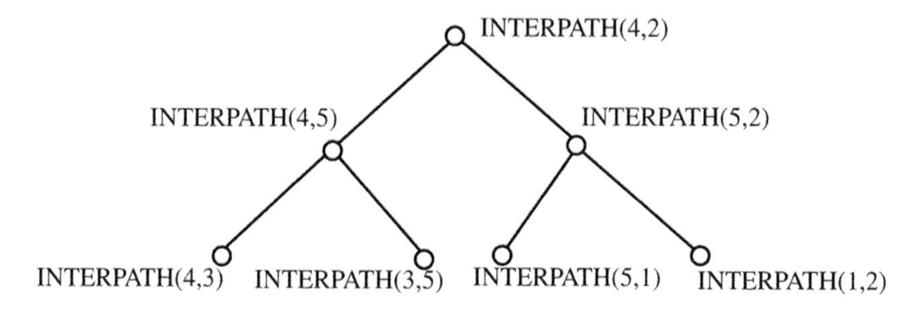

FIGURE 1.20: A tree illustrating different recursive call interpath(4,2).

1.5.1 An application of Floyd's algorithm

Radius, diameter, and center of a graph:
Consider a directed graph $G = (X, U)$ with a non-negative weight associated with each arc. The *distance* between two vertices x and y denoted by $d(x, y)$ is the *cost* of a shortest directed path from the vertex x to the vertex y. The cost of a directed path is the sum of the cost of its arcs. If there is no path from x to y, we set $d(x, y) = \infty$.

The *eccentricity* $e(v)$ of a vertex v is the *maximum* of the lengths of the paths from the vertex v to any vertex of the graph. Symbolically, $e(v) = \max(d(v, w) \mid w \in X)$.

The *radius* of the graph G, denoted by $r(G)$, is the *minimum* taken over all eccentricities of the vertices of G, that is, $r(G) = \min(e(v) \mid v \in X)$. The *diameter* $d(G)$ of G is the *maximum* of all eccentricities of the vertices, that is, $d(G) = \max(e(v) \mid v \in X)$. In other words, the diameter of a graph G is the maximum of the distances between any two vertices of G.

Finally, the *center* of the graph G is the set of all vertices whose eccentricities coincide with the radius of the graph G. These notions arise from the telecommunication problems. The graph of Figure 1.21 illustrates these notions.

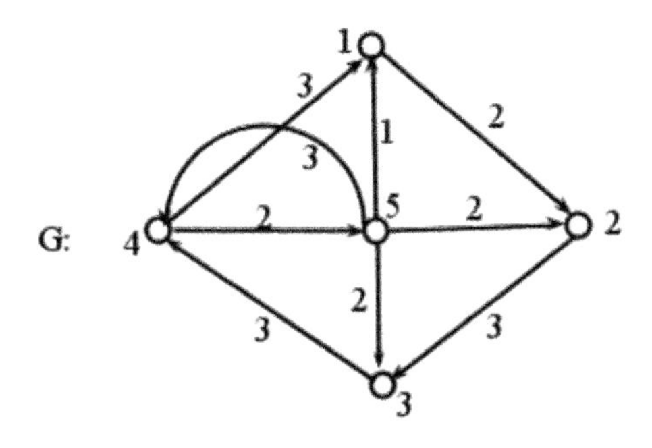

FIGURE 1.21: Graph illustrating eccentricities, radius and diameter.

TABLE 1.7: Eccentricity table of graph of Figure 1.21

	1	2	3	4	5
e	5	9	7	4	3

The eccentricities of different vertices of the graph of Figure 1.21 are given by the Table 1.7

Example 1.5:

Radius, Diameter and Center of graph of Figure 1.21.
The graph is represented by its weighted adjacency matrix W where

$$W = \begin{array}{c} \\ 1 \\ 2 \\ 3 \\ 4 \\ 5 \end{array} \begin{array}{c} \begin{array}{ccccc} 1 & 2 & 3 & 4 & 5 \end{array} \\ \begin{pmatrix} 0 & 2 & \infty & 3 & \infty \\ \infty & 0 & 3 & \infty & \infty \\ \infty & \infty & 0 & 3 & \infty \\ \infty & \infty & \infty & 0 & 2 \\ 1 & 2 & 2 & 3 & 0 \end{pmatrix} \end{array}$$

Then, we apply Floyd's algorithm to the matrix to obtain the matrix M where

$$M = \begin{array}{c} \\ 1 \\ 2 \\ 3 \\ 4 \\ 5 \end{array} \begin{array}{c} \begin{array}{ccccc} 1 & 2 & 3 & 4 & 5 \end{array} \\ \begin{pmatrix} 0 & 2 & 5 & 5 & \infty \\ 9 & 0 & 3 & 6 & 8 \\ 6 & 7 & 0 & 3 & 5 \\ 3 & 4 & 4 & 0 & 2 \\ 1 & 2 & 2 & 3 & 0 \end{pmatrix} \end{array}$$

Then, the eccentricity $e(i)$ of the vertex i is simply the *maximum* of the coefficients of the ith line of the matrix M. Finally, the radius of the graph is the *minimum* of the eccentricities and the diameter is the *maximum* of the eccentricities. In our example, the radius of the graph is 3 and the diameter is 9. The center of the graph is the set $\{5\}$.

Let us now write an algorithm to the radius, diameter and the center of a graph.

```
Algorithm to find the radius, diameter and the center of a given
graph G:
Floyd(W, M); (* call to Floyd's algorithm *)
(* now  M, the cheapest cost matrix of G *)
for i:= 1 to n do
begin
```

```
        (* find e[i] *)
        e[i] := M[i, 1];(* initialization of e[i] *)
        (* iteration *)
        for j := 2 to n do
            if e[i] <  M[i,j] then
                e[i] := M[i,j];
    end;
    (* We have found the eccentricity table e *)
    (* Initializing r and d *)
    r := e[1]; d := e[1];
    (* iteration *)
    for i:= 2 to n do
    begin
        if r > e[i] then
            r := e[i];
        if d < e[i] then
            d := e[i];
    end;
    (* finding the center C*)
    (* initialization of C*)
    for i: = 1 to n do
        C[i] := 0; (* C is empty *)
    (* iteration *)
    for i := 1 to n do
        if r = e[i] then
            C[i] := 1;
    (* print center *)
    (*C[i] = 1 if i in C, 0 otherwise *)
    for i := 1 to n do
        if C[i] =1 then write (i, ' ');
```

The complexity of the algorithm for finding the radius, diameter, and the center:
A call to Floyd requires $O(n^3)$ steps. We have to scan every entry of the matrix M to find the radius and diameter. This needs $O(n^2)$ steps. Finally finding the center requires only $O(n)$ steps as it scans the array e.

Hence the complexity is $O(n^3) + O(n^2) + O(n) = O(n^3)$.

1.6 Transitive Closure of a Directed Graph

Sometimes, we may be interested in knowing if there is a directed path from the vertex i to the vertex j of a directed graph. Transitive closure of the graph helps us answer this question directly.

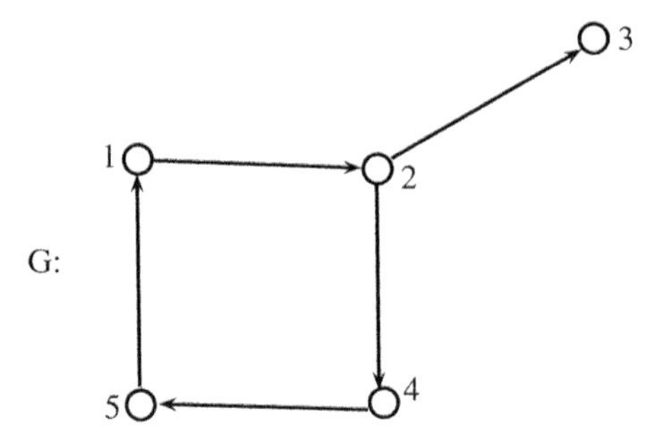

FIGURE 1.22: A graph to illustrate transitive closure.

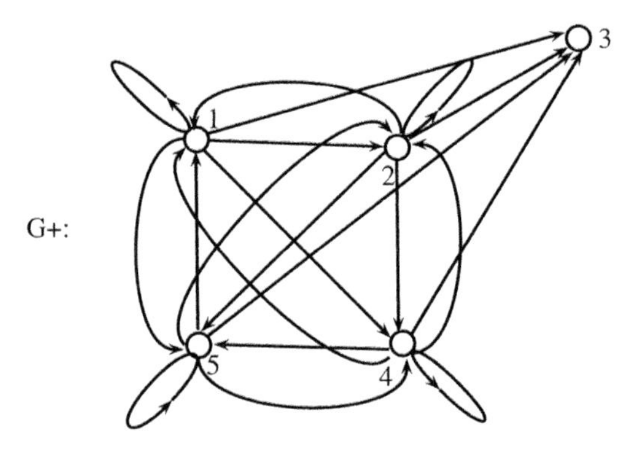

FIGURE 1.23: The transitive closure of the graph of Figure 1.22.

Consider a directed graph $G = (X, U)$. The transitive closure of the graph G denoted by G^+ is a graph whose vertex set is the same as that of the graph G, that is, X, and two vertices i and j (not necessarily distinct) of the transitive closure are joined by an arc (i, j) if there is a directed path of length at least *one* in the original graph G.

Example 1.6

Transtive closure of a graph: Consider the graph G of Figure 1.22.

This graph is equivalently represented by its adjacency matrix M where

$$M = \begin{array}{c} \\ 1 \\ 2 \\ 3 \\ 4 \\ 5 \end{array} \begin{array}{c} \begin{array}{ccccc} 1 & 2 & 3 & 4 & 5 \end{array} \\ \begin{pmatrix} 0 & 1 & 0 & 0 & 0 \\ 0 & 0 & 1 & 1 & 0 \\ 0 & 0 & 0 & 0 & 0 \\ 0 & 1 & 0 & 0 & 1 \\ 1 & 0 & 0 & 0 & 0 \end{pmatrix} \end{array}$$

The transitive closure of the graph of Figure 1.22 is drawn in Figure 1.23.

The adjacency matrix of the transitive closure of the graph G is the matrix TM where

$$TM = \begin{array}{c} \\ 1 \\ 2 \\ 3 \\ 4 \\ 5 \end{array} \begin{array}{c} \begin{array}{ccccc} 1 & 2 & 3 & 4 & 5 \end{array} \\ \begin{pmatrix} 1 & 1 & 1 & 1 & 1 \\ 1 & 1 & 1 & 1 & 1 \\ 0 & 0 & 0 & 0 & 0 \\ 1 & 1 & 1 & 1 & 1 \\ 1 & 1 & 1 & 1 & 1 \end{pmatrix} \end{array}$$

For example, the $(4,4)$ entry of TM is 1 because there is a directed path from the vertex 4 to itself, namely $(4,5,1,2,4)$, the $(5,3)$ entry of TM is one because of the directed path $(5,1,2,3)$, whereas the $(3,1)$ entry of TM stays as 0 because there is no directed path in the graph G from the vertex 3 to the vertex 1.

1.7 An $O(n^3)$ Transitive Closure Algorithm Due to Warshall

Input: A directed graph $G = (X, U)$ with $X = \{1, 2, \ldots, n\}$. The graph is represented by its adjacency matrix M.

Output: The adjacency matrix TM of the transitive closure of the graph G.

Algorithm: This algorithm is invented by Warshall. This algorithm is a special case of Floyd's algorithm but it predates Floyd's algorithm.

The algorithm constructs a sequence of matrices $M_0, M_1, \ldots, M_n$ where the matrix $M_k = m_{ij}^{(k)}$ for $1 \leq k \leq n$ is defined as follows:

$$m_{ij} = \begin{cases} 1 & \text{if there is a path from } i \text{ to } j \text{ not passing through any} \\ & \quad \text{intermediate vertex} > \text{k} \\ 0 & \text{otherwise} \end{cases}$$

Interpretation of $M_0 = (m_{ij}^{(0)})$:

By the above definition of the entries of the matrix M_0, $m_{ij}^{(0)}$ is equal to 1 if there is a directed path from i to j not passing through any intermediate vertex > 0 and is equal to 0, otherwise. Since the smallest numbered vertex is 1, this means that $m_{ij}^{(0)}$ is simply 1 if there is an arc from i to j, zero otherwise. This is the definition of the adjacency matrix of the graph G. Hence $M_0 = M$. This equation is used as initialization in Warshall's algorithm.

Interpretation of $M_n = (m_{ij}^{(n)})$:

By definition, $m_{ij}^{(n)}$ is equal to 1 if there is a directed path from i to j not going through any intermediate vertex $> n$ and is equal to 0, otherwise. Since no vertex of the graph is $> n$, this means that $m_{ij}^{(n)}$ is equal to 1 if there is a directed path from i to j in G and is equal to 0, otherwise.

This means that M_n is the desired output matrix TM. The subscript k in M_k represents the iteration number.

Induction leap:

How to find the matrix M_k given the matrix M_{k-1}?

We are given the n^2 entries $m_{ij}^{(k-1)}$ of the matrix M_{k-1}. We have to express $m_{ij}^{(k)}$ in terms of $m_{ik}^{(k-1)}$ and $m_{kj}^{(k-1)}$. This is because during the construction of M_k we have the right to use the vertex k as an intermediate vertex which is *not* the case with the matrix M_{k-1}.

We must not disturb the entries "one" of the matrix M_{k-1}, since a directed path from i to j not going through any intermediate vertex $> k-1$ is also a directed path not passing through any intermediate vertex $> k$. We are only interested in a possible reassignment of zero coefficients of the matrix M_{k-1} into 1. Since the concatenation of a path from i to k not going through any intermediate vertex $> k-1$ and a path from k to j not passing through any intermediate vertex $> k-1$ is a path from the vertex i to the vertex j not passing through any intermediate vertex $> k$, we have the following formula for $m_{ij}^{(k)}$.

$$m_{ij}^{(k)} = \max\left(m_{ij}^{(k-1)}, m_{ik}^{(k-1)} \times m_{kj}^{(k-1)}\right), \quad \text{for } 1 \leq i,\ j \leq n.$$

By setting $j = k$ in the above formula, we have $m_{ik}^{(k)} = \max(m_{ik}^{(k-1)}, m_{ik}^{(k-1)} \times m_{kk}^{(k-1)}) = m_{ik}^{(k-1)}$ and similarly by setting $i = k$ we get, $m_{kj}^{(k)} = m_{kj}^{(k-1)}$.

Hence, we can carry out our computation with only one copy of the matrix M. Let us describe the above formula in words:
There is a directed path not going through any intermediate vertex $> k$ if:

1. There is already a path from the vertex i to j not going through any intermediate vertex $> k-1$ (this is equivalent to saying: we must not modify the entries "one" of the matrix M_{k-1}) or

2. There is a path from i to k not passing through any intermediate vertex $> k-1$ and a path from k to j not going through any intermediate vertex $> k-1$.

We shall now write Warshall's algorithm. We use the type "Matrix" to represent $n \times n$ matrices of 0 and 1.

```
procedure Warshall( var M, TM : Matrix);
(* The input parameter is M, the adjacency matrix of the
```

```
given graph G. The output parameter is TM, the adjacency matrix
of the transitive closure of G. *)
var i, j, k : integer; (* i, j, k are for loops *)
begin
(* initialization of TM = M *)
    for i := 1 to n do
        for j := 1 to n do
            TM[i, j] := M[i, j];
(* iteration *)
    for k := 1 to n do (* k-th pass over TM *)
        for i : = 1 to n do
            for j := 1 to n do
                if TM[i, j] = 0 then TM[i, j] := TM[i, k] *
                 TM[k, j];
end;(* Warshall *)
```

The complexity of Warshall's algorithm:
The initialization of the matrix requires $O(n^2)$ steps, as there are two nested loops involved. The iteration step takes $O(n^3)$ steps, as there are three nested loops. Hence, the complexity of Warshall's algorithm is $O(n^2) + O(n^3) = O(n^3)$.

Let us execute Washall's algorithm on a given graph.

Example 1.7

Execution of Warshall's algorithm: Consider the graph which is an elementary circuit of length 4 with vertex set $\{1, 2, 3, 4\}$ and the arc set $\{(1, 2), (2, 3), (3, 4), (4, 1)\}$. The adjacency matrix of the above graph is

$$M = \begin{array}{c} \\ 1 \\ 2 \\ 3 \\ 4 \end{array} \begin{array}{c} \begin{array}{cccc} 1 & 2 & 3 & 4 \end{array} \\ \begin{pmatrix} 0 & 1 & 0 & 0 \\ 0 & 0 & 1 & 0 \\ 0 & 0 & 0 & 1 \\ 1 & 0 & 0 & 0 \end{pmatrix} \end{array}.$$

Initialization of TM : $TM = M$. Iteration $k = 1$. We use 1 as an intermediate vertex.

$$TM = \begin{array}{c} \\ 1 \\ 2 \\ 3 \\ 4 \end{array} \begin{array}{c} \begin{array}{cccc} 1 & 2 & 3 & 4 \end{array} \\ \begin{pmatrix} 0 & 1 & 0 & 0 \\ 0 & 0 & 1 & 0 \\ 0 & 0 & 0 & 1 \\ 1 & 1 & 0 & 0 \end{pmatrix} \end{array}$$

Iteration $k = 2$. We may use 1/2 as intermediate vertices.

$$TM = \begin{array}{c} \\ 1 \\ 2 \\ 3 \\ 4 \end{array} \begin{array}{c} \begin{array}{cccc} 1 & 2 & 3 & 4 \end{array} \\ \begin{pmatrix} 0 & 1 & 1 & 0 \\ 0 & 0 & 1 & 0 \\ 0 & 0 & 0 & 1 \\ 1 & 1 & 1 & 0 \end{pmatrix} \end{array}$$

Iteration $k = 3$. We may use 1/2/3 as intermediate vertices.

$$TM = \begin{array}{c} \\ 1 \\ 2 \\ 3 \\ 4 \end{array} \begin{array}{c} \begin{array}{cccc} 1 & 2 & 3 & 4 \end{array} \\ \begin{pmatrix} 0 & 1 & 1 & 1 \\ 0 & 0 & 1 & 1 \\ 0 & 0 & 0 & 1 \\ 1 & 1 & 1 & 0 \end{pmatrix} \end{array}$$

Iteration $k = 4$. We may use *any* vertices as intermediate vertices.

$$TM = \begin{array}{c} \\ 1 \\ 2 \\ 3 \\ 4 \end{array} \begin{array}{c} \begin{array}{cccc} 1 & 2 & 3 & 4 \end{array} \\ \begin{pmatrix} 1 & 1 & 1 & 1 \\ 1 & 1 & 1 & 1 \\ 1 & 1 & 1 & 1 \\ 1 & 1 & 1 & 1 \end{pmatrix} \end{array}$$

1.8 Navigation in Graphs

We are interested in traversing a multigraph in a systematic manner, that is, "visiting" the vertices and edges of a multigraph. The vertices are normally variables of type "struct" in the language C or the type "record" in Pascal. These vertices contain information and we are interested in reaching the information stored in the vertices to perform some kind of operation. For example, we may want to update the information in each vertex or simply write the contents of a particular field of each vertex. We shall study two important graph traversal methods: The *depth-first search* and *breadth-first search.*

Depth-first search:
The depth-first search of a graph can be viewed as a generalization of preorder traversal of a tree. If the graph does not possess a cycle, then the depth-first search (dfs) of a graph coincides with the preorder traversal of a tree.

Let us first explain the dfs in an intuitive manner. Consider a connected multigraph G. We start with a vertex, called the starting vertex, which will become the root of a spanning tree to be generated by dfs. From the starting

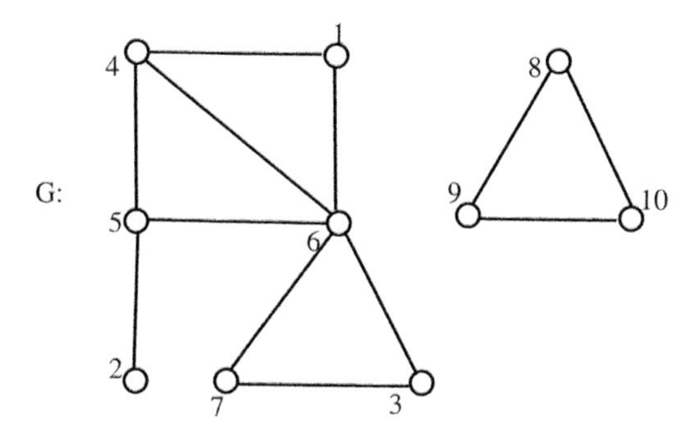

FIGURE 1.24: A simple graph to illustrate depth-first search.

vertex, we construct a path as long as possible. Then, we backtrack from the current vertex, that is, we climb up the tree which is a path at present. The climbing up continues until we find a vertex at which a side-path, that is, a new branch, may be constructed to a "new" vertex. If the graph is not connected, then we have to choose a new vertex of a new connected component as a starting vertex and perform the dfs etc., until all vertices of the graph are in some tree induced by dfs.

The dfs can be useful as a model around which one can design various efficient graph algorithms like finding: connected components of a graph, biconnected components, strongly connected components, finding triply connected components, testing planarity, etc.

Dfs may be used to perform "exhaustive search" to solve *TSP*, permutation generation, Knight's tour in a Chess board, etc.

Let us illustrate the dfs with an example.

Example 1.8

dfs: Consider the following simple graph G of Figure 1.24 consisting of two connected components.

The graph G is represented by the adjacency lists. In each adjacency list, the vertices are listed in increasing order. We use an array "mark" the vertices processed by the dfs. Initially, for each vertex i, we have $mark[i] = 0$. We set $mark[i] \leftarrow 1$ as soon as the vertex i is reached during the dfs. (0 means "unprocessed" or "unvisited" and 1 means "processed" or "visited.")

$L(1) = (4,6); L(2) = (5); L(3) = (6,7); L(4) = (1,5,6); L(5) = (2,4,6); L(6) = (1,3,4,5,7); L(8) = (9,10), L(9) = (8,10); L(10) = (8,9)$. Now let us take a vertex, say, the vertex 1, as the starting vertex and set $mark[1] \leftarrow 1$. The adjacency list of 1, $L(1)$ is scanned to find a vertex marked 0. The very first vertex 4 in $L(1)$ is chosen and we set $mark[4] \leftarrow 1$. The traversing of the list $L(1)$ is temporarily suspended and we start scanning the list $L(4)$ to find a vertex "unprocessed." The vertex 5 is found and we set $mark[5] \leftarrow 1$. Then, again the scanning of the list $L(4)$ is temporarily suspended and we start scanning the list $L(5)$ and we find the vertex 2 and we assign $mark[2] \leftarrow 1$. We now scan the list $L(2)$ in search of a unprocessed vertex. No new vertices are found by scanning $L(2)$ and we climb up the tree to find the vertex 5 which is the "dad" of vertex 2, that is, $dad[2] \leftarrow 5$. We now restart the scanning of the list $L(5)$, which was suspended previously. We find the vertex 6, then the vertex 3 and the vertex 7. Now all the vertices of the connected component containing the vertex 1 have been "processed." We now consider the second connected component and "visit" each vertex according to dfs. (See the forest F of Figure 1.25. The forest F is defined by the solid edges.)

The edges of the forest F are drawn as continuous edges, and the edges of the graph G *not* in the forest are depicted as dotted edges. The children/sons/daughters of a vertex are drawn from left to right. A vertex j is called a *descendant* of a vertex i, if there is a downward path from the vertex i to the vertex j in a tree of the forest generated by dfs. In this case, we also say that the vertex i is an *ancestor* of the vertex j. We consider a vertex x as an ancestor and descendant of itself. A descendant or ancestor of a vertex x other than itself is called a proper descendant or proper ancestor of x.

We associate to each vertex i of the graph, called the depth-first search number(dfsn), where $dfsn[i] = j$ if and only if i is the jth vertex visited during the dfs. Table 1.8 gives the dfsn of each vertex of the above forest F.

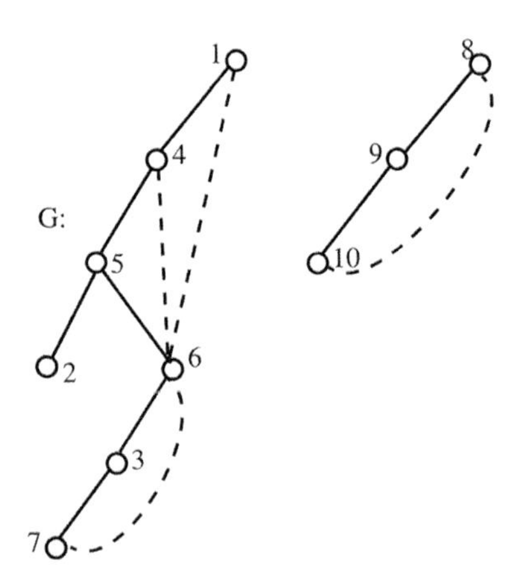

FIGURE 1.25: Illustration of depth-first search algorithm.

TABLE 1.8: Depth-first search number table of graph G of Figure 1.25

	1	2	3	4	5	6	7	8	9	10
dfsn	1	4	6	2	3	5	7	8	9	10

We now write the dfs algorithm in pseudo-code.

Input: A multigraph $G = (X, E)$ represented by n adjacency lists where the vertex set $X = \{1, 2, \ldots, n\}$.

Output: A partition of the edge set E into a set of edges belonging to the spanning tree/forest and a set of dotted edges, also called the *back edges*. The algorithm also assigns to each vertex a dfs number.

Algorithm: The algorithm is described *recursively*. We use an array "mark" of vertices where $mark[i] = 0$ if the vertex i is not yet visited by dfs and $mark[i] = 1$ if the vertex i is visited by the dfs. $dfsn[i] = j$ if i is the jth vertex visited during the search.

```
dfs algorithm:
(* initialization *)
for i := 1 to n do
    mark[i] := 0;
F : = empty;(* F, the forest under construction*)
counter : = 1; (* counter is used to assign the dfs number *)
(* end of initialization *)
procedure dfs( i : integer);
(* dfs visits each vertex of a connected component of G
containing i using depth-first search *)
var j : integer;(* vertex j scans L(i) *)
begin
    mark[i] := 1;
    dfsn[i] := counter;
    counter := counter + 1;
    (* scan L(i) *)
    for each vertex j in L(i) do
        if mark[j] = 0 then
        begin
            add the edge ij to F;
            dfs(j); (* recursive call *)
        end;
(* iteration *)
for i := 1 to n do
if mark[i] = 0 then
    dfs(i); (* invoking dfs *)
```

The edges of the graph G which are not in the spanning forest induced by dfs are the back edges. If the graph is connected, then the dfs induces a spanning tree, called the *dfs spanning tree.*

Let xy be an edge of the *dfs spanning forest*, that is, forest induced by dfs, with y, a son of x. Then, dfs(x) called dfs(y) *directly* during the dfs. In other words, the vertex y is marked zero, when we scan the list $L(x)$, the list of vertices adjacent to x.

We distinguish two types of calls:

1. Direct calls

2. Indirect calls

dfs(x) called directly dfs(y) if and only if the edge xy is an edge of the spanning forest constructed by dfs, that is, x is the father/mother of y in the spanning forest.

Note also that when we are at vertex x, scanning the list $L(x)$, and z in $L(x)$, that is, xz is an edge (and hence, x is also in the list $L(z)$), with $mark[z] = 1$, we cannot simply put the edge xz in the set of back edges, because the vertex z can be the father/parent of the vertex x.

If xy is a back edge, then neither dfs(x) nor dfs(y) called the other directly, but one called the other indirectly, that is, dfs(y) called dfs(z), which in turn called dfs(x), and hence y is an ancestor of x.

Remark 1.2. *The dfs spanning forest need not be unique. A dfs forest generated by the search procedure dfs depends on the starting vertices, that is, the roots of each tree in the spanning forest and the order of the vertices in each list $L(i)$.*

The complexity of the dfs:
Intuitively, during the traversal of the graph, each vertex is marked only once and each edge xy is traversed exactly twice – once from x and once from y. Hence the complexity of the dfs algorithm is $O(n + m) = O(\max(n, m))$ by the sum rule of the complexity. Here, m denotes the number of edges of the graph G.

In a formal manner, the initialization step takes $O(n)$ time. The iteration part is executed n times.

The number of steps in executing the call $dfs(i)$, without taking into account the recursive call to itself is the number of vertices in the list $L(i)$, which is equal to the degree of the vertex i. Thus, one call to dfs costs $O(d(i))$, where $d(i)$ is the degree of the vertex i. By Euler's theorem, the sum of the degrees of the vertices of a graph is twice the number of edges. Hence, the total costs of the "for" loop of the "procedure dfs" is $\sum_{i=1}^{n} d(i) = 2m$. Note that $dfs(i)$ is called only once, since as soon as we touch the vertex i, we mark i "visited" and we never call dfs on a vertex with its "mark" set to 1. Hence, the total number of steps required is $O(2n + 2m) = O(\max(n, m))$.

Property of a back edge:

Property 1.2. *Consider a connected graph and let T and B be the set of tree edges and back edges obtained by dfs. Then, each back edge xy joins an ancestor and a descendant in F. (A vertex is considered as an ancestor and a descendant of itself.)*

Proof. We may assume without loss of generality that the vertex x is "visited" before the vertex y. This means that the vertex y is marked "unvisited" when we reach the first time the vertex x. Since xy is an edge of the graph, we have the vertex y in the list $L(x)$. But then the call $dfs(x)$ would not be complete until all the vertices of $L(x)$ are "visited." Hence, the vertex y is marked "visited" between the initiation of the call dfs(x) and the completion of the call $dfs(x)$. All vertices between the initiation of dfs(x) and the end of the call dfs(x) will become descendants of the vertex x in the dfs spanning tree constructed by the search. In particular, the vertex y is a *proper* descendant of x which is *not* a son of x in the dfs spanning tree. Hence the property. □

Relation "x is to the left of y":
Let us note that the children of a vertex are in the left-to-right order in the dfs spanning forest and the different trees of the spanning forest are also drawn in the left-to-right order. Consider a dfs spanning tree T of a connected graph and consider two vertices x and y which are not related by the relation descendant-ancestor. Such vertices can be put in relation called, "is to the left of." We say that the vertex x lies to the left of the vertex y or y lies to the right of x if the following condition is satisfied:

We draw the unique path from the root to the vertex y. Then, all vertices to the "left" of this path are treated as to the left of the vertex x and all vertices to the "right" of this path are considered as to the right of the vertex y.

If the vertex i is a proper ancestor of the vertex j in a dfs spanning tree, then $dfsn[i] < dfsn[j]$ and if i is to the left of j, then also $dfsn[i] < dfsn[j]$.

Remark 1.3. *The order in which the different calls terminate in the dfs is the* postorder traversal *of the dfs tree.*

1.9 Applications of Depth-First Search

1.9.1 Application 1: Finding connected components

Input: A multigraph $G = (X, E)$ with $X = \{1, 2, \ldots, n\}$.

Output: List of vertices of each connected component and the number of connected components.

Algorithm: Of course, the connected components of a graph can be easily found by "eyeballing" if the graph is drawn in the geometric representation and if it is sufficiently small. On the other hand, if we are given a complex electrical circuit, the question of whether the circuit is connected is not at all obvious "visually."

Our problem is to instruct a computer so that it writes different connected components of a multigraph. The algorithm can be easily written using the dfs. We use an integer variable "nc" to calculate the number of connected components of G. Then, we add the following statement in the dfs algorithm.

After the assignment $mark[i] := 0$ at the beginning of the "procedure dfs," add the following print statement: write(' ',i); which means that as soon as we "touch" or "visit" a vertex i, we immediately print the corresponding vertex. Add the statement nc := 0; in the initialization part of the algorithm.

Finally we rewrite the "iteration part" of the dfs algorithm as follows:

```
for i := 1 to n do
     if mark[i] = 0 then
     begin
          nc := nc + 1;(* update nc *)
          dfs(i); (* call to dfs. i is a root of a sub-tree
                   in dfs spanning forest which is under
                   construction *)
                   writeln;(* print a new line character to separate
                   components*)
     end;
writeln(nc);(* print the number of components *)
```

Justification of the algorithm: Each call dfs(i) in the "iteration part" corresponds to the root i of a subtree in the dfs spanning forest. Once the call dfs(i) is initiated, all vertices connected to the vertex i by a "downward" path in dfs forest are marked with 1. Hence, the validity of the algorithm.

The complexity of finding connected components:
The time needed to find the connected components is the same as that of dfs, since the instructions added are of time complexity $O(1)$. (Recall that $O(1)$ represents a constant.)

1.9.2 Application 2: Testing acyclic graph

Input: A multigraph $G = (X, E)$ with vertex set $\{1, 2, \ldots, n\}$.

Output: "Yes" if the graph is acyclic, that is, if G has no cycles;
"No" otherwise. In this case, print the cycles the dfs encounters.

Algorithm: We shall use the following fact: Every back edge we meet during the dfs search defines a cycle and conversely if a graph possesses a cycle, then we must meet a back edge in the course of our dfs search. Why? This is because of Property 1.2. Note that a back edge ij is encountered the first time in the course of the dfs search if we are at vertex i and $dfsn[i] \geq dfsn[j]$. Of course, the same back edge $ij = ji$ is traversed the second time when we are again at vertex j. A loop is considered as a back edge. Let us now write the algorithm. We shall use an array "dad" where $dad[j] = i$ if the vertex i is the father/parent of the vertex j in the spanning dfs forest.

```
dfs algorithm for cycles:
(* initialization *)
for i := 1 to n do
    mark[i] := 0;
counter := 1;(* counter to assign dfs number *)
acyclic : = 0; (* acyclic is 0 if the graph has no cycles.
                  acyclic is 1, otherwise *)
(* end of initialization *)
procedure dfs(i:integer);
(* dfs visits each vertex of a connected component of G
containing i using depth-first search *)
var j, k : integer;(* vertex j scans L(i) *)
begin
    mark[i] := 1;
    dfsn[i] := counter;
    counter := counter + 1;
    (* scan L(i) *)
    for each vertex j in L(i) do
        if mark[j] = 0 then
        begin
            dad[j] := i;(* ij is an edge of dfs forest *)
            dfs(j); (* recursive call *)
        end
        else (* j is marked ''visited'' *)
            if ((dad[i] <> j) and (dfsn[i] > dfsn[j]))or(i = j)
            (* ij is a back edge possibly a loop *)
            then
            begin
                acyclic := 1;
                (* print the cycle *)
                if i = j
                then
                write(i,i,' is a loop')
                else
                begin
                    k := i;
                    repeat (* climb up the tree
                              till we meet the vertex j *)
                        write(k,' ');
                        k := dad[k];
                    until k = j;
                end
```

```
                writeln;(* new line to separate the cycles *)
            end;
end;(* dfs *)
(* iteration *)
for i := 1 to n do
if mark[i] = 0 then
    dfs(i); (* invoking dfs *)
if acyclic = 0
then
    writeln ('yes')
else
    wtiteln('no');
```

The complexity of the algorithm testing for cycles:
The complexity is $O(n + m)$, which is the same as that of dfs, since we have only added the statements to the "else part" in the "procedure dfs" of the "dfs algorithm." In the worst case, the added statements cost $O(n)$ steps. By the sum rule [6] of big Oh notation, we may write the complexity as $O(n+m) = O(\max(m, n))$. (The sum rule says that, in a sum, it is enough to take into account the "dominating term" of the sum, ignoring constants and constant coefficients.)

1.9.3 Application 3: Finding biconnected components of a connected multigraph

Hopcroft's algorithm:
Consider a connected multigraph $G = (X, E)$ with $X = \{1, 2, \ldots, n\}$. We assume that the graph G has no loops. A vertex x is a *cut vertex or articulation vertex* of the graph G if the removal of the vertex x from G results in a disconnected graph. Note that the removal of the vertex x also results in the removal of all edges incident with the vertex x.

An edge with the analogous cohesive property is called a *bridge* or *cut edge* or isthmus of the graph G, that is an edge e is a bridge of the graph G if the removal of the edge e from the graph results in a disconnected graph. Note that by removing an edge $e = xy$, the end vertices of the edge e will *still* be in the graph $G - e$. It can be easily seen (by constructing examples) that the removal of an edge results in a graph with at most two connected components, whereas the removal of a vertex may result in a graph with several connected components.

Remark 1.4. *If we allow loops in the multigraph, then a vertex x incident with a loop and another edge which is not a loop is considered as a cut vertex.*

In the following graph of Figure 1.26, the vertices 3, 4, and 6 are cut vertices and the bridges are edges 47 and 36. Note that an end vertex of a bridge is a cut vertex if its degree is strictly more than one.

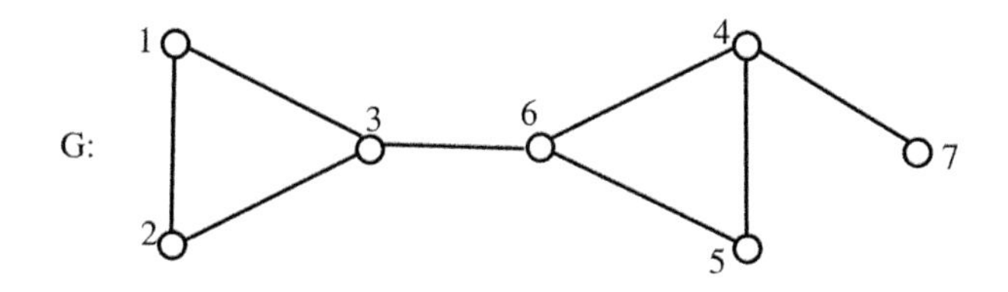

FIGURE 1.26: Illustration of cut points and bridges.

Example 1.9: Cut vertices and bridges

Consider a tree T with at least two vertices. Each vertex of degree one of this tree T (there must be at least two such vertices, otherwise, $\delta(T) \geq 2$ and hence the graph has a cycle [6], which is impossible) is not a cut vertex of T and all other vertices are cut vertices of T.

On the other hand, every edge of T is a bridge.

A connected graph with at least two vertices is a *biconnected graph* or a *block* if it does not contain a cut vertex. A *biconnected component* or *block* of a connected graph is a *maximal* induced subgraph without any cut vertex. In other words, the graph H is a biconnected component of the graph G if H is biconnected in its own right and for any vertex x lying outside of H, the subgraph of G induced by the vertices of $H \cup x$ is either disconnected or contains a cut vertex.

A graph can be partitioned into mutually disjoint connected components. In an analogous manner, a given connected graph can be decomposed into mutually *edge-disjoint* biconnected components. Note that a graph is biconnected if and only if its unique biconnected component is itself.

The nomenclature "biconnected" is justified because of the following property enjoyed by a biconnected graph with at least three vertices:

Theorem 1.2. *A connected graph with at least three vertices is biconnected if and only if between any two distinct vertices of the graph there are at least two internally vertex-disjoint elementary paths.*

In the following graph of Figure 1.27, we have a connected graph, with four biconnected components.

The list of edges of the three biconnected components of the graph of Figure 1.27 are:

1. 12, 14, 24, 29, 19, 49
2. 23
3. 36, 37, 67
4. 59, 58, 89

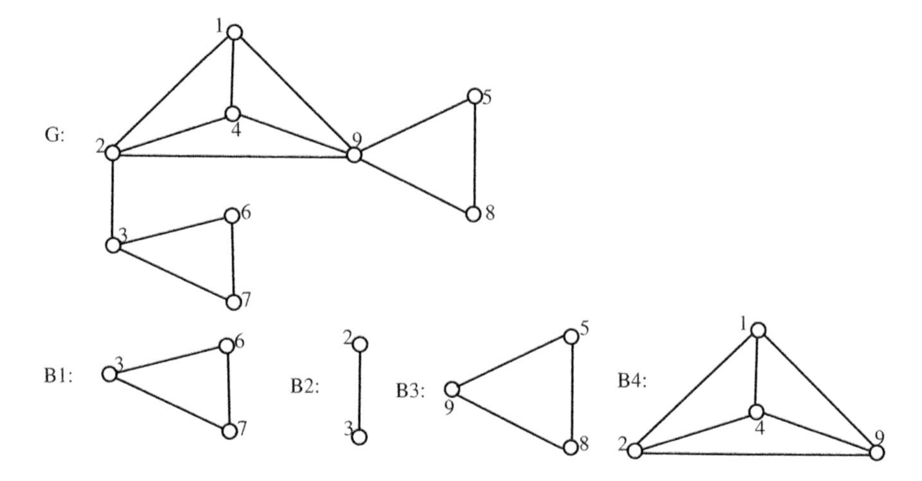

FIGURE 1.27: Illustration of biconnected components of a graph.

The biconnected components listed above satisfy the following properties:

1. Two distinct biconnected components intersect in at most one vertex.
2. If two distinct biconnected components intersect in exactly one vertex x, then this vertex must be a cut vertex of the graph.

The above two properties are in general true for any connected graph with at least two vertices. Intuitively, this means that the different cut vertices of a graph are the "frontiers" between the biconnected components.

We shall now describe the biconnected components algorithm.

Input: A connected multigraph $G = (X, E)$ with $X = \{1, 2, \ldots, n\}$.

Output: A list of edges of each biconnected component of the graph G.

Algorithm: We shall first illustrate the algorithm on the graph of Figure 1.27. Step 1: The graph is represented by its adjacency lists. Let us assume that the vertices are listed in the increasing order in each list $L(i)$, that is, $L(1) = (2, 4, 9)$, $L(2) = (1, 3, 4, 9)$, etc., till $L(9) = (1, 2, 4, 5, 8)$. We choose a starting vertex, say, the vertex 1.

Step 2: The graph is searched according to the algorithm dfs. This search gives us a partition of the edge set into T and B, where T is the set of edges of the spanning tree induced by the dfs and B is the set of back edges which is equal to $E \setminus T$. During the search, each vertex is assigned a number called the dfs number. Recall that dfsn$[i] = j$ if and only if the vertex i is the jth vertex "visited" in the course of the dfs (see the graph of Figure 1.28 below obtained by dfs).

The following Table 1.9 gives us the dfsn of different vertices.

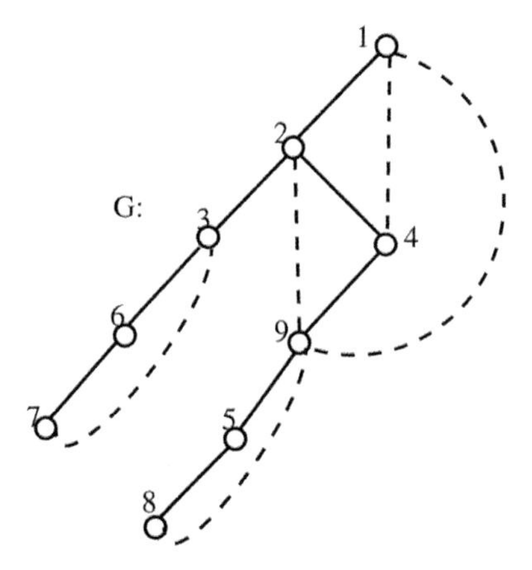

FIGURE 1.28: Depth-first search of the graph of Figure 1.27.

TABLE 1.9: Depth-first search number (dfsn) table of the graph of Figure 1.28

	1	2	3	4	5	6	7	8	9
dfsn	1	2	3	6	8	4	5	9	7

To help in finding the biconnected components, we need to define the following function "LOW" on the set of vertices of the graph.

The function LOW:
LOW$[i]$ is the *smallest* dfsn less than or equal to *dfsn*$[i]$ one can reach either from i or from one of the descendants of i in the dfs spanning tree using *only once* a back edge. If all such back edges lead to vertices with dfsn $> dfsn[i]$, or if there is no back edge from i or from one of its descendants of i, then we define LOW$[i]$ = *dfsn*$[i]$.

Mathematically, we write

$$\text{LOW}[i] = \min(\{\text{dfsn}[i]\} \cup \{\text{dfsn}[j] \mid \text{there is a back edge from a descendant of } i \text{ to } j\}).$$

The function LOW will be computed by the algorithm in postorder traversal of the depth-first search spanning tree obtained. In our example, the postorder traversal of the dfs tree is: $(7, 6, 3, 8, 5, 9, 4, 2, 1)$. So, the first vertex for which LOW is computed is the vertex 7.

LOW$[7] = \min(dfsn[7], dfsn[3]) = \min(5, 3) = 3$ because of the back edge 73 from the vertex 7. LOW$[6] = \min(dfsn[6], dfsn[3]) = \min(5, 3) = 3$ because of the back edge 73 from the son 7 of the vertex 6.

LOW$[3] = 3$.

$\text{LOW}[2] = \min(dfsn[2], dfsn[1], dfsn[1], dfsn[2]) = 1$ because of the back edges $41, 91, 92$.

Note that a vertex is considered as a descendant of itself. The following table gives the LOW function of different vertices of our example graph (Table 1.10).

As we have already remarked, the different cut vertices of the graph G serve as "frontiers" between biconnected components. So we are first interested in characterizing the cut vertices of the graph G. Let us again refer to the graph of Figure 1.28. The vertex 3 is a cut vertex, whereas the vertex 6 is not. Why is vertex 6 not a cut vertex? This is because of the following fact: There is a back edge, namely, 73, from the son of 6, that is, from a proper descendant of 6, to a proper ancestor of 6. This back edge gives us a second path from 3 to 7 and the first path from 3 to 7 is obtained via the tree edges. On the other hand, there is no back edge from a proper descendant of the vertex 3 to a proper ancestor of 6. Hence, there are no two vertex-disjoint paths from a proper descendant of 3 to a proper ancestor of 3. This makes the vertex 3, a cut vertex of the graph G. In a similar manner we can conclude that the vertices 2 and 9 are cut vertices.

We now state the following theorem whose proof is intuitively clear from the above discussion. For a formal proof, the reader can consult the book [3].

Theorem 1.3. *Consider a connected multigraph G and consider a partition of the edge set of the graph G into T, a set of tree edges and B, a set of back edges induced by a search dfs. Then, a vertex i is a cut vertex of G if and only if it satisfies exactly one of the following properties:*

1. *i is the root of the tree T and the root i has at least two sons in the spanning dfs tree T.*

2. *i is not the root and i possesses at least one son s in the spanning dfs tree with* $\text{LOW}[s] \geq dfsn[i]$.

Back to our example illustrating Hopcroft's biconnected components algorithm: With this characterization of cut vertices, we continue our illustration. We shall use a stack S of edges. Initially our stack is empty. Each edge traversed the *first* time in the course of the dfs is pushed into the stack. Recall that each edge of the graph is encountered twice during the search procedure. Note that a tree edge xy is pushed into the stack in the order (father, son), where x is the father of y, whereas a back edge da is pushed into the stack with the vertex a an ancestor of the vertex d. Note that a loop is considered as a back edge.

TABLE 1.10: LOW function table of the graph of Figure 1.28

	1	2	3	4	5	6	7	8	9
LOW	1	1	3	1	7	3	3	7	1

In our example, the stack will look as below when we are at vertex 7 before backtracking from 7 to the father of 7 which is 6 (the stack grows from left right).

$$S = (12, 23, 36, 67, 73).$$

The edge at the top of the stack, namely, edge 73, is a back edge whereas all other edges in the stack are tree edges. Tree edges occur as a result of direct recursive calls, that is, xy is a tree edge if and only if $dfs(x)$ calls $dfs(y)$ directly during the search.

When we backtrack from the vertex 7, that is, when we climb up the tree from vertex 7 to his father 6, the following test is performed.

$\text{LOW}[7] \geq dfsn[6]$? If the test is true, then we pop up the stack till the edge 67 (including the edge 67) to emit a biconnected component. In our example, the test is false, since $\text{LOW}[7] = 3 < dfsn[6] = 4$. So no pop-up occurs. We are at the vertex 6 and the search finds no more new vertices from 6. So we climb up the tree from 6 to the father of 6 which is 3. The following test is performed.

$\text{LOW}[6] \geq dfsn[3]$? The inequality is true, since $\text{LOW}[6] = 3$ and $dfsn[3] = 3$. Hence, we pop up the stack till and including the edge 36.

Thus, our first biconnected component is

$$B_1 = (73, 67, 36)$$

and the stack will look as follows:

$$S = (12, 23).$$

We are at vertex 2 and no new vertices are found from the vertex 2. Hence we climb up the tree from 2 to his father 1. The test $LOW[2] \geq dfsn[1]$ turns out to be true and we pop up the stack till the edge 23.

The second biconnected component found is

$$B_2 = 23$$

and the stack

$$S = (12).$$

We are now at vertex 2. We find a new vertex 4 by scanning the list $L(2)$ and the edge 24 is pushed. Then, the following edges are pushed into the stack: $41, 49, 91, 92, 95, 58, 89$. No new vertices are found from the vertex 8. At this point, the stack S from left to right is

$$S = (12, 24, 41, 49, 91, 92, 95, 58, 89).$$

We backtrack from the vertex 8 to his father 5 and the test $\text{LOW}[8] \geq dfsn[5]$ is false. We further backtrack from 5 to his father 9 and the test $\text{LOW}[5] \geq dfsn[9]$ turns out to be true. Hence, the stack is popped till we arrive at the edge 95.

The third biconnected component found is

$$B_3 = (89, 58, 95)$$

and the stack

$$S = (12, 24, 41, 49, 91, 92).$$

We climb up the tree from 9 to 4 (the test is false) and then from 4 to 2 (the test is false) and finally from 2 to 1. The test LOW$[2] \geq dfsn[1]$? is true. Hence, the stack is popped till the edge 12. This gives us the fourth and last biconnected component B_4 where

$$B_4 = (12, 24, 41, 49, 91, 92).$$

Note that the stack is finally empty.

With this hand simulation as our model, we are now almost ready to write the Hopcroft's biconnected components algorithm in pseudo-code. One more final point remains. We have to express the definition of the function LOW in a recursive manner. Since a son j of a vertex i in a dfs spanning tree T is a proper descendant i, we rewrite LOW as follows:

$$\mathrm{LOW}[i] = \min(\overbrace{\{dfsn[i]\}}^{\text{part1}} \cup \overbrace{\{LOW[j] \mid j \text{ is a son of } i \text{ in T }\}}^{\text{part2}} \cup \overbrace{\{dfsn[k] \mid ik \in B\}}^{\text{part3}},$$

where B is the set of back edges. The search procedure dfs is augmented with suitable statements to print different biconnected components. We use the following variables: the arrays $mark[1..n]$ of 0..1, $dfsn[1..n]$ of integer, LOW$[1..n]$ of integer, $dad[1..n]$ of integer. We also use a stack S whose objects are edges. The vertices are $1, 2, \ldots, n$.

```
Biconnected components algorithm:
procedure dfs_biconn( i : integer);
var j : integer; (* j is used to scan L(i)*)
  LOW, dad, dfsn : array[1..n] of integer;
begin
    mark[i] := 1;
    dfsn[i] := counter;
    counter := counter + 1;
    (* initialization of LOW[i]  *)
    LOW[i] := dfsn[i];(* part 1 of the def. of LOW *)
    (* scan L(i) for a new vertex *)
    for each j in L(i) do
    if mark[j] = 0
    then
    begin
        dad[j] := i;(* add edge ij to dfs tree *)
        push ij into S;
```

```
            dfs_biconn(j);(* recursive call *)
            (* at this point we have computed LOW[j] *)
            (* update LOW[i], the part 2 of the def. of LOW *)
            LOW[i] := min(LOW[i], LOW[j]);
            if LOW[j] >= dfsn[i]
            then(* a biconnected component is found *)
            begin (*write the edges of biconnected component*)
            pop S till(including) the edge ij;
            writeln;(* new line *)
        end
        else(* j already visited*)
            (* test if ij is a back edge *)
            if (dad[i] <> j) and (dfsn[i] > dfsn[j])
            then (* update LOW. part 3 of the def. of LOW *)
            begin
                LOW[i] := min(LOW[i], dfsn[j]);
                push ij into S;
            end;
    end;(* dfs_biconn *)

    (* initialization *)
    for i := 1 to n do
        mark[i] := 0;
    counter := 1:
    S := empty;(* empty stack *)
    dfs_biconn(1); (* procedure call *)
```

Complete program in Pascal:
Let us write the complete program for finding biconnected components in Pascal.

```
    Biconnected components in pascal:
    program biconnected;
    const maxn = 50; (* maximum number of vertices *)
          maxm = 250;(* maximum number of edges *)
    type pointer = ^node;
           node  = record
                info : integer;
                next : pointer;
           end;

   var n, m, i, j, x, y, counter : integer;
       L : array[1..maxn ] of pointer;
       S : array[1..maxm ] of integer;
   (*S, stack of edges and an edge
        is a pair of vertices *)
```

```
        top :integer; (* top of the stack S *)
    procedure input_graph;
    var i, j, x, y :integer;(* xy is an edge *)
                 t :pointer;
    begin(* input_graph *)
        write(' enter the number of vertices ');
        readln(n);
        write(' enter the number of edges ');
        readln(m);

        (* initialization. graph with n vertices and 0 edges *)
        for i := 1 to n do
            L[i] := nil;
        (* read the edges *)
        for i := 1 to m do
        begin
            write('enter the edge ', i);
            readln(x,y);
            (* add y at the head of the list L[x] *)
            new(t); (* create a node pointed by t *)
            t^.info := y;
            t^.next := L[x];
            L[x] := t;(* attach t at the head of L[x] *)
            (* add x at the head of list L[y] *)
            new(t);
            t^.info := x;
            t^.next := L[y];
            L[y] := t;
        end;
    end;(* input_graph *)
    procedure output_graph;
    var t : pointer;
        i : integer;
    begin
        for i := 1 to n do
        begin
            t := L[i];
            if t = nil
            then
                write('no vertices joined to ', i)
            else
            begin
                write('the vertices joined to ',i, ' are :'),
                (* scan the list L[i] *)
                while t <> nil do
```

```
            begin
                write(t^.info,'  ');
                t := t^.next;
            end;
            writeln;
        end;
end;(* output_graph *)

procedure dfs_biconnected;
var t : pointer; j:integer;
    dfsn : array[1..maxn] of integer;
    LOW : array[1..maxn ] of integer;
    dad : array[1..maxn ] of integer;
begin
    mark[i] := 1;
    dfsn[i] := counter; counter := counter + 1;
    (* initialization of LOW: part 1 *)
    LOW[i] := dfsn[i];
    (* scan L[i] *)
    t := L[i];
    while t <> nil do
    begin
        j := t^.info;
        if mark[j] = 0
        then
        begin
            dad[t^.info] := i;(* add (i,j) to dfs tree *)
            (* push (i, j) into the stack *)
            S[top+1] := i;
            S[top+2] := j;
            top := top +2; (* update top *)
            dfs_biconnected(j);(* recursive call *)
            (* at this point we have found LOW[j] *)
            (* update LOW[i]: part 2 *)
            if LOW[i] > LOW[j] then
                LOW[i] := LOW[j];
            if LOW[j] >= dfsn[i] then
            begin
                (* a biconnected component is found *)
                (* pop S till the edge ij *)
                repeat
                    write('(',S[top],',',S[top-1],')');
                    top := top - 2;
                until S[top+1] = i;
                writeln;
```

```
            end;
        end(* then*)
        else
            (* test if ij is a back edge *)
            if (dad[i] <> j) and (dfsn[i] > dfsn[j])
            then
            begin
                S[top + 1] := i;
                S[top + 2] := j;
                top := top + 2;
                (* update LOW[i]: part 3 *)
                if LOW[i] > dfsn[j] then
                    LOW[i] := dfsn[j];
            end;
        t := t^.next;
    end;(* while*)
end;(* dfs_biconnected *)

begin(* main program*)
(* initialization *)
for i := 1 to n do
mark[i] := 0;
counter := 1;
top := 0;(* empty stack S *)
(* end of initialization *)
input_graph;
output_graph;
dfs_biconnected(1);(* call *)
end.
```

The complexity of the biconnected components algorithm:
The complexity of the algorithm is the same as that of the dfs search, since the added instructions, namely, calculating the function LOW, pushing each edge once into the stack and popping each edge out of the stack takes $O(m)$ steps. Hence, the complexity is $O(n+m)$. But the input is a connected graph and for a connected graph we always have $m \geq n-1$. Hence the complexity is $O(m)$ by *the rule of sum.* (We can ignore lower-order terms while computing complexity.)

1.10 Depth-First Search for Directed Graphs

The algorithm of the depth-first search for un undirected graph holds equally well for a directed graph also. In a directed graph if (i,j) is an arc and (j,i) is not an arc, then we place the vertex j in the list $succ[i]$ but the

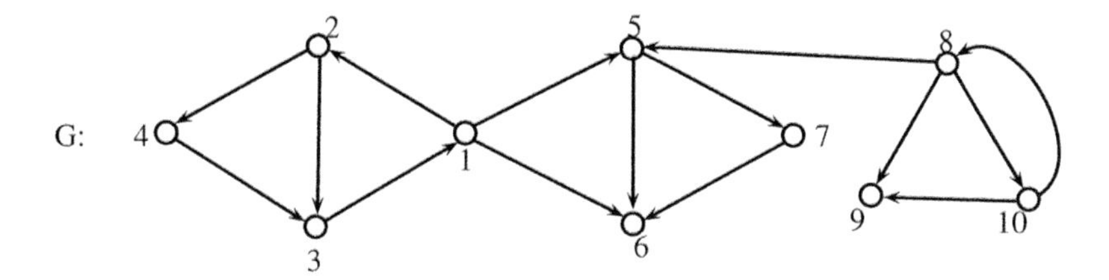

FIGURE 1.29: Digraph to illustrate the depth-first search.

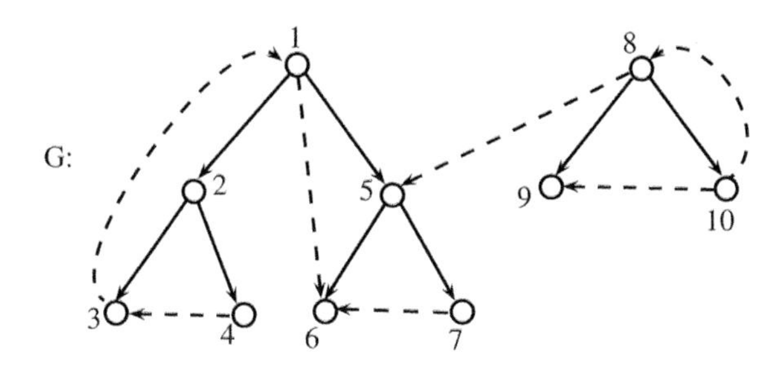

FIGURE 1.30: Depth-first search of graph of Figure 1.29.

vertex i will not be in the list $succ[j]$. In the case of an undirected graph, if ij is an edge, then we include the vertex j in the list $L[i]$ and *also* include the vertex i in the list $L[j]$.

The search dfs for a directed graph is illustrated by the following example. Consider the graph of Figure 1.29.

As in the case of an undirected graph, the graph is represented by list of successors of each vertex.

$succ[1] = (2, 5, 6)$; $succ[2] = (3, 4)$; $succ[3] = (1)$; $succ[4] = (3)$; $succ[5] = (6, 7)$; $succ[6] = ()$; $succ[7] = (6)$; $succ[8] = (5, 9, 10)$; $succ[9] = ()$; $succ[10] = (8, 9)$.

Let us now perform the dfs on the graph of Figure 1.30. The first call dfs(1) "visits" the vertices $1, 2, \ldots, 7$ and gives us a tree with the root 1. Then, the second call dfs(8) "visits" the remaining vertices $8, 9, 10$ and gives us a second tree of the dfs spanning forest. The original graph is restructured as in the following graph Figure 1.30. In this figure, the solid arcs define a spanning arborescence.

We observe the following properties:

Property 1.3. *There are* four *types of arcs:*

1. *The tree arcs which are drawn as continuous arcs. If (i, j) is an arc in the dfs spanning forest induced by the algorithm, then $dfs(i)$ called* directly *$dfs(j)$ during the search (e.g., $(2, 4)$).*

2. *Dotted arcs like* $(1, 6)$ *which go from an ancestor towards a decendant. These arcs are called* descending dotted arcs. *If* (i, j) *is a descending dotted arc, then* $dfsn[i] < dfsn[j]$.

3. *Dotted arcs like* $(3, 1)$ *which go from an ancestor towards a descendant. These arcs are called* mounting dotted arcs. *Note that a loop is considered as a mounting dotted arc. If* (i, j) *is a mounting dotted arc, then* $dfsn[i] > dfs[j]$.

4. *Dotted arcs like* $(4, 3)$ *and* $(8, 5)$ *which join two vertices that are not in the relation ancestor-descendant or descendant-ancestor. Such arcs are called* dotted cross arcs.

We shall now observe and prove the following key property of a dotted cross arc.

Property 1.4. *A dotted cross arc always goes from the right to the left. (We assume that the children of each vertex are drawn from left to right and the different trees of the dfs spanning forest are also drawn in the left to right fashion.) In other words, if* (i, j) *is a dotted cross arc, then* $dfsn[i] > dfsn[j]$.

Proof. Consider a dotted cross arc (i, j). This means that the vertex j is "visited" before the vertex i is "visited" during the search. Otherwise, suppose the vertex i is "visited" before the vertex j. That is, when we reach the vertex i for the first time, the vertex j is *still* marked "unvisited." Since (i, j) is an edge of the graph, we have j in the list of $succ[i]$. But then the "visit" of i would not be complete unless we touch the vertex j. Hence, j must be a descendant of i in the dfs spanning forest. Since (i, j) is a dotted edge, we conclude that (i, j) must be a descending dotted edge, a contradiction.

Hence j is "visited" before the vertex i. This means that when we reach i for the first time, j is marked "visited." Hence, $dfsn[i] > dfsn[j]$. (Geometrically, the vertex i is to the right of the vertex j.) □

1.11 Applications of Depth-First Search for Directed Graphs

1.11.1 Application 1: Finding the roots of a directed graph

Consider a directed graph $G = (X, U)$ where the vertex set $X = \{1, 2, \ldots, n\}$. A root of the directed graph G is a vertex r, from which we can reach all other vertices of the graph by a directed path from r. A root does not always exist. A graph may have more than one root. With the help of the dfs, one can find all the roots of a given directed graph.

Input: A directed graph $G = (X, U)$ with the vertex set $X = \{1, 2, \ldots, n\}$.

The graph is represented by n successor lists, $succ[1], succ[2], \ldots, succ[n]$.

Output: Find all the roots of the graph G.

Algorithm: We mark all vertices as "unvisited," set the "counter" used to assign dfs number to 1. Then, we call the procedure dfs(1). If we mark as "visited" all the vertices of the graph with this call, then the vertex 1 is a root. Then, we repeat the calls with $dfs(2), dfs(3), \ldots, dfs(n)$. Let us now write this algorithm.

```
procedure dfs( i : integer);
(* dfs visits each vertex of a connected component of G
containing i using depth-first search *)
var j : integer;(* vertex j scans succ[i] *)
begin
   mark[i] := 1;
   dfsn[i] := counter;
   counter := counter + 1;
   (* scan succ[i] *)
   for each vertex j in succ[i] do
       if mark[j] = 0 then
       begin
           add the arc ij to F;
           dfs(j); (* recursive call *)
       end;

(* initialization *)
for i := 1 to n do
    mark[i] := 0;
counter := 1;
F := empty; (* empty forest *)
(* iteration *)
for i := 1 to n do
begin (* test if i is a root *)
    dfs(i): (* call to the procedure dfs *)
    if counter = n+1
    then write(i,' is a root');
    (* reinitialize: mark all vertices ''unvisited'' *)
     for i := 1 to n do
        mark[i] := 0;
     counter := 1;
end;
```

The complexity of finding all the roots of a graph:
A call to dfs costs $O(n+m)$ and we make n calls, one call with each vertex of the graph. Hence the time complexity of the algorithm is $O(n(n+m))$.

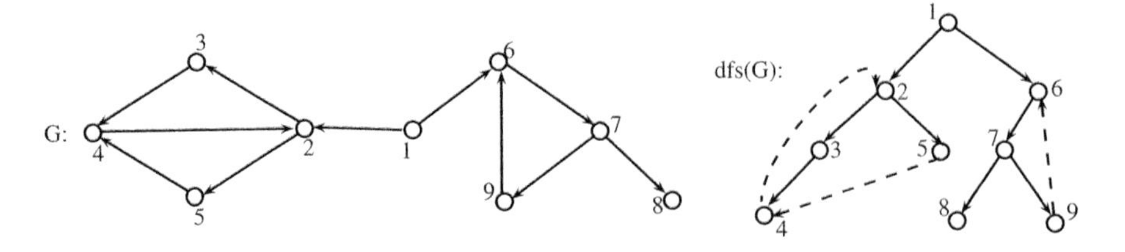

FIGURE 1.31: Illustration to test if a graph is acyclic.

1.11.2 Application 2: Testing if a digraph is without circuits

The existence of a directed circuit in a given digraph can be easily tested using the dfs.

Input: A digraph G.

Output:
"yes" if the graph has no directed circuit.
"no" if the graph possesses a directed circuit.

Algorithm: By definition, each mounting dotted arc encountered during the search clearly defines a circuit. Conversely, if the given digraph possesses a circuit, then one must encounter a mounting dotted arc in the course of the search.

To prove the converse part, consider a graph with at least one circuit and perform the depth-first search on the graph. Now choose the vertex x whose dfsn is the smallest among the vertices belonging to a circuit. Consider an arc (y, x) on some circuit containing the vertex x. Since x belongs to a circuit, such an arc must exist ($d^-(x) \geq 1$.) Since x and y are on the same circuit, y must be a descendant of x in the dfs spanning forest. Thus, the arc (y, x) cannot be a dotted cross arc. Because the dfsn of $y >$ the dfsn of x, (y, x) cannot be a tree arc or dotted descending arc. Hence, (y, x) must be a dotted mounting arc as in the figure below. Hence the converse part of the assertion (see the graph of Figure 1.31).

In the above figure, the vertex x in the above argument is the vertex 2, and the vertex y is the vertex 4 and the circuit in discussion is the circuit $(2, 3, 4, 2)$.

1.11.3 Application 3: Topological sort

Consider a directed graph without circuits. Geometrically, a *topological sort* means laying the vertices of a directed graph without circuits in a horizontal manner from left to right such that each arc of the graph goes from left to right. In other words, if the graph has n vertices, we are interested in assigning the integers $1, 2, \ldots, n$ to the vertices of the graph such that if there is an arc from the vertex i to the vertex j, then we have the inequality $i < j$.

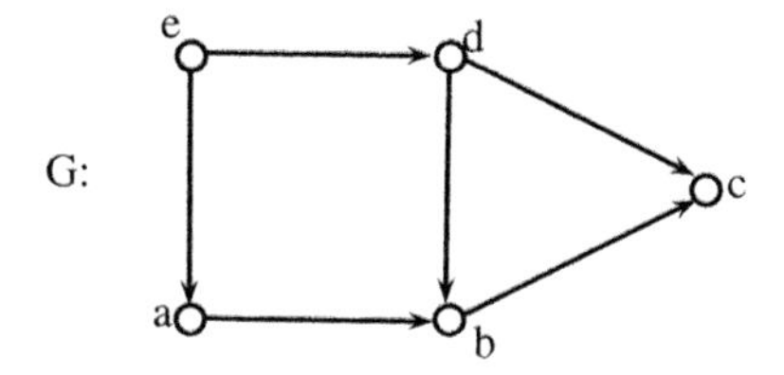

FIGURE 1.32: A graph to illustrate topological sort.

We shall shortly show that a topological sort is possible if and only if the graph is without circuits. Before proving the assertion, let us look at an example.

Example 1.10

Consider the following graph of Figure 1.32 without circuits.

Let us assign the integer 1 to the vertex e, 2 to a, 3 to d, 4 to b and finally 5 to the vertex c. This assignment defines a topological sorting of the vertices of the graph. The same graph is redrawn below in Figure 1.33 to reflect the property of "topological sorting."

Theorem 1.4. *A directed graph G admits a topological sort if and only if the graph is without circuits.*

Proof. One part of the theorem is obvious. If the graph admits a topological sort, then by definition we can assign the integers $1, 2, \ldots, n$ as new labels to its vertices in such a way that if (i, j) is an arc of the graph, then $i < j$. If the graph possesses a circuit $(v_1, v_2, \ldots, v_s, v_1)$, then because of the property of the new labeling of the vertices, we have $v_1 < v_2 < \cdots < v_s$. But then, (v_s, v_1) is an arc of the circuit with $v_1 < v_s$, a contradiction.

Now for the second part, consider a *longest* directed path $(x_0, x_1, \ldots, x_d)$ in the given graph. Its length is d. We shall first prove that the initial vertex x_0 of our longest path satisfies $d^-(x_0) = 0$. If not, there is an arc (x, x_0) in the graph. We shall distinguish two cases.

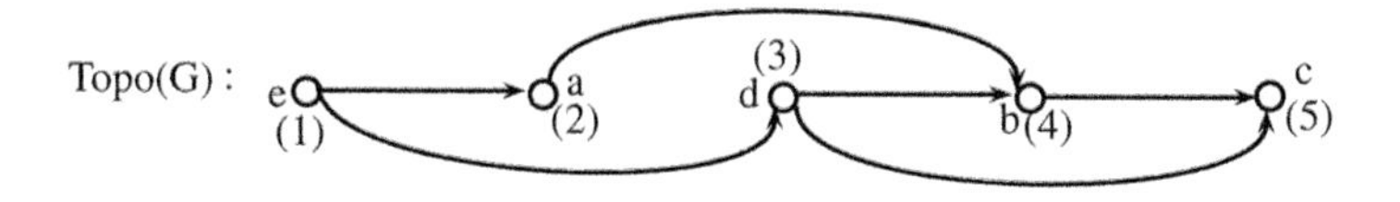

FIGURE 1.33: Topological sorting of the graph of Figure 1.32.

Case 1. $x \neq x_i$ for $1 \leq i \leq d$.
We then have a directed path $(x, x_0, x_1, \ldots, x_d)$ of length $d+1$, a contradiction.

Case 2. $x = x_i$ for some i with $1 \leq i \leq d$.
Then, $(x_0, \ldots, x_i, x_0)$ is a circuit of the graph, a contradiction.

Hence there is a vertex, say 1, in the graph with no arc entering into it. Now consider the vertex deleted subgraph graph $G-1$. It has still no circuits. Hence there is a vertex, say, 2, with no arcs entering into it in the graph $G-1$. Then, we consider the subgraph $G-1-2$, etc. till we arrive at the graph with exactly one vertex which will be labeled as n.

These vertices $1, 2, \ldots, n$ define a topological ordering of the vertices of the graph G. □

We shall now prove the following result concerning the adjacency matrix of a digraph without circuits. This result is an immediate consequence of the topological sort.

Theorem 1.5. *Let G be a directed graph on n vertices without circuits. Then, its vertices can be labeled by the integers $1, 2, \ldots, n$ so that its adjacency matrix is an upper triangular.*

Proof. Since the graph is without circuits, it admits a topological sort. Let this order be $1, 2 \ldots, n$. We now relabel the vertices of the graph in this order. According to this labeling, if (i, j) is an arc, then $i < j$. This means that, if $i > j$, then (i, j) is *not* an arc of the graph. This means that in the adjacency matrix M of the graph G, we have $M[i, j] = 0$ if $i > j$. Hence the result. □

Remark 1.5. *According to Theorem 1.5, a graph without circuits can be represented by an upper triangular matrix. Since the graph has no loops, the diagonal entries of this matrix are also 0. Hence a graph on n vertices and without circuits may be represented economically by a one-dimensional array using only $(n^2 - n)/2$ memory space, (by omitting 0 entries on and below the main diagonal) instead of n^2 space needed to store an $n \times n$ matrix.*

Algorithm for the topological sort:

Input: A directed graph G without circuits.

Output: Assigning the integers $1, 2, \ldots, n$ to the vertices of the graph in such a way that if (i, j) is an arc of the graph, then we have $i < j$.

Algorithm 1.4. *Choose a vertex x such that the in-degree of the vertex x is 0. According to Theorem 1.4 such a vertex exists. Label this vertex as 1. Then, consider the vertex deleted subgraph $G_1 = G - 1$ and choose a vertex in G_1 for which the in-degree is 0. Call this vertex 2. Then, consider $G_2 = G - 1 - 2$ and choose a vertex in G_2 with in-degree 0. Continue like this till we arrive at the graph with only one vertex. This last vertex will be labeled as n. The sequence $1, 2, \ldots, n$ is the required topological sort.*

Let us write the algorithm in pseudo-code. We use the following variables: The graph is represented by n successor lists, $succ[1], succ[2], \ldots, succ[n]$. We use an array $id[1..n]$ where $id[i]$ is the in-degree of the vertex i, that is, the number of arcs entering into the vertex i. Once a vertex i is processed by the algorithm, we set $id[i]$ equal to -1 or some value which cannot be a legal in-degree of a vertex. The variable "counter" is used to count the number of vertices printed in topological order.

```
Topological sort1:
(* initialization of counter *)
counter := 0;
while counter < n do
begin
(* scan the array id to find vertices of in-degree zero *)
    for i := 1 to n do
        if id[i] = 0 then
        begin
            write(i,' ');
            counter := counter + 1;
            id[i] := -1; (* vertex i is processed *)
            (* scan succ[i] to reduce the in-degree by one *)
            for each j in succ[i] do
                id[j] := id[j] - 1;
        end;
end;
```

The complexity of the topological sort:
The worst case arises when the graph is a directed path $(1, 2, \ldots, n)$. The array id is defined as $id[i] = 1$ for $0 \leq i < n$ and $id[n] = 0$. The algorithm prints the vertex n first and the variable counter is set to 1. During the next pass through the while loop, the algorithm prints the vertex $n-1$ and the counter is assigned the value 2, etc. Hence, the while loop is executed n times for this graph. Each execution of while loop costs $O(n)$ steps, because of the for loop inside the while loop. Hence the complexity is $O(n^2)$ steps. The desired topological sort is $(n, n-1, \ldots, 1)$.

Algorithm 1.5. *This algorithm uses dfs. It is based on the following property: A directed graph is without circuits if and only if there is no dotted mounting arc during the dfs. A simple write statement at the end of the dfs procedure gives us a* reverse *topological sort. We use a stack S of vertices. We will give the procedure below:*

```
Topological sort algorithm using dfs:
(* initialization *)
for i := 1 to n do
    mark[i] := 0;
```

```
top := 0; (* empty stack *)
(* end of initialization *)

procedure toposort( i : integer);
begin
    mark[i] := 1;
    (* scan the list succ(i) *)
    for each vertex j in succ(i) do
        if mark[j] = 0 then
            dfs(j);(* recursive call *)
    (* push i into the stack *)
    top := top +1;
    S[top] := i;
end;

(* iteration *)
for i := 1 to n do
if mark[i] = 0 then
    dfs(i); (* invoking toposort *)
(* print the vertices in topological sort *)
for i: = n downto 1 do
write (S[i],' ');
```

Let us justify the algorithm:

Proof. When the procedure "toposort" finishes scanning entirely the list $succ(i)$, it pushes the vertex i into the stack. Since the graph is without circuits, there are are no dotted mounting arcs from i. The only possible arcs going out of i are the arcs of the forest, dotted descending arcs, and dotted cross arcs. But these arcs are directed toward vertices whose successor lists have been completely scanned and therefore already pushed into the stack. Hence the proof. □

The complexity of the topological sort:
We have added only two elementary statements to the procedure dfs, namely, incrementing the variable "top" and pushing a vertex into the stack. Hence the complexity is the same as that of the complexity of the dfs procedure which is $O(m + n)$.

1.11.3.1 An application of topological sort: PERT

PERT and the critical path method: Imagine a project like constructing a big house. A project like this is often divided into a number of smaller subprojects like clearing land, building foundation, constructing walls, ceilings, carpentry works, plumbing, electricity work, landscaping, etc. In order to complete well the entire project within a given time limit, these subprojects should be carried out in a certain order. For example, land clearing should be

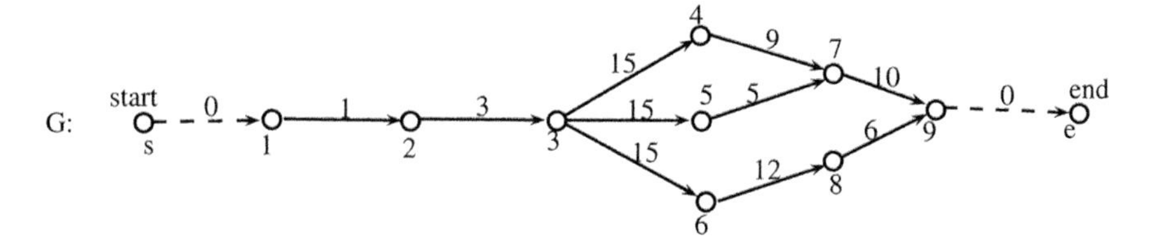

FIGURE 1.34: An activity digraph.

completed first before the foundation work starts, and plumbing and electricity may be taken later once the walls and ceilings will have been completed, etc. These subprojects are called the *activities of tasks*; that is, a project is constituted of activities. These activities are interrelated heavily by the constraints like a certain activity may not start until other activities have been completed.

Let us illustrate this by an example.

Example 1.11

The vertices of the graph (see Figure 1.34) without circuits represent different activities and the graph itself represents a project. An arc from the vertex i to the vertex j means the activity i must have been completed before initiating the activity j. The weight $w(i, j)$ associated with an arc (i, j) represents the time needed (e.g., measured in weeks) to finish the task i. In our example, the task 1 has no predecessor and the task 9 has no successor. Such vertices always exist in a directed graph without circuits. For example, the first vertex in a topological sort has no predecessor, and the last vertex of the sort has no successor. We introduce two new vertices s and e to the activity digraph. The vertex s corresponds to the starting of the project and the vertex e to the end of the entire project. The vertex s is joined *to* all vertices of the graph for which the in-degree is 0 and all vertices of the graph for which the out-degree is 0 are all joined *to* the vertex e. Finally, we associate the weight to all arcs of the form (s, i) and all arcs of the form (j, e). In our example, there is only one vertex of in-degree 0, namely the vertex 1 and the only vertex of out-degree 0 is 9.

We are interested in finding the *minimum* number of weeks needed to finish the entire project. This problem is equivalent to finding a directed path of *maximum* weight in a graph without circuits. In our example, a directed path of maximum weight is $(1, 2, 3, 4, 7, 9)$ and its weight is 38. Note that task 2 can start only after completing task 1, and 3 can be initiated after finishing 3, etc. This means that to complete the entire project we need a *minimum* of 38 weeks. Why? The time required to complete the entire project cannot be less than 38 weeks, since otherwise not all of the tasks of the graph could be completed.

To see this, let us write the vertices of the activity digraph in topological order. The desired order is $(1, 2, 3, 4, 5, 6, 7, 8, 9)$. We first start and finish the task 1; then we start and finish the task 2 and then 3. Now we can start *simultaneously* the tasks 4, 5 and 6 since these three activities are not related by the constraints of precedence (technically these tasks 4, 5, and 6 form a stable set, that is, there are no arcs joining these three vertices in the graph. After finishing 4, we can initiate 7. Note that after finishing 5, we cannot immediately start the task 7, since we have to wait for the completion of the task 4 which takes longer than the task 5. (The task 7 has two predecessors: 4 and 5.) After finishing 6, we start 8. The task 7 starts before the task 8 but is completed only after the task 8, because the weight of a longest path to 7 is 28, whereas the one to 8 is 31. That is, 7 starts two weeks later than 8 but is completed one week earlier.

A directed path of maximum weight is called a *critical path.* The different tasks on a longest weighted path are called *critical tasks.* In our example, the critical tasks are $1, 2, 3, 4, 7, 9$. If the completion of one of the tasks on a critical path is delayed, say, by one week, then the entire project will be delayed by one week. Techniques referred to as CPM and PERT use weighted directed graphs without circuits as models.

Longest path in a directed path without circuits:

We shall now present a *polynomial time* algorithm to find the *maximum* weight of a directed path (also called longest directed path) in a given directed graph *without* circuits and with a weight associated to each arc.

There is *no* known polynomial time algorithm for finding the *maximum* weight of a directed path in an arbitrary weighted directed graph. This problem belongs to a large class of problems called *NP-Complete problems.*

Input: A weighted directed graph without circuits $w(i, j)$ is the weight of the arc (i, j).

Output: $t[i]$, the *maximum* weight of a directed path arriving at the vertex i for each vertex i.

Algorithm: Let us first perform a topological sort of the vertices of the given graph. Let this order be $1, 2, \ldots, n$. The algorithm processes the vertices in this order. For each vertex i, we list all the predecessors of the vertex i in the list *pred*(i). Let us write the algorithm in pseudo-code.

```
Maximum weight of a directed path arriving at each vertex:
The vertices are ordered as 1,2,...,n according to the topological sort.
(* initialization *)
for i := 1 to n do
    t[i] := 0;
    (* iteration *)
```

```
for i := 1 to n do
begin
    for each vertex j in the list pred(i) do
        if t[j] + w(j,i) > t[i] then
            t[i] := t[j] + w(j,i);
end;
```

The complexity of finding a longest dipath in a digraph without circuits:
A call to topological sort requires $(n+m)$ steps. The initialization of the table t takes $O(n)$ steps. Now consider the iteration step: It has two nested loops. For each vertex i, the interior loop is executed exactly $d^-(i)$ times where $d^-(i)$ is in-degree of the vertex i, that is, the number of arcs entering into the vertex i. Since the sum of the in-degrees of all vertices of a graph is the number of arcs of the graph, the cost of the iteration is $O(m)$. Hence the complexity is $O(n+m)+O(m)+O(m) = O(\max(n,m))$, by the sum rule of the complexity.

1.11.4 Application 4: Strongly connected components algorithm

Consider a directed graph $G = (X, U)$ with vertex set $X = \{1, 2, \ldots, n\}$. The graph G is *strongly connected* if between any two vertices x and y there is a directed path from x to y and a directed path from y to x. In other words, a graph is strongly connected, if any two distinct vertices lie on a circuit (not necessarily elementary). By convention, the singleton sequence (x) is considered as a directed path of length 0 from x to itself. In this subsection, edge means directed arc/edge.

Example 1.12: Strongly connected graph

The graph G of Figure 1.35 is strongly connected. We see that any two vertices lie on a circuit.

The vertices 2 and 5 lie on the circuit $(2, 1, 3, 4, 5, 1, 3, 2)$ which is not even simple, because the arc $(1, 3)$ is repeated twice.

A graph which is not strongly connected can be decomposed into *strongly connected components.* A strongly connected component of a directed graph G is a *maximal* induced subgraph H which is strongly connected. That is, H is an induced subgraph of G with the following properties:

1. The graph H is strongly connected.
2. For any vertex x of G but not in H, the subgraph of G induced by $H \cup \{x\}$ is *not* strongly connected.

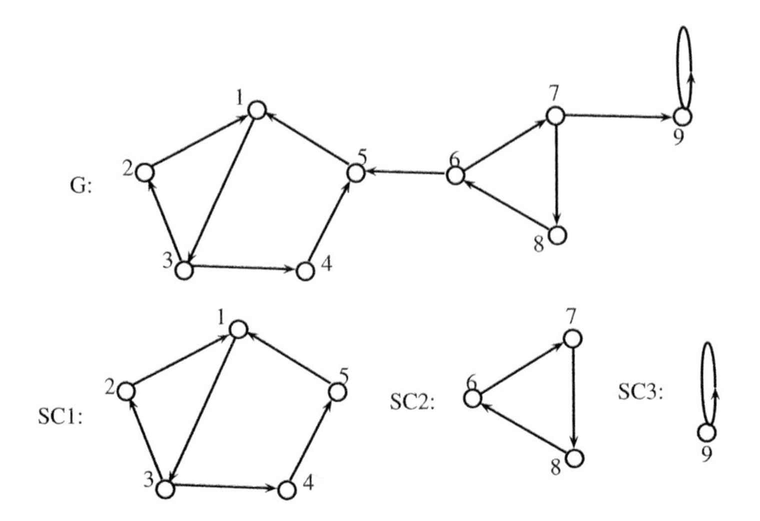

FIGURE 1.35: A graph and its strongly connected components.

Let us illustrate this notion by an example. Note that a graph is strongly connected if and only if its only strongly connected component is itself.

Example 1.13

Consider the graph of Figure 1.35.

The graph of Figure 1.35 is not strongly connected because there is no directed path from the vertex 1 to the vertex 6 even though there is a directed path from the vertex 6 to the vertex 1. The graph possesses *three* strongly connected components. These strongly connected components are induced by the vertex sets $\{1, 2, 3, 4, 5\}$ and $\{6, 7, 8\}$ and $\{9\}$.

Note that every vertex of the graph G lies in exactly one strongly connected component, but there are edges like $(6, 5)$ and $(7, 9)$ which are not in any strongly connected components. In other words, the different strongly connected components of a graph define a partition of the vertex set of X of G into a union of mutually disjoint subsets of X.

Tarjan's algorithm for strong components:

We shall now present an algorithm for finding strong components of a directed graph. This algorithm is based on the dfs of the given graph G. There is some *similarity* between the strong components algorithm and the biconnected components algorithm we have already studied. Recall that we have defined a function LOW which helped us to find the biconnected components of a connected graph using dfs. The function LOW used the dfs spanning tree, the dfsn of the vertices and the back edges. In a similar manner, we will be defining a function LOWLINK using dfs spanning forest, dfsn, dotted mounting arcs and dotted cross arcs.

In the biconnected components algorithm, we have pushed and popped the edges of the undirected connected graph, because the different biconnected components form a partition of the edge set. In the strongly connected components algorithm, we will be pushing and popping the vertices of the graph because the different strong components of a graph form a partition of the vertex set.

In short, the strong components algorithm and the biconnected components are *"dual" algorithms.*

Même quand l'oiseau marche, on sent qu'il a des ailes.

Lemierre (1733–1793)

This duality will be shortly explained.

With this analogy in mind, we shall study the strongly connected components algorithm.

Input: A directed graph $G = (X, U)$ with vertex set $X = \{1, 2, \ldots, n\}$.

Output: The list of vertices of each strongly connected component.

Algorithm: Let us first illustrate this algorithm by means of an example. Consider the following graph G of Figure 1.36.

To find the strong components of the graph, we search the graph according to dfs. The graph is represented by successor list of each vertex: $succ[1] = (2,4); succ[2] = (3,4); succ[3] = (2); succ[4] = (3); succ[5] = (1,6,9); succ[6] = (7,8); succ[8] = (7); succ[9] = (10); succ[10] = (9)$.

The dfs structure of the graph G of Figure 1.36 is given in Figure 1.37:

The following Table 1.11 gives the dfsn of different vertices.

The function LOWLINK: Like the function "LOW" of the biconnected components algorithm, we shall define the function "LOWLINK" on the set of vertices of the graph.

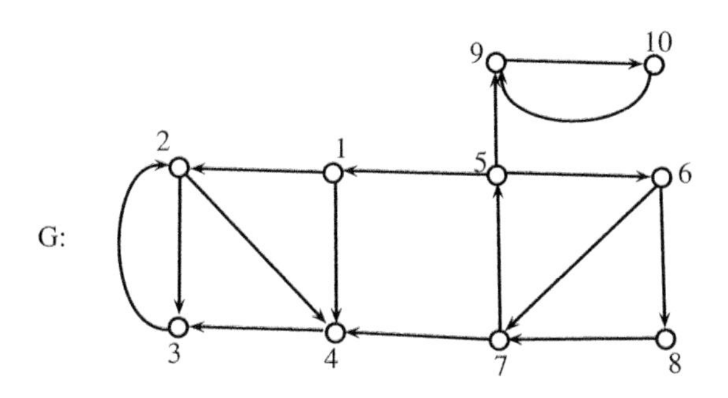

FIGURE 1.36: Illustration of strong components algorithm.

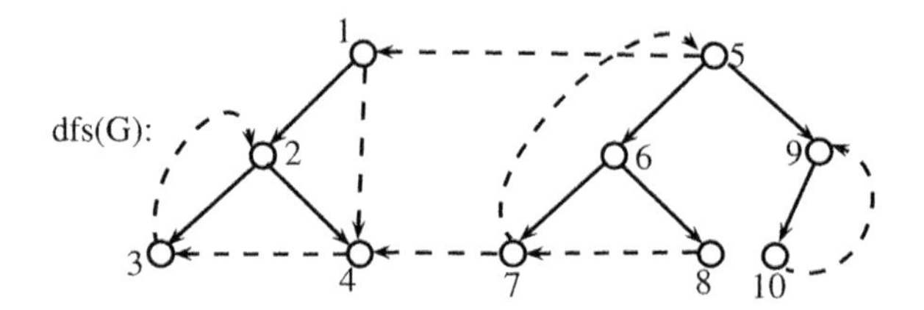

FIGURE 1.37: dfs: strong components algorithm.

TABLE 1.11: Depth-first search number (dfsn) table of graph of Figure 1.37

	1	2	3	4	5	6	7	8	9	10
dfsn	1	2	3	4	5	6	7	8	9	10

LOWLINK[i] is the smallest dfsn $\leq dfs[i]$ one can reach, either from the vertex i or from one of the descendants of i in the dfs spanning forest, using only one dotted mounting or cross arc. In the case of utilization of a dotted cross arc in the calculation of LOWLINK, this dotted cross arc *must* lie in some circuit of the graph. If such dotted arcs lead to vertices whose depth-first search number $> dfsn[i]$ or if no such dotted arcs exist from i or from one of its descendants, then we define LOWLINK$[i] = dfsn[i]$.

Mathematically, we write LOWLINK$[i] = \min(\{dfsn[i]\} \cup B)$, where

$$B = \{dfsn[j] \mid \exists \text{ an arc} \in M \cup C \text{ lying in a circuit from a descendant of } i \text{ to } j\},$$

where M is the set of mounting arcs and C is the set of cross arcs. Let us calculate the function LOWLINK in our example. In the algorithm, the different LOWLINKS are computed according to the postorder traversal of the vertices of the spanning dfs forest. The postorder of our dfs forest is $(3, 4, 2, 1, 7, 8, 6, 10, 9, 5)$. Hence the first vertex for which LOWLINK is computed is the vertex 3.

LOWLINK[3] is $\min(dfsn[3], dfsn[2]) = \min(3, 2) = 2$, because of the mounting dotted edge $(3, 2)$ from the vertex 3. LOWLINK[4] is $\min(dfsn[4], dfsn[3]) = \min(4, 3) = 3$ because of the dotted cross edge $(4, 3)$ from the vertex 4 and this arc lies in a circuit $(4, 3, 2, 4)$. LOWLINK$[2] = \min(dfsn[2], dfsn[2]) = 2$ because of the dotted mounting arc from the son 3 of the vertex 2. LOWLINK$[1] = 1$. LOWLINK[7] is $\min(dfsn[7], dfsn[5] = \min(7, 5) = 5$. Note that the dotted cross arc $(7, 4)$ from the vertex 7 is not taken into account for the calculation of LOWLINK[7] because this cross arc does *not* lie in any circuit of the given graph. Similarly, while computing LOWLINK[5], the dotted cross arc $(5, 1)$ will not be considered because of the absence of a circuit passing through this arc. On the other hand, while computing LOWLINK[8], we have to take into consideration the dotted cross arc $(8, 7)$ because this arc lies in a circuit $(8, 7, 5, 6, 8)$ of the given graph.

TABLE 1.12: LOWLINK function table of graph of Figure 1.37

	1	2	3	4	5	6	7	8	9	10
LOWLINK	1	2	2	3	5	5	5	7	9	9

The following table gives the function LOWLINK (Table 1.12).

In our example, there are *four* strong components. These components are induced subgraphs of the graph G generated by the vertex set:

$$\{2, 3, 4\};\ \{1\};\ \{5, 6, 7, 8\} \text{ and } \{9, 10\}.$$

We make the following important observation from the example:

The intersection of the *arcs* of the strong component generated by the set $\{2, 3, 4\}$ and the arcs of the dfs spanning forest is a *tree* of the spanning forest. This is true for the other three strong components as well. In fact this is true in general. For a proof, see [3].

The root of the tree obtained by the intersection of the arcs of a strong component and the dfs spanning forest is called the root of the strongly connected component.

Putting it differently, in each strong component, choose the vertex r whose dfsn is *minimum* among all vertices of the strong component. This vertex is called the *root* of the strongly connected component.

Back to our example, the root of the strong component induced by the set $\{2, 3, 4\}$ is the vertex 2 and the root of the strong component generated by $\{1\}$ is 1, $\{5, 6, 7, 8\}$ is 8 and the one generated by $\{9, 10\}$ is the vertex 9.

Once we have identified the *roots* of the strong components, we can easily find the set of vertices of each strong components. We use a stack S of vertices. Initially, the stack S is *empty*. Recall that we have used a stack of edges in the biconnected components algorithm.

As we traverse the graph by means of the dfs, we push the vertices into the stack as soon as we "visit" them. When should we pop up the stack to get the strong components? As in the biconnected components algorithm, when we *backtrack*, that is, when we *climb up* the dfs forest under construction, if a *root* r of a strong component is encountered, then we pop up the stack till (including) the vertex r, to obtain the strong component containing the vertex r.

In our example, we push into the stack the vertices 1 and 2 and finally the vertex 3. No more new vertices are found from the vertex 3. So we prepare to backtrack to the father of the vertex 3 which is 2. At this point, the stack will look as (stack grows from left to right)

$$S = (1, 2, 3).$$

Now we test if the vertex 3 is a root of a strong component. The answer is "no." So no pop-up occurs. From the vertex 2, we find a new vertex 4 and hence we push into the stack the vertex 4. No new vertices are found from the vertex 4 and we are ready to backtrack from 4 to his father 2. At this point the stack will look as

$$S = (1, 2, 3, 4).$$

We perform the test: Is 4 a root of a strong component? Since this is not the case, no pop-up occurs. We climb up to the vertex 2. From 2, no new vertices are found and we perform the test if 2 is a root of a strong component. Since the answer to the test is "yes," we pop up the stack S till the vertex 2. This is the vertex set of the first strong component found by the algorithm.

The first strong component is induced by the set $\{4, 3, 2\}$.

We go back to the vertex 1. Since no new vertices are found from 1, we perform the test if 1 is a root of a strong component. Since 1 is a root, we pop the stack till the vertex 1.

The second strong component found is induced by the set $\{1\}$.

At this point the stack is again empty. The dfs takes a new vertex 5 as a root of a tree in the spanning forest under construction and initiates the dfs search from this vertex 5. From 5 we find 6 and from 6 we find the vertex 7. No new vertices are found from 7. So we backtrack. At this point, the stack will be

$$S = (5, 6, 7).$$

Test fails for 7 and we move to 6. From 6, we find 8. No new vertices are "visited" from 8. We prepare to climb up the tree from 8 to the father of 8 which is 6. At this point, the stack will be $S = (5, 6, 7, 8)$. 8 is not the root. From 6 no new vertices are "visited." Since 6 is not a root, we reach the vertex 5. From 5, we find the new vertex 9, from 9 we find the vertex 10. At this point, the stack is $S = (5, 6, 7, 8, 9, 10)$. 10 is not a root and we reach back to 9. No new vertices are found from 9. 9 is a root and we pop up the stack till the vertex 9.

Hence the third strong component is induced by the set $\{10, 9\}$.

The stack $S = (5, 6, 7, 8)$. Finally, we climb up to the vertex 5. Since no new vertices are found from 5, it is a root. Hence we pop up the stack till the vertex 5.

Hence the fourth and final strong component is induced by the set $\{8, 7, 6, 5\}$.

Note that the order in which the strong components are found is the same as that of the *termination* of their dfs calls. In our example, this order is $2, 1, 9, 5$. The call dfs(2) ends before the call dfs(1) ends. The call dfs(1) ends before the call dfs(9) ends. Finally, the call dfsn(9) ends before the call dfs(5) ends.

It remains to recognize the different vertices which are the roots of strong components. From the example, we observe the following property of the roots of strong components.

The root r of a strong component satisfies the equality LOWLINK$[r]$ = $dfsn[r]$ and conversely if a vertex x satisfies the equality LOWLINK$[x]$ = $dfsn[x]$, then x is a root of some strong component.

Before proceeding to write the algorithm, we shall express the function LOWLINK in a recursive manner. In this recursive definition, LOWLINK of a vertex x is expressed in terms of the LOWLINK of the children of the vertices of x and dotted mounting arcs and cross arcs from a descendant of the vertex x.

$$\text{LOWLINK}[i] = \min(\overbrace{\{dfsn[i]\}}^{\text{part1}} \cup \overbrace{\{\text{LOWLINK}[s] \mid \text{ s, a son of i in dfs forest}\}}^{\text{part2}} \cup C')$$

where

$$C' = \overbrace{\{dfsn[j] \mid \exists \text{ an arc} \in M \cup C \text{ lying in a circuit from a descendant of i to j}\}}^{\text{part3}},$$

where M is the set of mounting arcs and C is the set of cross arcs. We are now ready to describe the algorithm in pseudo-code.

We use the following variables: The array "mark" (mark[i] = 0 if the vertex i is "unvisited" and mark[i] = 1 if the vertex i has been "visited during the dfs). The array "dfsn" where dfsn[i] is the depth-first search number of the vertex i and the array "LOWLINK" to compute the LOWLINK of each vertex. The variable "counter" is used to assign the dfsn. The stack S elements are the vertices of the graph.

```
Strongly connected components algorithm:
procedure dfs_strong_comp( i : integer);
var j : integer;
LOWLINK : array[1..n] of integer; (* n is the number of vertices *)
begin
    mark[i] := 1;
    dfsn[i] := counter; counter := counter + 1;
    (* initialization of LOWLINK: part 1 of recursive definition *)
    LOWLINK[i] := dfsn[i];
    top := top +1; S[top] := i; (* push i into the stack S *)
    (* scan the list succ(i) to find a new vertex j *)
    for each j in succ(i) do
        if mark[j] = 0
        then (* j is a son of i in dfs forest *)
        begin
            dfs_strong_comp(j); (* recursive call *)
            (* at this point  we have computed LOWLINK[j] *)
            (* update LOWLINK: part 2 of recursive definition *)
            LOWLINK[i] := min(LOWLINK[i],LOWLINK[j]);
        end
```

```
        else(* j is "visited" *)
        (* test if arc (j,i) is a mounting  or cross arc *)
            if (dfsn[i] > dfsn[j]) and ((j,i) in a circuit)
            then
                (* update LOWLINK: part 3 of LOWLINK definition *)
                LOWLINK[i] := min(LOWLINK[i],dfsn[j]);
    (* test if the vertex i is a root of a strong component *)
    (* if so pop up the stack till the vertex i *)
    if LOWLINK[i] = dfsn[i]
    then
        repeat
            write(S[top],' ');
            top := top - 1;
        until S[top+1] = i;
        writeln;(*new line to separate strong components *)
    end;

    (* initialization *)
    for i := 1 to n do
        mark[i] := 0;
    top := 0; (* empty stack *)
    counter := 1; (* to assign dfs number *)
    for i := 1 to n do
        if mark[i] = 0
        then dfs_strong_comp(i); (* call to procedure dfs_
        strong_comp*)
```

The above algorithm is simply an implementation of the function LOWLINK. In order to write a working program, we have to implement "part 3" of the recursive definition of LOWLINK. In this part 3 of LOWLINK, we take into account a dotted cross arc (i, j) to compute LOWLINK if this arc lies in a circuit of the graph (see part 3 of the recursive definition of LOWLINK).

There is a circuit passing through the dotted cross arc (i, j) if and only if the vertex j is *still* in the stack S.

To illustrate this, let us again refer to our example illustrating strong components algorithm.

In the example, the order of the roots of the strong components according to the *termination* of their dfs call is $(2, 1, 9, 5)$. Note that we have taken into account the cross arc $(4, 3)$ in the computation of LOWLINK[4], since this arc lies in a circuit or equivalently the terminal vertex 3 of the cross arc $(4, 3)$ is in the stack S while we examine this cross arc. Similarly, we have taken into consideration the cross arc $(8, 7)$ in the computation of LOWLINK[8], when we are at vertex 8 traversing this cross arc, since the end vertex of the arc, namely, the vertex 7 is in the stack S.

On the other hand, the cross arc $(5, 1)$ is *not* taken into consideration in the computation of LOWLINK[5], since the end vertex 1 of this cross arc does not belong to the stack S. Note that the strong component containing the vertex 1 has already been emitted by the algorithm when we are at vertex 5 examining the cross arc $(5, 1)$. For a similar reason, we discard the cross arc $(7, 4)$ in the computation of LOWLINK[7].

```
Strong components program in Pascal:
program strongcomp;
   const maxn = 50; (* maximum number of vertices *)
         maxm = 250;(* maximum number of edges *)
   type pointer = ^node;
          node  = record
               info : integer;
               next : pointer;
          end;

  var n, m, i, j, x, y, counter : integer;
      succ : array[1..maxn ] of pointer;
      S : array[1..maxn ] of integer;(* stack of vertices *)
      top :integer; (* top of the stack S *)
 procedure input_graph;
 var i, j, x, y :integer;(* (x,y) is an arc *)
             t :pointer;
 begin(* input_graph *)
     write(' enter the number of vertices ');
     readln(n);
     write(' enter the number of arcs ');
     readln(m);

     (* initialization. graph with n vertices and 0 arcs *)
     for i := 1 to n do
         succ[i] := nil;
     (* read the arcs *)
     for i := 1 to m do
     begin
         write('enter the arc ', i);
         readln(x,y);
         (* add y at the head of the list succ[x] *)
         new(t); (* create a node pointed by t *)
         t^.info := y;
         t^.next := succ[x];
         succ[x] := t;(* attach t at the head of L[x] *)
     end;
 end;(* input_graph *)
 procedure output_graph;
```

```
 var t : pointer;
     i : integer;
 begin
     for i := 1 to n do
     begin
         t := succ[i];
         if t = nil
         then
             write('no successor to ', i)
         else
         begin
             write('the successor to ',i, ' are :');
             (* scan the list succ[i] *)
             while t <> nil do
             begin
                 write(t^.info,'  ');
                 t := t^.next;
             end;
             writeln;
         end;
 end;(* output_graph *)
procedure dfs_strong_comp( i : integer);
var t : pointer; j : integer;
    dfsn : array[1..maxn] of integer;
    LOWLINK : array[1..maxn] of integer;
    instack : array[1..maxn] of boolean;
    (* instack[i] is true if i is in stack, false otherwise *)
begin
    mark[i] := 1;
    dfsn[i] := counter; counter := counter + 1;
    (* push i into stack *)
    top := top + 1; S[top] := i;
    isstack[i] := true;
    (* initialization of LOWLINK[i] using part 1 *)
    LOWLINK[i] := dfsn[i];
    (* scan the list succ[i] *)
    t := succ[i];
    while t < > nil do
    begin
        j := t^.info;
        if mark[j] = 0
        then (* j is a new vertex *)
        begin
            dfs_strong_comp(j); (* recursive call *)
             (* at this point we have computed LOWLINK[j]*)
             (* update LOWLINK[i] using part 2*)
             if LOWLINK[i] > LOWLINK[j]
             then LOWLINK[i] := LOWLINK[j];
```

```
            end
            else
                if (isstack[j] = true) and (dfsn[i] > dfsn[j])
                then (* (i,j) is cross arc in a circuit *)
                    (* update LOWLINK[i]: part 3 *)
                    if LOWLINK[i] > dfsn[j]
                    then LOWLINK[i] := dfsn[j];
            t := t^.next;
        end;
        (* test if i is a root of a strong component *)
        if LOWLINK[i] = dfsn[i]
        then(* i is a root of a strong component *)
        begin
            repeat
                write(S(top],' ');
                isstack[S[top]] := false;
                top := top - 1;
            until S[top +1] = i;
            writeln;(* new line *)
        end;
    end;
    begin (* strongcomp *)
     (* initialization *)
        for i := 1 to n do
        begin
            mark[i] := 0;
            isstack[i] := false;
        end;
        top := 0; (* empty stack *)
        counter := 1;
      (* end of initialization *)
         input_graph;
         output_graph;
         for i := 1 to n do
             if mark[i] = 0
             then dfs_strong_comp(i); (* call *)
    end.
```

Complexity of strong components algorithm:
The complexity of the strong components algorithm is the same as that of the dfs procedure. Intuitively, each vertex is "visited" only once and pushed/popped only once. Each edge is traversed only once. Hence the complexity of Tarjan's algorithm is $O(\max(n, m))$.

Remark 1.6. *We have already said that the biconnected components algorithm and the strongly connected components algorithm are dual algorithms. This duality actually comes from the following observation:*

In a graph, vertices and edges can be considered as dual concepts. The different biconnected components of an undirected connected graph may be obtained as equivalence classes induced by a suitable equivalence on the set of edges E of the graph G. This relation R_1 is defined as below.

Two edges e_1 and e_2 are related by the relation R_1 if $e_1 = e_2$ or there is an elementary cycle passing through these two edges. This is an equivalence relation on the set of edges of the graph G. Reflexivity of R_1 is evident by the definition. Symmetry of R_1 is also clear, since the order of appearance of the edges e_1 and e_2 on an elementary cycle is not relevant according to the definition. Now let us prove transitivity of R_1. If $e_1R_1e_2$ and $e_2R_1e_3$, then there is an elementary cycle C_1 passing through e_1 and e_2 and an elementary cycle C_2 passing through e_2 and e_3.

Then, the symmetric difference (considering the cycles C_1 and C_2 as sets of edges) $C_1 \Delta C_2 = (C_1 \cup C_2) \backslash (C_1 \cap C_2)$ is an elementary cycle passing through the edges e_1 and e_3.

The subgraphs induced by the edges of the equivalence classes of this relation R_1 are the biconnected components *of the graph.*

Let us define an equivalence relation R_2 on the vertex set of a directed graph so that the different classes of this equivalence relation induces strong components of the directed graph. Two vertices x and y are in relation under R_2 if either $x = y$ or there is a directed elementary path from the vertex x to the vertex y and *an elementary directed path from the vertex y to the vertex x. Then, we can prove easily that the relation R_2 is an equivalence relation on the vertex set of the graph. The subgraphs induced by the vertices of the equivalence classes of R_2 are the* strong components *of the graph.*

1.12 Traveling Salesman Problem

One of the famous NP-complete problems is the traveling salesman problem (TSP). Let us state this problem.

Input: There are n cities $1, 2, \ldots, n$ and the cost of traveling between any two cities $i, j, (i \neq j)$ is $c_{ij} = c(i, j)$, $(1 \leq i, j \leq n)$. We assume that the cost function is a symmetric function, that is, $c(i, j) = c(j, i)$.

Output: Find a tour of a salesman starting from the city 1, visiting each city exactly once and returning to the starting point 1, by *minimizing* the total cost of the travel. The cost of a tour is the sum of the costs of the travel between the cities.

This problem can be cast into a graph problem by representing the cities by vertices and the intercity routes by the corresponding edges, with the cost of traveling between the vertices i and j as a weight of the edge ij.

Input: A complete simple graph K_n on n vertices and a weight attached to each edge ij, $1 \leq i, j \leq n$.

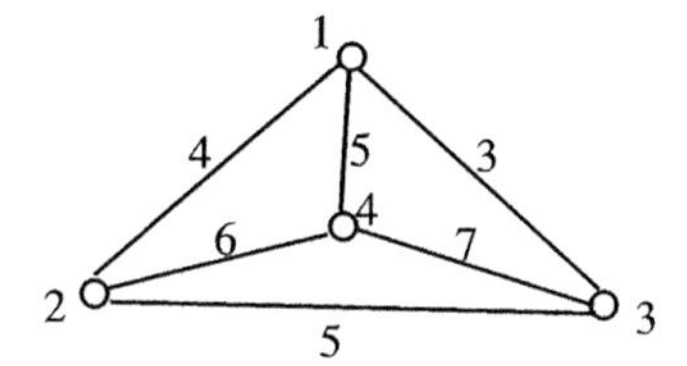

FIGURE 1.38: Illustrating the traveling salesman problem.

TABLE 1.13: Three tours of Figure 1.38

Tour number	Tour	Cost of tour
1	$(1, 2, 3, 4, 1)$	21
2	$(1, 3, 2, 4, 1)$	19
3	$(1, 3, 4, 2, 1)$	20

Output: Find a Hamiltonian cycle of the complete graph K_n with the sum of the weights of the edges of the cycle as a minimum, that is, find an elementary cycle passing through each vertex exactly once with the sum of the weights of the edges as a minimum.

Let us illustrate this problem with an example (see graph of Figure 1.38). In the graph of Figure 1.38, the different tours and their corresponding costs are given in Table 1.13. In Table 1.13, there are three tours and the minimum tour is the tour number 2 whose cost is 19. Note that the tours $(1, 2, 3, 4, 1)$ and the reverse tour $(1, 4, 3, 2, 1)$ give the same cost. Hence such duplicated tours are not listed in the table. The following lemma gives the number of Hamiltonian tours in a complete graph K_n.

Lemma 1.2. *The number of Hamiltonian tours in K_n is $(n-1)!/2$ (by identifying the reverse tours, that is, for $n = 3$, $(1, 2, 3, 1) = (1, 3, 2, 1)$, because they have the same cost).*

Proof. Every tour of K_n defines a cyclic permutation and conversely every cyclic permutation of the set $\{1, 2 \ldots, n\}$ defines a tour of K_n. But the number of cyclic permutation of the n set $[n]$ is $(n-1)!$ (see [6]). By identifying a tour and its reverse tour (because they have the same total cost), the desired number is $(n-1)!/2$. □

Brute-force algorithm to solve TSP:
The first algorithm which comes to our mind is the brute-force algorithm: Generate all the possible tours of the salesman and find one for which the total cost is a minimum. But this algorithm takes exponential time, because there are essentially $(n-1)!/2$ tours and the following inequality shows that factorial n is an exponential function (for a proof, see [6]).

$$n^{n/2} \leq n! \leq ((n+1)/2)^n .$$

For a proof of this inequality, see [6]. The following program implements the brute-force algorithm. This program does not check for equality between a tour and its reverse tour. There are four cities 1, 2, 3 and 4. 0 is not a city. The reader is requested to see the recursive permutation generation program in [6] and appreciate the similarity with TSP brute-force program.

```
Brute-force C program for tsp:
#include <stdio.h>        //load input-output library
#include <stdlib.h>
#include <limits.h>       //INT_MAX is defined here
#define N 4              // N, the number of cities

int  v[N+1]={0,1,2,3,4} ; // v= ville, means city in French
int  c[N+1][N+1]={{0,0,0,0,0},{0,0,4,3,5},{0,4,0,5,6},{0,3,5,0,7},
      {0,5,6,7,0}};
//c, the cost matrix
int  t[N+1]; //t, the Hamiltonian tour
int  mc=INT_MAX ; //mc, minimum cost

void tsp(int k)

{/*tsp finds the cost of a minimum cost Hamiltonian cycle   */
int i,temp,s; // i for the loop, temp for swap and s for the sum

if (k==1) //basis of recurrence



{s=0;
    for (i=1;i<N;i++)
       s = s+c[v[i]][v[i+1]];
         s = s+c[v[N]][v[1]];
    if (s<mc)
    {mc=s;
    //save the Hamiltonian cycle
     for (i=1;i<=N;i++)
     t[i]=v[i];
    }
}
else // recurrence
    { tsp(k-1); //recursive call
        for (i=1;i<k;i++)
        {//swap v[i] et v[k]
        temp=v[i];
        v[i]=v[k];
        v[k]=temp;
```

```
            tsp(k-1); //recursive call
            //restore array v

            //swap again v[i] et v[k]
            temp=v[i];
            v[i]=v[k];
            v[k]=temp;
        }
    }
}

int main ( )

{ int i ;
  tsp ( N ) ; //call with k = N
//print the result
printf("the minimum cost tour is %d \n", mc );
//print the tour
for (i=1;i<=N;i++)
printf("%3d",t[i]);
printf("%3d",t[1]);
system ("pause");
return 0;
}
```

1.12.1 Approximate algorithms for traveling salesman problem

Definition 1.2 (Approximate algorithm). *An algorithm A is said to be an* approximate algorithm *for the traveling salesman problem (TSP), if for any instance I of the TSP (an* instance *of a problem is obtained by specifying particular values of the parameters of the problem, for example, the graph of Figure 1.38 is an instance of TSP) if the ratio*

$$r(A) = A(I)/E(I)$$

is bounded above by a constant c, where $A(I)$ is the value found by the algorithm A on the input I and $E(I)$ is the exact value of the instance. Theoretically, we may obtain $E(I)$ by executing the brute-force algorithm on the instance I. Note that $c \geq 1$. If the constant c is 1, then A is an exact algorithm.

We shall now study three approximate algorithms for TSP [7].

Nearest neighborhood (NN) algorithm:
This is a heuristic. A *heuristic* for TSP is an algorithm which gives an *acceptable solution* but not necessarily optimal solution for TSP. We shall describe the heuristic.

TABLE 1.14: Execution of NN algorithm on Figure 1.38

Partial tour	Vertex chosen
1	3
$(1, 3)$	2
$(1, 3, 2)$	4
$(1, 3, 2, 4)$	1
$(1, 3, 2, 4, 1)$	

Input: A complete graph K_n with vertices $1, 2, \ldots, n$ and a cost function c associated to each of the $n(n-1)/2$ edges of K_n. We further assume the *triangle inequality*, that is, for any three cities i, j, k, we have $c(i, k) \leq c(i, j) + c(j, k)$.

Output: A solution for this instance I such that the ratio

$$NN(I)/E(I) \leq 1/2\,(\lceil \log_2 n \rceil + 1)$$

where $NN(I)$ is the value obtained by the NN algorithm on the instance I and $E(I)$ is the exact value on the instance I.

Algorithm: We shall describe in an informal way this algorithm. Choose the starting city 1 and let $c_1 = 1$. Choose a city $c_2 \neq c_1$ $(2 < n)$ such that $c(c_1, c_2)$ is a minimum (ties may be broken by choosing a smallest city c_2). Next choose a city c_3 $(3 < n$ and $c_3 \neq c_1, c_2)$ such that $c(c_2, c_3)$ is a minimum. More generally, having thus constructed a partial tour $c_1, c_2, \ldots, c_p$, $(p < n)$, we choose c_{p+1} which is different from c_i for $i = 1, 2, \ldots, p$ and $c(c_{p+1}, c_p)$ is a minimum. Finally, having chosen c_n, we choose c_1, the starting city to close the path. We have thus constructed the *complete* tour $(c_1, c_2, \ldots, c_n, c_1)$. An algorithm must be seen to be believed. Let us execute the NN algorithm on an example. Let us refer to the graph of Figure 1.38.

Example 1.14: Execution of NN algorithm

Table 1.14 illustrates the execution.

The complete tour obtained is $(1, 3, 2, 4, 1)$, whose cost is 19, which is exact in this case. We shall now present a C program to implement the NN algorithm for TSP.

```
C program implementing the NN algorithm:
#include <stdio.h>        //load the input/output library
#include <stdlib.h>
#include <limits.h>       //INT_MAX is defined in limits.h
#define N 4              // N is the number if cities
```

```
//int  v[N+1]={0,1,2,3,4} ; // v= villes, means cities in French
int  c[N+1][N+1]={{0,0,0,0,0},{0,0,4,3,5},{0,4,0,5,6},{0,3,5,0,7},
      {0,5,6,7,0}};
// c, the cost matrix
int  t[N+1]; //t, the resulting tour
int  mark[N+1]={0,0,0,0,0};
// mark[i]=0 if city i is not yet visited
//mark[i]=1, otherwise

void tsp_nn( )

int i,j,k,min,counter; //counter is the number of cities visited
//min, the nearest cost neighborhood
// k is the name of the next city to include
t[1]=1; //start with 1
mark[1]=1;
counter=1;
i=1; //i, current city

while (counter<N)
        {
           min=INT_MAX; //initialisation
            for ( j=1;j<=N;j++)
                if ((mark[j]==0)&&(i!=j)&&(c[i][j]<min))
                    {
                       min=c[i][j];
                       k=j;
                    }
            mark[k]=1; //visit city k
            counter=counter+1;
            i=k;
            t[counter]=k;
        }//end while
} //end tsp_nn

int main ()
{
int i,mc=0;//mc, minimum cost tour
tsp_nn() ; //function call
for (i=1;i<N;i++)
    cm=cm+c[t[i]][t[i+1]];
    cm=cm+c[t[N]][t[1]];

//print the result
printf("the minimum cost is %d\n",cm);
//print tour
for (i=1;i<=N;i++)
    printf("%3d",t[i]);
    printf("%3d",t[1]);
system("pause");
return 0
}
```

TABLE 1.15: An approximate algorithm for TSP using MST

Algorithm: We shall describe the algorithm in an informal way.

Step 1: Find a minimum cost spanning tree for the graph K_n, by applying Prim's algorithm (see Section 1.2.1)

Step 2: To each edge of the minimum cost tree T, add one more additional edge. This operation converts the T into an Eulerian graph ET, that is a connected graph in which the degree of each vertex is even.

Step 3: Find a Eulerian cycle of the graph ET. (A Eulerian cycle is a cycle passing through each edge exactly once, but a vertex may be visited more than once. In an eulerian cycle, each vertex x is visited exactly $d(x)$ times).

Step 4: Convert the Eulerian cycle into a Hamiltonian cycle, that is, traveling salesman tour, by taking *"shortcuts."*

An approximate algorithm for TSP using minimum spanning tree and Eulerian cycle (MSTEULER):

Input: A complete graph K_n with vertices $1, 2, \ldots, n$ and a cost function c associated to each of the $n(n-1)/2$ edges of K_n. We further assume the *triangle inequality*, that is, for any three cities i, j, k, we have $c(i,k) \leq c(i,j) + c(j,k)$.

Output: A solution for this instance I such that the ratio

$$\text{MSTEULER}(I)/E(I) < 2$$

where MSTEULER(I) is the value obtained by the MSTEULER algorithm on the instance I and $E(I)$ is the exact value on the instance I.

Before illustrating the algorithm with an example, let us see an algorithm to find a Eulerian cycle in a Eulerian graph. We shall recall the definition of a *bridge*. In any graph G, not necessarily connected, a bridge is an edge e of G (e is not a loop), such that the number of connected components of $G - e$ is strictly greater than the number of components of G.

Algorithm to find a Eulerian cycle in a Eulerian graph:

This algorithm is invented by Fleury.

Fleury's algorithm is given in Table 1.16.

Execution of Fleury's algorithm:

Let us execute Fleury's algorithm on the graph of Figure 1.39.

Start with any vertex, say, the vertex 1. Choose any edge incident with the vertex 1, say, the edge e_3. Now $E_1 = \{e_3\}$ and $E_2 = E \setminus E_1$. Now we are at vertex 3 and we must not choose the edge e_2 as our next edge as the edge e_2 is a bridge of the edge induced subgraph E_2, because we have an alternative edge e_5 or e_5 is incident with 3 which are not bridges of the edge induced subgraph E_2. We choose the edge, say, e_5 as our next edge of the cycle. We are now at the vertex 4. Again, we must not choose the edge e_4 for the same

TABLE 1.16: Informal Fleury algorithm

Input: A connected graph (loops and multiple edges allowed) $G = (X, E)$ in which the degree of each vertex is even.

Output: A Eulerian cycle.

Algorithm: Start with any vertex x_1. Traverse any edge incident with the vertex x_1 (this edge may be a loop). Having reached a vertex x_k, $(1 \leq k \leq n)$, we distinguish two types of edges:

Type 1: The edges already in the cycle being constructed by Fleury's algorithm.

Type 2: The edges not yet included in the partial Eulerian cycle already constructed by the algorithm.

Let us denote the set of edges of type 1 by E_1 and the set of edges of type 2 by E_2. Note that $E = E_1 \cup E_2$ and $E_1 \cap E_2 = \emptyset$. Initially, $E_1 = \emptyset$ and $E_2 = E$, the entire edge set of G.

At the vertex x_k, *unless there is no alternative*, we select an edge incident with x_k *in the edge induced subgraph* E_2, which is *not* a bridge of the edge induced subgraph E_2. The algorithm terminates as soon as $E_2 = \emptyset$.

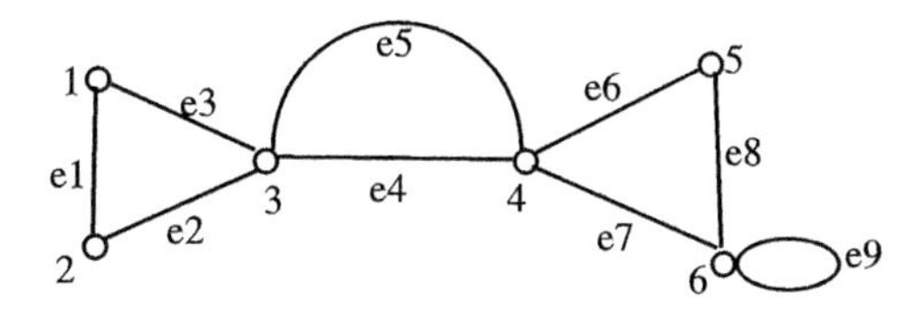

FIGURE 1.39: Illustration of Fleury's algorithm.

reason and we choose an edge, say, e_7 incident with 4. At vertex 6, we must choose the loop e_9. Now again at vertex 6, since there is no alternative, we choose the edge e_8 even though it is a bridge of the edge induced subgraph $E_2 = \{e_8, e_6, e_4, e_2, e_1\}$. Continuing like this, we choose the sequence of edges e_8, e_4, e_2, e_1(note that there are alternative choices left for these edges even though they are bridges). Now $E_2 = \emptyset$ and the algorithm stops. We have thus the Eulerian cycle,

$$(e_3, e_5, e_7, e_9, e_8, e_6, e_4, e_2, e_1).$$

Now we are ready to execute the MSTEULER algorithm for TSP. Consider the graph of Figure 1.38. We shall execute the algorithm on this graph of Figure 1.38.

Step 1. Execution of Prim's algorithm on the graph of Figure 1.38 gives us the spanning tree Figure 1.40.

Step 2. Doubling the edges of the graph of the spanning tree Figure 1.40 gives us the graph of Figure 1.41.

Step 3. A Eulerian cycle of the graph of Figure 1.41 is: $(e_1, e_2, e_5, e_6, e_3, e_4)$.

Step 4. Taking shortcuts gives the Hamiltonian cycle: $(1, 2, 4, 3, 1)$.

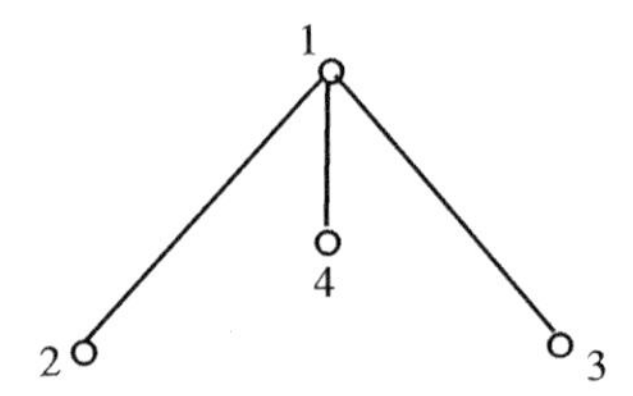

FIGURE 1.40: Spanning tree of Figure 1.38.

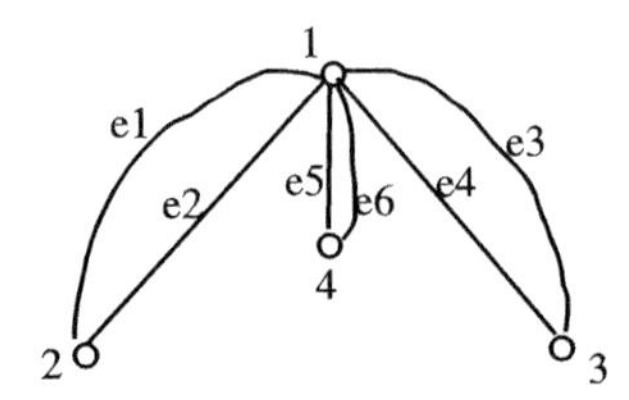

FIGURE 1.41: Doubling the edges of the tree of Figure 1.40.

How to take the shortcuts:
We follow the vertices visited by the Eulerian cycle already constructed. The first edge is e_1 whose ends are 1 and 2. Hence the Hamiltonian cycle starts with the vertex 1 followed by the vertex 2. Now the next edge in the Eulerian cycle is e_2 whose end vertices are 2 and 1 (in the order appearing in the cycle), but these vertices are already in the Hamiltonian cycle under construction. Hence we do not include them. We now examine next edge in the Eulerian cycle which is e_5 whose ends are 1 and 4 (in the order) and the vertex 1 is already in the cycle under construction but the vertex 4 is a new vertex (which is not in the cycle under construction). Hence we take the *shortcut* from the vertex 2 directly to the vertex 4. The partial cycle constructed is $(1, 2, 4)$. The next edge in the Eulerian cycle is e_6 whose ends are 4 and 1. Both are already in the partial Hamiltonian cycle. So we do not include them. The next edge we consider is e_3 whose ends are 1 and 3. But 3 is not in the cycle. Hence we take the *shortcut* from the vertex 4 to the vertex 3. The cycle thus constructed is $(1, 2, 4, 3)$. Since all the vertices are in the cycle, we complete the cycle by adding once more initial vertex 1 of the cycle. The cycle obtained is $(1, 2, 4, 3, 1)$. The cost of the cycle is 20.

Theorem 1.6. *For any instance of the TSP, we have*

$$\text{MSTEULER}(I)/E(I) < 2$$

where the MSTEULER(I) is the cost obtained by the MSTEULER algorithm and $E(I)$ is the exact cost.

Proof. Let us observe that if C is a Hamiltonian cycle and e is an edge of C, then $C \setminus e$ is a spanning tree of the graph G. Hence, if T is a minimum cost spanning tree of G, then $w(T) \geq w(C \setminus e)$ where $w(T)$ is the sum of the weights of the edges of T and $w(C \setminus e)$ is the sum of the weights of the edges of $C \setminus e$. By doubling each edge of T, we obtain the graph ET, with $w(ET) = 2w(T)$. Since a Eulerian cycle traverses each edge of ET exactly once, its total weight is also $w(ET)$. Taking shortcuts will not increase the cost $w(ET)$ as:

$$c(x_1, x_2) + c(x_2, x_3) + \cdots + c(x_{k-1}, x_k) \leq c(x_1, c_k)$$

by triangle inequality, where c is the cost function associated with each edge. Hence the theorem. □

Another approximate algorithm for TSP using a minimum matching (tspmm):

Input: A complete graph K_n with vertices $1, 2, \ldots, n$ and a cost function c associated to each of the $n(n-1)/2$ edges of K_n. We further assume the *triangle inequality*, that is, for any three cities i, j, k, we have $c(i,k) \leq c(i,j) + c(j,k)$.

Output: A solution for this instance I such that the ratio

$$tspmm(I)/E(I) < 3/2$$

where $tspmm(I)$ is the value obtained by the MSTEULER algorithm on the instance I and $E(I)$ is the exact value on the instance I.

Algorithm: The algorithm is given in Table 1.17.

TABLE 1.17: An approximate algorithm for TSP using a minimum matching

Algorithm: We shall describe the algorithm in an informal way.

Step 1: Find a minimum cost spanning tree T of K_n using Prim's algorithm in Section 1.2.1. Hence this step is the same as the first step of the previous algorithm.

Step 2: Recall that in any graph, the number of vertices whose degrees are odd integer is always an even integer (see [6]). We find a *perfect matching* (a *matching* is a collection of edges no two of which are adjacent. A perfect matching is a matching with exactly $n/2$ edges for a n vertex graph. Of course, n must be even.) among the vertices of odd degree of T whose total weight is a minimum.

Step 3: Add the edges of a minimum weight matching found in step 2. The graph thus obtained MT is a connected graph whose degrees are even (because of the addition of a matching). Hence MT is an eulerian graph. Find an eulerian cycle using Fleury's algorithm 1.16.

Step 4: As in the previous algorithm (see step 4 of Table 1.15), take shortcuts of the Eulerian cycle obtained in step 3.

1.13 Exercises

1. Consider a directed graph G represented by its adjacency matrix. Write a program in C to find the vertex-deleted subgraph $G - k$ where k is any vertex of the graph G. Find the complexity of your program.

2. Consider a directed graph G represented as linked lists. Write a program in C to find the vertex-deleted subgraph $G - x$ where x is any vertex of the graph G. Find the complexity of your program.

3. Consider a directed graph G represented by its adjacency matrix. Write a program in C to find the arc-deleted subgraph $G-(x, y)$ where (x, y) is any directed edge of the graph G. Find the complexity of your program.

4. Consider a directed graph G represented as linked lists. Write a program in C to find the arc-deleted subgraph $G-(i, j)$ where (i, j) is any directed edge of the graph G. Find the complexity of your program.

5. Write a program in C to represent a graph in the form of its incidence matrix. What is the complexity of your program?

6. Let G be a connected simple graph containing at least one bridge, with weights associated to each edge of the graph G. Let T be a minimum cost spanning tree found by Prim's algorithm/Kruskal's algorithm.

   Then, T contains all the bridges of the graph G. True or false? Justify your answer.

7. Let $G = (X, E)$ be a simple graph on n vertices with weights associated to each edge of G and let T be a minimum cost spanning tree obtained by applying Prim's algorithm on the graph G. Let $c_1, c_2, \ldots, c_{n-1}$ be the edges of the tree T arranged in non-decreasing order; that is, $c_1 \leq c_2 \leq \cdots \leq c_{n-1}$. Consider any spanning tree T' of G with edge costs $c'_1 \leq c'_2 \leq \cdots \leq c'_{n-1}$.

   Show that $c_i \leq c'_i$ for all i with $1 \leq i \leq n-1$.

8. Apply Prim's algorithm on the graph of Figure 1.42. The weights of edges are written on each edge. Give the weight of the spanning tree thus obtained.

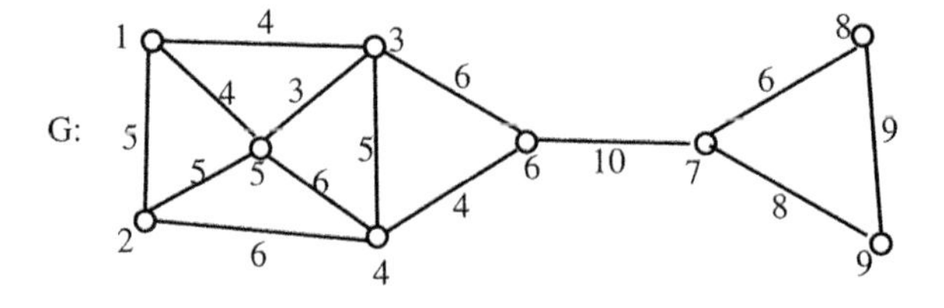

FIGURE 1.42: A weighted simple graph.

9. Apply Kruskal's algorithm on the graph of Figure 1.42. The weights of edges are indicated on each edge. Give the weight of the spanning tree obtained.

10. Apply Dijkstra's algorithm on the graph of Figure 1.43 by taking the vertex 1 as the source vertex. The weights of arcs are indicated on each directed edge. Give the array "dad." Draw the spanning arborescence obtained by the algorithm.

11. Execute Warshall's algorithm on the graph of Figure 1.43 by ignoring the weights associated to each arc.

12. Execute Floyd's algorithm on the graph of Figure 1.43. Give the matrix $INTER_{5\times 5}$ obtained at the end of the execution. Execute the call inter-path(1,4) and write the vertices printed by the call.

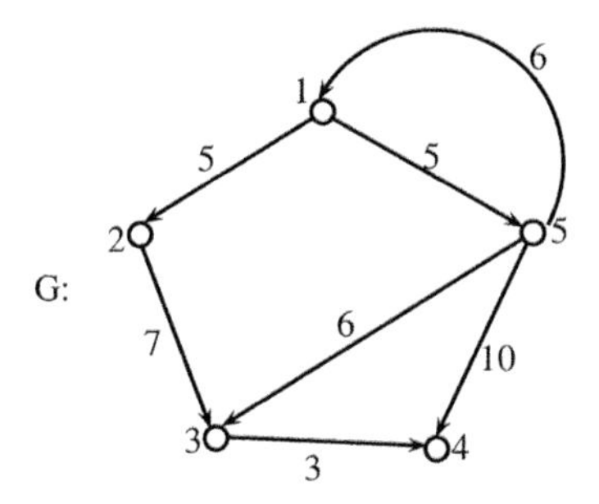

FIGURE 1.43: A weighted directed graph.

13. Prove that in a directed graph, the number of directed walks of length k from a vertex i to a vertex j is the (i, j) entry of the matrix M^k where M is the adjacency matrix of the graph. Using this result, design a $O(n^4)$ algorithm for transitive closure of a graph.

14. Execute Hopcroft's algorithm to find the biconnected components on the graph of Figure 1.44. The vertices are arranged in increasing order in each adjacency list. Process the vertices in increasing order. Draw clearly the dfs tree together with the back edges. Give the arrays a dfs number and LOW.

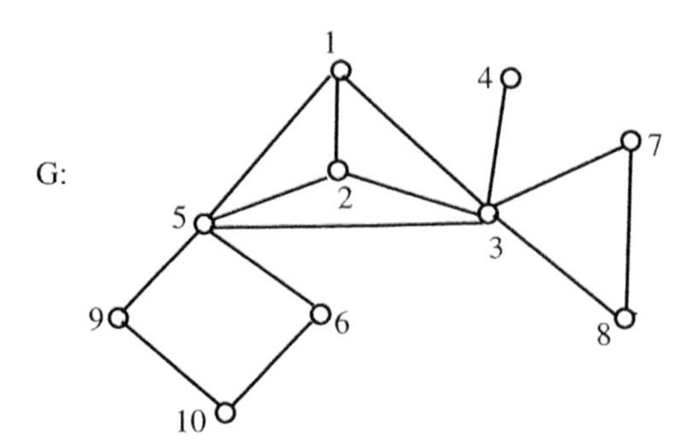

FIGURE 1.44: A simple connected graph.

15. Add suitable instructions in Hopcroft's biconnected components algorithm to print the cut vertices and bridges of a connected graph with at least two vertices.

16. Execute Tarjan's strongly connected components algorithm on the graph of Figure 1.44 after giving following directions/orientations to the edges of the graph:

    Orient the edges as follows: $(1,2), (2,3), (3,1), (3,7), (7,8), (8,3),$ $(4,3), (1,5), (2,5), (3,5), (5,6), (6,10), (10,9), (9,5)$.

    Write the arrays dfsn and LOWLINK. Draw the dfs spanning forest with forward, backward, and cross dotted arcs.

17. Execute Dijkstra's negative cost algorithm on the following graph of five vertices. The graph is represented by its 5×5 costs matrix W.

$$W = \begin{pmatrix} 0 & 7 & 8 & 9 & 10 \\ 0 & 0 & 8 & 9 & 10 \\ 0 & -2 & 0 & 9 & 10 \\ 0 & -4 & -3 & 0 & 10 \\ 0 & -7 & -6 & -5 & 0 \end{pmatrix}$$

    Can you guess the complexity of Dijkstra's negative cost algorithm from this example?

18. Execute the topological sort algorithm on the graph of Figure 1.44 after assigning the following orientations to the edges of the graph:

    Orient the edges as follows: $(1,2), (1,3), (1,5), (2,3), (2,5), (3,5), (4,3),$ $(5,6), (6,10), (10,9), (5,9), (3,7), (7,8), (3,8)$.

19. Write a program in Pascal for Dijkstra's algorithm.

20. Write a program in C for Dijkstra's algorithm.

21. Write a program in Pascal for Prim's algorithm.

22. Write a program in C for Prim's algorithm.

Chapter 2

Graph Algorithms II

> The things we have understood in such a way that they can be instructed to a computer could be called science. Otherwise they are only art.
>
> **D. E. Knuth**

In this chapter on graph algorithms II, we introduce another systematic way of searching a graph known as breadth-first search (BFS). Testing if a given graph is geodetic and finding a bipartition of a bipartite graph are given as applications. Next, matching theory is studied in detail. Berge's characterization of a maximum matching using alternating chain and König-Hall theorem for bipartite graphs are proved. We consider matrices and bipartite graphs: Birkhoff-von Neumann theorem concerning doubly stochastic matrices is proved. Then, the bipartite matching algorithm using a tree growing procedure (Hungarian method) is studied. The Kuhn-Munkres algorithm concerning maximum weighted bipartite matching is presented. Flow in transportation networks is studied. The Ford-Fulkerson algorithm to find a maximum flow and a minimum cut capacity in a transportation network is treated. As applications of bipartite networks, we give an algorithm to determine if a given sequence of pairs of integers is realizable by a p-graph (that is, a graph in which the number of parallel edges/directed arcs from a vertex a to a vertex b is at most p). Greedy algorithm and approximate algorithm for a minimum transversal in a graph are studied. Havel-Hakimi's algorithm concerning degree sequence of a simple graph is studied. Exact exponential algorithm and polynomial greedy algorithm for graph coloring are studied. Finally, the relation between the chromatic number and the examination time tabling problem is discussed.

2.1 Breadth-First Search

In Chapter 1 on Graph Algorithms I, we have discussed the depth-first search (dfs). Another systematic way of processing/visiting the vertices of a graph is the breadth-first search (bfs). In dfs, we go as deeply as possible into the graph before fanning out to other vertices. On the other hand, during the bfs, we "fan" out to as many vertices as possible, before going "deeper" into the graph (see [3]).

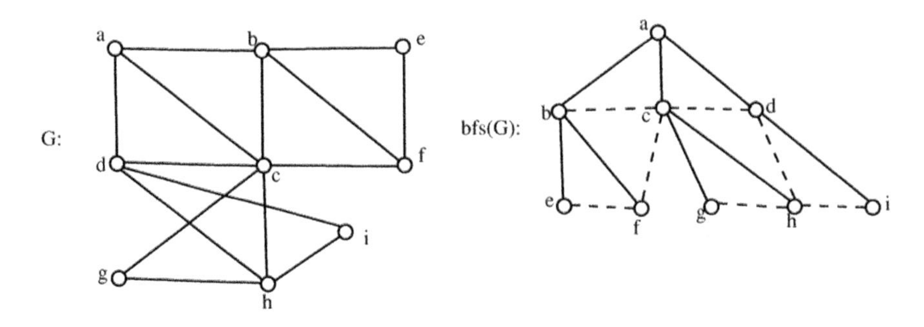

FIGURE 2.1: A graph G and its breadth-first search, bfs(G).

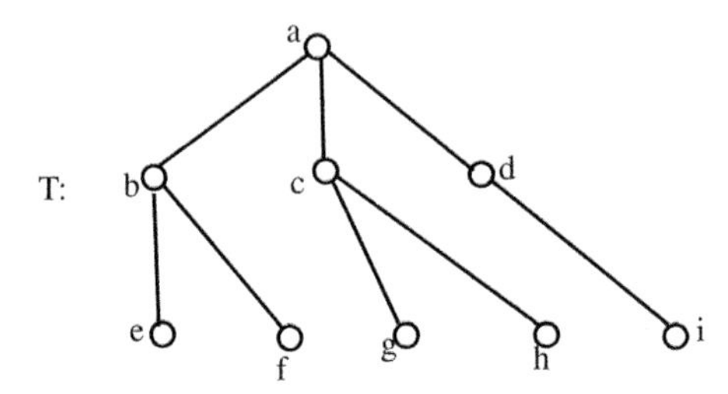

FIGURE 2.2: Breadth-first search tree of graph G of Figure 2.1.

We shall illustrate this procedure by an example. Let us consider the connected simple graph G of Figure 2.1.

The graph is represented by the adjacency lists: $L(a) = (b, c, d)$; $L(b) = (a, c, e, f)$; $L(c) = (a, b, d, f, g, h)$; $L(d) = (a, c, h, i)$; $L(e) = (b, f)$; $L(f) = (b, c, e)$; $L(g) = (c, h)$; $L(h) = (c, d, g, i)$; $L(i) = (d, h)$.

Initially all the vertices are marked "unvisited." We start the search with any vertex as the starting vertex, say, the vertex a. This vertex a will be the root of the bfs tree under construction. We mark the vertex a "visited." Then, we scan completely the adjacency list of the vertex a looking for vertices marked "unvisited." We mark each of the vertices b, c, d "visited" and add the edges ab, ac, ad to the bfs tree. We now scan the list $L(b)$ looking for vertices marked "unvisited." Two new vertices e, f are found and they are marked "visited" and the edges be, bf are added to the bfs tree.

Now by scanning the list $L(c)$, we find two new vertices g, h and the edges cg, ch are added to the tree. Next by scanning the list $L(d)$, only one new vertex i is found and the edge di is added to the tree. The tree T obtained by the search is defined by the solid edges in Figure 2.2 and all other edges of the graph G which are not in the tree T are drawn as dotted edges.

We observe the following properties of the dotted edges.

Observation 2.1. *A dotted edge, which is not a loop, is always a cross edge, that is, a dotted edge joins two vertices x, y where neither is an ancestor or descendant of the other. This property is just the opposite of the property of the dotted edges of the dfs of an undirected graph.*

Every cross edge defines an elementary cycle as in the dfs. More precisely, consider a cross edge xy, and let the vertex z be the closest common ancestor of the vertices x and y in the bfs tree. Then, the unique path from z to x in the tree and the unique path from z to y in the tree together with the cross edge gives an elementary cycle of the graph.

For example, in Figure 2.1, the cross edge hi defines the cycle (a, c, h, i, d, a), the common ancestor being the root vertex a.

Having seen an example, let us now describe the bfs algorithm in pseudo-code.

Breadth-first or level-order search algorithm:
Input: An undirected graph $G = (X, E)$ with $X = \{1, 2, \ldots, n\}$. The graph is represented by the n lists $L(i), i = 1, 2 \ldots, n$.

Output: A partition of the edge set E into the edge F, where F is a spanning forest of G, with F being a vertex disjoint union of trees and the set D of dotted edges.

Algorithm: An array $mark[1..n]$ is used where $mark[i] = 0$ if and only if the vertex i is not yet "visited" during the search. Initially, $mark[i] := 0$ for each vertex i. We use a "queue of vertices" Q. Initially, the queue is empty. The bfs tree is represented by an array of "dad," that is, $dad[j] = i$ if the vertex i is the father of the vertex j in the tree, that is, ij is an edge of the bfs tree under construction. Initially, the set of edges of the bfs tree T is empty. We use an array $bfsn[1..n]$ where bfsn stands for the breadth-first search number. More precisely, $bfsn[i] = k$ if i is the kth vertex "visited" during the bfs. Thus, the bfsn of the root vertex of the first tree in F is 1. The integer variable counter is initialized to 1. The algorithm is given below in pseudo-code.

```
procedure bfs(x : vertex);
(* bfs marks by breadth first search, all the vertices
  accessible from the vertex x by an elementary path  *)
var y,z:vertex; Q: queue of vertex;
begin(*bfs*)
  (* initialization of Q *)
  Q:= empty;
  mark[x]:=1; bfsn[x]:=counter; counter;= counter +1;
  enqueue x in Q;
  while Q is nonempty do
  begin
    y:= head of the queue Q;
    dequeue;(*remove the front vertex of Q*)
    for each vertex z in L(y) do
      if mark[z]=0
      then begin
        mark[z] :=1; bfsn[z]:=counter; counter: counter+1;
        enqueue z in the queue Q;
```

```
        dad[z]:=y;(* add the edge yz to the forest F*)
        F:=F + {yz};
     end;(*while*)
  end;(*bfs*)

  begin(* main program*)
    (*initialization of array mark , F and counter*)
    for each vertex x do
      mark[x]:=0;
    F:=empty;counter:=1;
    for each vertex x do
      if mark[x] = 0 then
      bfs(x);(* call*)
  end;
```

Note that if the graph is not connected, bfs will be called from the main program on a vertex of each connected component, that is, there will be as many calls to bfs as the number of connected components of the graph. Let us now discuss the complexity of bfs.

Complexity of bfs algorithm:
As in the dfs, each vertex is processed at most a constant amount of time (e.g., marking, enqueueing, dequeueing) and each edge xy is traversed exactly twice, once from x and once from y. Hence, the time taken by bfs is $O(n+2m)$ which is $O(\max(n,m))$ by the sum rule of "big-oh" notation.

Property 2.1 (Level sets: Partition of the vertex set induced by bfs). *Bfs induces a partition of the vertex set of a connected graph with the following properties:*

The vertex set

$$X = L_0(a) \cup L_1(a) \cup \cdots \cup L_k(a)$$

where a is the root vertex, k is the eccentricity of the vertex a and the set $L_i(a)$ consists of all vertices at distance i from the vertex a. Symbolically,

$$L_i(a) = \{\, x \mid d(a,x) = i \,\}.$$

The sets $L_i(a)$ are called level sets with respect to the vertex a. By the definition of level sets, there cannot be an edge of the graph between vertices of the level sets $L_i(a)$ and the level sets $L_j(a)$ if $|i-j| > 1$.

Example 2.1: Level sets induced by bfs

Let us refer to graph G of Figure 2.1. In this graph, the level sets are $L_0(a) = \{\, a \,\}$ then $L_1(a) = \{\, b,c,d \,\}$ and finally $L_2(a) = \{\, e,f,g,h,i \,\}$. Note that $L_k(a) = \emptyset$ for all $k \geq 3$.

This partition can be easily obtained from the bfs algorithm as follows:

An array variable $L[1..n]$ of integer is declared together with an integer variable k to assign levels to vertices. k is initialized to 1. In the beginning of the bfs procedure, we add the instruction $L[a] = 0$ and write the statement $L[z] := k$ just after the instruction $mark[z] := 1$. Finally, the variable k is updated by the statement $k := k + 1$ just before the end of the while loop.

Once the array L is available, to find the vertices at level i from the root vertex a, we simply scan the L array to print the vertices x for which $L[x] = i$.

2.2 Applications of bfs Algorithm

As in the case of the dfs algorithm, bfs can be used directly to find the connected components of a graph and to test the existence of cycles in graphs.

We shall now see few applications of bfs: Testing if a given graph is geodetic, testing if a graph is bipartite and finding a maximum matching in a graph.

Geodetic graph:
An undirected graph is geodetic if between any two vertices there is a unique elementary path of shortest length. Let us recall that the length of an elementary path is the number of edges in the path.

By the definition, a geodetic graph must be a simple connected graph. For example, a tree is a geodetic graph. Other examples are the famous Petersen graph of Figure 2.3 and any elementary cycle of odd length.

An algorithm to test if a graph is geodetic:
The algorithm is based on the following simple characterization of geodetic graphs due to Parthasarathy and Srinivasan (see [11]).

Theorem 2.1 (Characterization of geodetic graphs). *Consider a simple connected graph G and a partition of the vertex set X generated by bfs from an arbitrary root vertex x. Then, the graph is geodetic if and only if for every root*

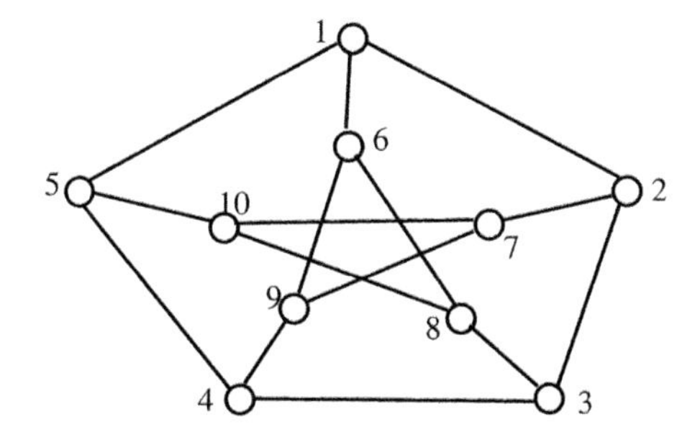

FIGURE 2.3: An example of a geodetic graph.

vertex x, a vertex y of the ith level set $L_i(x)$ is joined by an edge to exactly one vertex z in the $(i-1)$th level set $L_{i-1}(x)$.

Proof. Let G be a geodetic graph. Then, each vertex y in $L_i(x)$ is adjacent to exactly one vertex z of $L_{i-1}(x)$. For, if there is a vertex $y \in L_i(x)$ adjacent to two vertices z_1, z_2 belonging to L_{i-1}, then by the definition of level sets, there are two paths of shortest length i from the root vertex x to the vertex z, one going through z_1 and the other passing via z_2, a contradiction.

For the converse part, suppose that every vertex of the ith level set is joined to exactly one vertex of the $(i-1)$th level set, for all root vertex $x \in X$. If G is not geodetic, then there are vertices x and y such that there are two shortest paths P_1 and P_2 from x to y. Let $z \neq y$ be the last vertex common to both P_1 and P_2 while going from x to y. Then, clearly, $z \neq x$ for, otherwise the two paths will be identical. Now, if the distance between x and z is i, then in the bfs with x as the root vertex, the vertex z in the ith level set has two adjacent vertices in the $(i-1)$th level set, a contradiction. □

We are now ready to describe the algorithm:
Input: A connected simple graph G.

Output: "1" if G is geodetic, "0" if not.

Algorithm: By the above characterization of geodetic graphs, G is geodetic if and only if there is no dotted edge between two consecutive level sets, for any root vertex $x \in X$.

The algorithm is given below in pseudo-code: We declare an integer variable "geo" which is initialized to 1. If there is a dotted edge between two consecutive level sets, then the variable "geo" is set to 0 and the procedure "bfs" terminates, thanks to the statement "return" and the control falls back to the statement which activated the procedure "bfs." $L[s] = L[f] + 1$ if and only if the vertex f is the father of the vertex s in the bfs tree. Note that the square brackets in $L[x]$ denote the level number of the vertex x, whereas the parenthesis in $L(x)$ denotes the list of vertices adjacent to the vertex x.

```
procedure bfs_goedetic(x : vertex);
  var y,z:vertex; Q: queue of vertex;
begin(*bfs_geodetic*)
  (* initialization of Q *)
  Q:= empty;
  mark[x]:=1; L[x]:=0;
  enqueue x in Q;
  while Q is nonempty do
  begin
    y:= head of the queue Q;
    dequeue;(*remove the front vertex of Q*)
    for each vertex z in L(y) do
      if mark[z]=0
```

```
      then begin
        mark[z] :=1;
        enqueue z in the queue Q;
        dad[z]:=y;(* add the edge yz to the T*)
        L[z]:= L[y]+1;(*assign level number to z*)
      end
      else (*z already visited*)
        if L[y]=L[z]+1
        then
          if dad[y] is not z
          (* yz is a dotted edge joining consecutive level sets*)
          then begin geo:=0; return; end;
  end;(*while*)
end;(*bfs*)

begin(* main program*)
  (*initialization of mark ,T,geo*)
  for each vertex x do
    mark[x]:=0;
    geo:=1; T:=empty;
  for each vertex x do
  begin
    bfs_geodetic(x);(*call*)
    if geo = 1
    then (* reinitialize the array mark and tree T*)
      for each vertex x do begin mark[x]:=0; T:= empty; end
    else break;(* quit the for loop, not geodetic*)
  end;
  write(geo);
end;
```

Remark 2.1. *During the bfs, each edge is traversed twice. A dotted edge joining two vertices of the same level is first traversed from left to right and a dotted edge xy joining two vertices of consecutive levels with x in level i and y in level $i+1$ is first traversed from x to y, that is, geometrically, assuming that the sons of a vertex are drawn in the left to right manner, the vertex x is to the right of the vertex y. Hence in the above algorithm, we can simply replace the "else" part of the "for loop" in the procedure "bfsgeodetic" by the following statement:*

if $L[y] = L[z] - 1$ *then* $geo{:=}0$;

Complexity of the geodetic graph algorithm:
The "for loop" of the main program activates the procedure "bfs_geodetic" at most n times for an n vertex graph. The procedure "bfs_geodetic" is activated exactly n times if and only if the graph is geodetic. Since the cost of each activation is $O(\max(n,m))$, the total time spent is $O(n\max(n,m))$.

Algorithm for testing if a graph is bipartite using bfs:
Input: An undirected connected graph $G = (X, E)$. If the graph is not connected, we can consider its connected components separately.

Output: "yes" if the graph G is bipartite, "no" otherwise.

Algorithm: The algorithm is given in Figure 2.4. The algorithm is based on the following property:

Property 2.2. *A graph is bipartite if and only if there are no dotted edges joining two vertices of the same level sets.*

Proof. By König's theorem (see the chapter on Introduction to Graph Theory in [6]), a graph G is bipartite if and only if G does not contain an odd cycle.

Let us first apply the bfs procedure on G from a root vertex a. Suppose there is a dotted edge xy joining two vertices of the same level, say, $x, y \in L_i[a]$. Now consider the closest common ancestor of the vertices x and y in the bfs tree. Let s be such an ancestor. Then, the shortest path from s to x and the shortest path from s to y together with the dotted edge xy will be an elementary cycle of length $d(s, x) + d(s, y) + 1 = 2d(s, x) + 1$, which is odd, a contradiction.

Conversely, suppose G contains an odd elementary cycle. We shall show that there is a dotted edge between two vertices of the same level, when we perform bfs of the graph G. Let $C = (x_1, x_2, \ldots, x_{2p+1})$ be an elementary odd cycle of shortest possible length. First of all, we shall prove that the distance $d(x_1, x_{p+1}) = p$ in the graph G (evidently, the distance between x_1 and x_{p+1} along the edges of the cycle C is p). Otherwise, suppose there is a path P in the graph G between x_1 and x_{p+1} with length strictly less than p. Let the distance path from x_1 to x_{p+1} along the edges of the cycle be Q. Since the paths P and Q have the same initial and final vertices and $P \neq Q$, their union $P \cup Q$ contains at least one elementary cycle. Since the length of the path $Q <$ the length of the path P, there must be an elementary cycle C' in $P \cup Q$ with the following property.

Exactly two vertices x_i and x_j of the cycle C are in $C' \cap Q$ and the length of the subpath of P from x_i to $x_j >$ the length of the subpath of Q from x_i to x_j. In fact, the cycle C' is the union of the subpath of P from x_i to x_j and the subpath of Q from x_i to x_j.

Since p is the length of a smallest possible odd cycle and the length of the cycle $C >$ the length of the cycle C', the length of the cycle C' must be even. This means that the subpaths of P from x_i to x_j and the subpath of Q from x_i to x_j are of the same parity. This implies that the removal of the edges of the subpath of P from x_i to x_j and adding the edges of the subpath of Q from x_i to x_j gives us an odd cycle whose length is $<$ the length of the cycle C, a contradiction. Hence, the distance $d(x_1, x_{p+1}) = p$ in the graph G. By a similar argument, it can be proved that the distance between the vertices x_1 and x_{p+2} is p in the graph G.

Now let us perform the bfs of the graph G starting from the root vertex x_1 which belongs to the cycle C. By the definition of the level set $L_p(x_1)$, the

```
procedure bfs_bipartite(x : vertex);
  var y,z:vertex; Q: queue of vertex;
begin(*bfs_bipartite*)
  (* initialization of Q *)
  Q:= empty;
  mark[x]:=1; L[x]:=0;
  enqueue x in Q;
  while Q is nonempty do
  begin
    y:= head of the queue Q;
    dequeue;(*remove the front vertex of Q*)
    for each vertex z in L(y) do
      if mark[z]=0
      then begin
        mark[z] :=1;
        enqueue z in the queue Q;
        dad[z]:=y;(* add the edge yz to the T*)
        L[z]:= L[y]+1;(*assign level number to z*)
      end
      else(*z already visited*)
        if L[y]=L[z]
        then (*yz is a dotted edge in the same level set*)
          begin bip:=0; return; end;
  end;(*while*)
end;(*bfs_bipartite*)

begin(* main program*)
  (*initialization of mark ,T,bip*)
  for each vertex x do
    mark[x]:=0;
  bip:=1; T:=empty;
  bfs_bipartite(x);(*call with any vertex x*)
  if bip = 1
  then
    begin write("yes");
      (*write the bipartition of G*)
      for i:=1 to n do
        if L[i] mod 2 = 0
        then write(i,' ');
      writeln;(*newline to separate bipartition*)
      for i:=1 to n do
      if L[i] mod 2 =1 write(i,' ');
    end
  else write("no");
end;
```

FIGURE 2.4: Algorithm to test if a connected graph is bipartite.

vertices x_{p+1} and x_{p+2} are in the level set $L_p(x_1)$. But then, the vertices x_{p+1} and x_{p+2} are joined by a dotted edge (since they are consecutive vertices of the cycle C).

Hence the property. □

Property 2.3. *Consider a connected bipartite graph G. Perform a level order search of the graph G starting from any vertex x. Then, the bipartition of the graph G is given by the union of evenly subscripted level sets and the union of oddly subscripted level sets. Symbolically, the bipartition of the vertex set X is $L_0(x) \cup L_2(x) \cup L_4(x) \cup \cdots$ and $L_1(x) \cup L_3(x) \cup \cdots$.*

Proof. Since graph G is bipartite, by Property 2.2 there are no cross edges between two vertices of the same level. By Property 2.2 of the level sets, there can be no edge between any two oddly/evenly subscripted level sets. This implies that there can be no edge between any two vertices of the union $\cup_{i=0}^{e(x)} L_{2i}(x)$ and no edge joining two vertices of $\cup_{j=1}^{e(x)} L_{2j-1}$, where $e(x)$ denotes the eccentricity of the vertex x. □

Example 2.2: Bipartition of a bipartite graph

Consider graph G of Figure 2.5.

Let us perform a level order search from the vertex 1. Vertices are listed according to the increasing order in each adjacency list. The level order search gives us the bfs tree of Figure 2.5.

$L_0(1) = \{1\}$ $L_1(1) = \{2, 6, 7\}$ $L_2(1) = \{3, 8, 5, 12\}$ $L_3(1) = \{4, 9, 13, 11\}$ $L_4(1) = \{10\}$. Hence the bipartition of the graph is $L_0(1) \cup L_2(1) \cup L_4(1) = \{1, 3, 8, 5, 12, 10\}$ and $L_1(1) \cup L_3(1) = \{2, 6, 7, 4, 9, 13, 11\}$.

Having seen Properties 2.2, 2.3 and Example 2.2, we can now write the algorithm to test if a given graph is bipartite.

We perform a level order search from an arbitrary vertex. We use a Boolean variable "bip" which is initialized to 1 (means true). If we find a dotted edge joining two vertices in the same level set, then the variable "bip" is set 0 (means false) and the procedure "bfs_bipartite" terminates thanks to the instruction "return." If the graph is bipartite, the algorithm writes the bipartition by scanning through the array $L[1..n]$ printing the evenly subscripted

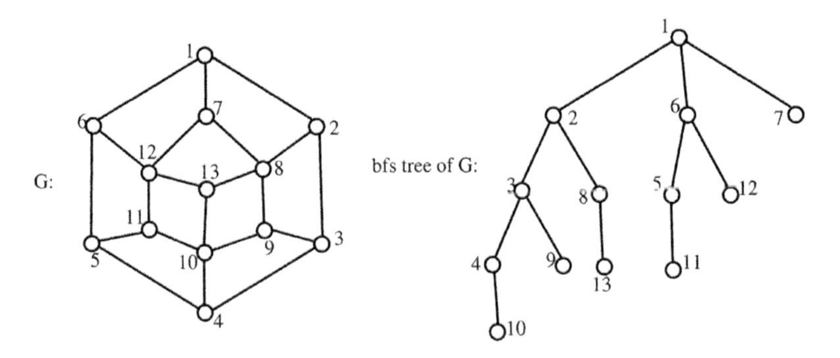

FIGURE 2.5: Finding a bipartition of a bipartite graph G of Figure 2.5.

vertices in one line and oddly subscripted vertices in the next line (n is the number of vertices of the graph).

Complexity of the algorithm "bfs_bipartite":
Clearly the complexity is the same as that of the bfs which is $O(\max(m, n))$.

2.3 Matchings in Graphs

Consider an undirected graph $G = (X, E)$. A matching M in G is a set of edges of G satisfying the following two conditions: 1. No loop belongs to M, 2. No two of the edges of M are adjacent, that is, no two of the edges of M share a common vertex. The following example illustrates the concept.

Example 2.3

$M = \{fa, gb, hc, id, je\}$ is a matching of graph G of Figure 2.6. Another matching is $M' = \{ab, cd, fj, gh\}$.

A vertex x of the graph G is saturated by the matching M or $M-$ saturates the vertex x, if there is an edge in M incident with the vertex x. The set of all saturated vertices under the matching M is denoted by $S(M)$. If the edge xy is in the matching M, we say that the vertex x is matched under M with the vertex y.

A matching M of the graph G is a maximum matching, if there is no matching M' of G such that $|M'| > |M|$. Here, $|M|$ stands for the number of edges in M. A perfect matching M is a matching which saturates every vertex of the graph G. Perfect matchings are also called *1*-factors. Every perfect matching is a maximum matching but the converse is not always true. Note that the number of edges in a perfect matching is exactly $n/2$ where n is the number of vertices. Hence the minimal condition for the existence of a perfect matching is that the number of vertices of the graph must be even.

A perfect matching in a graph of n vertices may be viewed as a "bijection along the edges" from a suitable set of $n/2$ vertices onto its complement set of vertices.

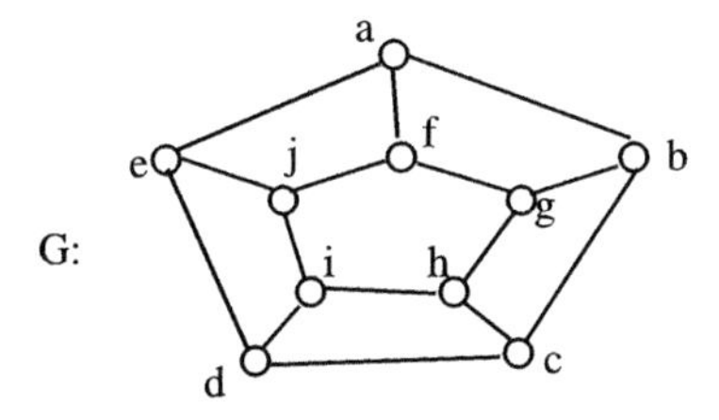

FIGURE 2.6: Illustration of matchings.

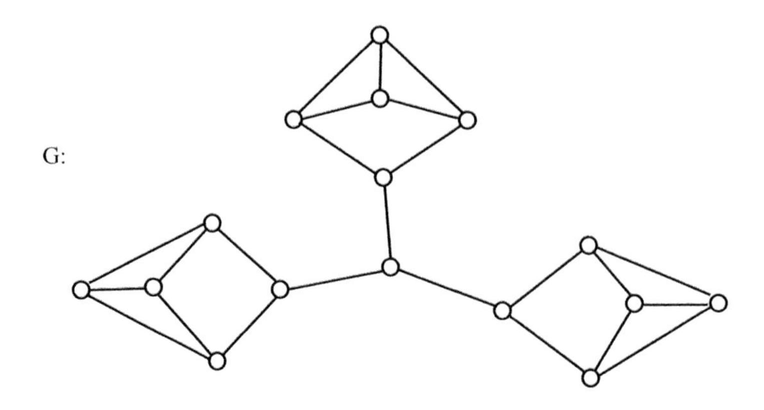

FIGURE 2.7: A 3-regular graph without a perfect matching.

When we discuss the maximum matching problem, we may assume without loss of generality that the graph is a simple graph, that is, it has no loops and no multiple edges.

Let us refer to Example 2.3 (see the graph G of Figure 2.6). M is a perfect matching, whereas M' is not. The graph of Figure 2.7 is an example of a 3-regular graph without a perfect matching.

Example 2.4

We shall see below a few real-world situations in which matchings arise naturally.

We would like to distribute 64 square chocolates among 32 children. Let us imagine that the 64 pieces are arranged in the form of an 8×8 square, like a Chess board. The first child gets two square pieces which are at the diagonally opposite corners of the 8×8 square.

Question: Is it possible to distribute the remaining 62 square pieces (corresponding to the truncated Chess board) among the remaining 31 children in such a way that each child gets a 1×2 rectangular piece?

The problem is equivalent to finding the existence of a perfect matching in the following graph: The vertices correspond to the 62 squares of the truncated Chess board and two vertices are joined by an edge if the squares corresponding to vertices share a common side, not merely a single point. We can demonstrate that the graph thus defined does not possess a perfect matching. We will use the following parity argument (see Tables 2.1 and 2.2).

Color the squares of an 8×8 Chess board alternately black and white, that is, if a square s is colored black, then a square sharing a common side with the square s should be colored white and vice versa. Note that the two diagonally opposite corner squares are colored with the same color. Hence the truncated Chess board has either 32 black squares and 30 white squares or 32 white squares and 30 black squares. But an edge of a matching saturates exactly one black square and one white square. In order to obtain a perfect matching, we should have the same number of black and white squares. Hence we cannot distribute the 62 pieces among the 32 children in the form of a 1×2 rectangle.

TABLE 2.1: Checker board

×	B	W	B	W	B	W	B
B	W	B	W	B	W	B	W
W	B	W	B	W	B	W	B
B	W	B	W	B	W	B	W
W	B	W	B	W	B	W	B
B	W	B	W	B	W	B	W
W	B	W	B	W	B	W	B
B	W	B	W	B	W	B	×

Example 2.5: Domino

In Table 2.2, the shape of a dominoe is drawn. The dominoe is drawn horizontally in the table. It can be also drawn vertically.

TABLE 2.2: Domino

W	B

Example 2.6: Personnel assignment problem

In a certain company, there are p technicians $t_1, t_2, \ldots, t_p$ and q jobs $j_1, j_2, \ldots, j_q$. Each technician is qualified to do one or more jobs.

Question: Is it possible to assign each technician one and only one job for which he/she is qualified?

We construct a bipartite graph with vertices $\{t_1, t_2, \ldots, t_p\} \cup \{j_1, j_2, \ldots, j_q\}$. Two vertices t_k and j_r are joined by an edge if and only if the technician t_k is qualified for the job j_r. Now the problem is to find existence of a matching saturating each vertex t_k for $1 \leq k \leq p$ in the bipartite graph thus constructed.

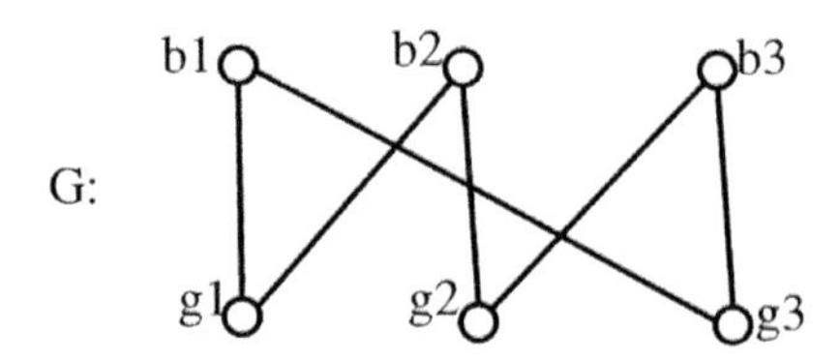

FIGURE 2.8: Illustration of the dancing problem.

Example 2.7: Dancing problem

In a certain college, each girl student has exactly $k > 0$ boyfriends and each boy student has exactly $k > 0$ girlfriends.

Question: Is it possible to organize a weekend dance in such a way that each boy dances with exactly one of his girlfriends and each girl dances with exactly one of her boyfriends? (See Figure 2.8.)

In the graph of Figure 2.8, each boy b_i has exactly 2 girlfriends and each girl g_i has exactly 2 boyfriends. Thus, the bipartite graph is regular of degree 2. The perfect matching $M = \{ b_1g_3, b_2g_1, b_3g_2 \}$ corresponds to a dance in which the boy b_1 dances with the girl g_3, b_2 with g_1 and b_3 with g_2.

Such a dance is always possible. We construct a bipartite graph in which each vertex corresponds to a student and two vertices are joined by an edge if one vertex corresponds to a boy b and the other to a girl g with b a boyfriend of g, g a girlfriend of b. We can show that the bipartite graph thus constructed possesses a perfect matching. This will be proved later.

Alternating and augmenting paths with respect to a matching: Consider a matching M in a graph $G = (X, E)$. An M-alternating path or an alternating path with respect to M is an elementary path in G whose edges are alternately in M and $E \setminus M$, that is, for any two consecutive edges, one of the edges should be in M and other must be in $E \setminus M$. An M-augmenting path is an M-alternating open path whose initial and final vertices are unsaturated by the matching M. The following example clarifies the notion (see Figure 2.9). Note that an edge xy of G where the vertices x and y are unsaturated by a matching M is an augmenting path relative to the matching M. A path of length zero, that is, a path consisting of the single vertex (x) is regarded as an alternating path.

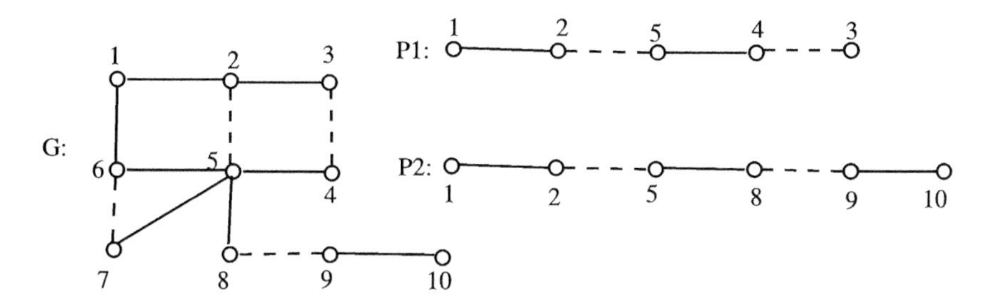

FIGURE 2.9: A graph and its alternating paths.

Example 2.8: Alternating path

Consider a matching $M = \{25, 34, 67, 89\}$ of the graph G of Figure 2.9. The path $P_1 = (1, 2, 5, 4, 3)$ is an alternating path with respect to the matching M. The path P_1 is not an M-augmenting path because the initial vertex 1 of P_1 is not saturated by the matching M.

On the other hand, the path $P_2 = (1, 2, 5, 8, 9, 10)$ is an M-augmenting path, because the initial vertex 1 and the final vertex 10 of P_2 are saturated by the matching M.

We make the following observation: Consider an elementary cycle $C_{2p} = (x_1, x_2, \ldots, x_{2p}, x_1)$ of length $2p$. Then, the set of alternating edges of this cycle $M = \{x_1x_2, x_3x_4, \ldots, x_{2p-1}x_{2p}\}$ is a matching, as well as the set $M' = \{x_2x_3, x_4x_5, \ldots, x_{2p}x_1\}$. In the same manner, alternating edges of an elementary open path form a matching.

Property 2.4. *Consider a matching M in a graph G. Let P be an augmenting path with respect to the matching M. Then, the length of the path P is an odd integer.*

Proof. By the definition of an augmenting path, the initial and final vertices of P are distinct and are unsaturated by the matching. Hence the path starts with an initial edge not in M and ends with an edge also not in M. Hence the number of edges in the path P must be odd. □

Let us recall the operation of symmetric difference of two sets: For any two sets A and B, the symmetric difference of A and B denoted by $A \Delta B$ is the set $(A \cup B) \setminus (A \cap B)$ which is also equal to $(A \setminus B) \cup (B \setminus A)$.

Lemma 2.1 (Berge). *Consider a graph $G = (X, E)$ and two matchings M_1 and M_2 in the graph G. Let $H = (X, M_1 \Delta M_2)$ be a spanning subgraph of G where $M_1 \Delta M_2$ is the symmetric difference $(M_1 \cup M_2) \setminus (M_1 \cap M_2)$. Then, the* *connected components of the graph H can only be one of the following three types:*

Type 1: An isolated vertex x, that is, the degree of the vertex x in the graph H is zero.

Type 2: An elementary cycle of even length whose edges are alternately in M_1 and M_2.

Type 3: An elementary open path whose edges are alternately in M_1 and M_2 whose initial and final vertices are unsaturated by exactly one of the matchings M_1 and M_2.

Proof. Let us first observe that the degree of each vertex of the spanning subgraph (X, M_1) is either 0 or 1, since M_1 is a matching. Similar observation holds for the spanning subgraph (X, M_2). Hence the degree of each vertex x in the spanning subgraph $H = (M_1 \Delta M_2)$ is at most 2, because x can be incident with at most one edge of M_1 and at most one edge of M_2. This means that the only possible components of the spanning subgraph are isolated vertices, an elementary path, and an elementary cycle.

(See Figure 2.10. The spanning subgraph induced by the symmetric difference of matchings $M_1 \Delta M_2$ has five connected components of which three are isolated vertices, one is an elementary path of length 2 and the other is an elementary cycle of length 4.)

By the definition of $M_1 \Delta M_2 = (M_1 \cup M_2) \setminus (M_1 \cap M_2)$, the edges alternate in M_1 and M_2 in each of the component containing at least one edge of H. Hence the length of the cycle component must be even, otherwise either M_1 or M_2 will not be a matching.

Finally, for a path component, if the initial edge belongs to the matching M_1, then the initial vertex of the path component cannot be saturated by M_2, since otherwise either the degree of the initial vertex will be 2 in H or the initial edge belongs to both M_1 and M_2, a contradiction to the symmetric difference $M_1 \Delta M_2$. A similar argument holds for the final vertex of any path component of H. □

An efficient maximum matching algorithm in a graph hinges on the following theorem due to the eminent graph theorist: Claude Berge.

Theorem 2.2 (Berge). *A matching M in a graph $G = (X, E)$ is a maximum matching if and only if the graph G possesses no M-augmenting path relative to M.*

Proof. Consider a matching M and an augmenting path P relative to M. We shall show that the matching M is not a maximum matching. By Property 2.4, the length of such a path must be an odd integer. Let $(x_1, x_2, \ldots, x_{2p+2})$ be an M-augmenting path of length $2p+1$. By the definition of augmenting path relative to M, this means that the edges $x_i x_{i+1}$ are in the matching M for

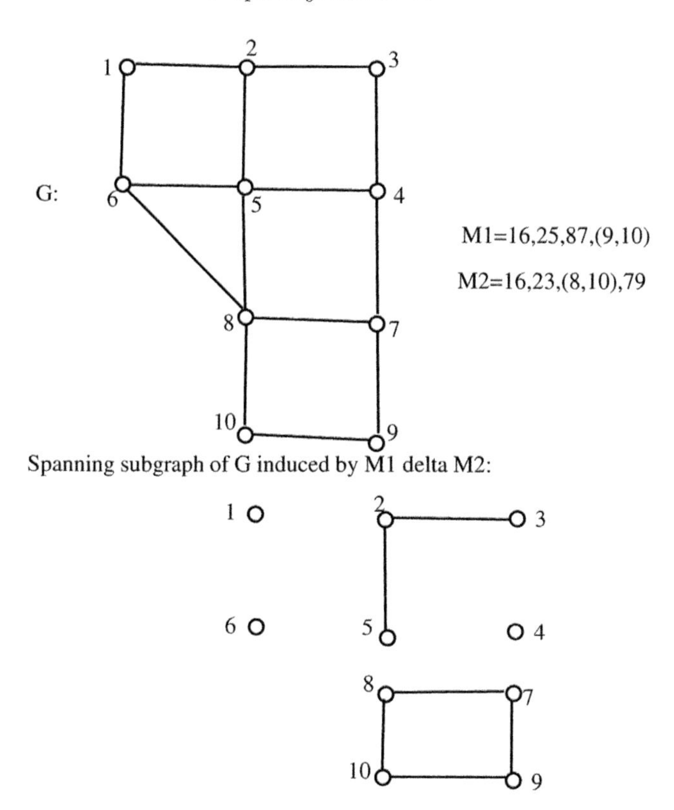

FIGURE 2.10: A graph G and the spanning subgraph induced by $M_1 \Delta M_2$.

$2 \leq i \leq 2p$ and the other edges of the path are in $E \setminus M$ with x_1 and x_{2p+1} unsaturated by M. Now define a new matching M' which consists of the edges of the symmetric difference

$$(M \cup P) \setminus (M \cap P) = \{ M \cup \{ x_1x_2, x_2x_3, \ldots, x_{2p}x_{2p+1} \} \setminus \\ \{ x_2x_3, x_4x_5, \ldots, x_{2p}x_{2p+1} \}.$$

(Note that P represents the set of edges of the path P.) Then, M' is a matching and $|M'| = |M| + 1$, since we have removed p alternate edges of P from M and added $p + 1$ alternate edges of P to M to obtain the new matching M'. Hence M is not a maximum matching.

For the converse part, consider a matching M which is not maximum. We shall prove the existence of an M-augmenting path. Let M' be a maximum matching in G. Then, $|M'| > |M|$.

Consider the spanning subgraph $H = (X, M'\Delta M)$ induced by the edge set $(M' \setminus M) \cup (M \setminus M')$. By Lemma 2.1, the connected components of H are isolated vertices or elementary cycles of even length with edges alternately in M and M' or an elementary path with edges alternately in M and M' with initial and final vertices saturated by exactly one of the matchings M and M'.

Since, $|M' \setminus M| > |M \setminus M'|$, the spanning subgraph H contains at least one more edge of M' than of M. But the (even) cycle components of H account for the equal number of edges of M and M'. Hence there must be some path component P of H such that P contains more edges of M' than of M. Such a path must start with an edge of M' and end with an edge of M'. The initial and final vertices of P are M'-saturated in H, and hence must be M-unsaturated in G. Thus, P is an M augmenting path in G. □

We are now ready to present an outline of a maximum matching algorithm:

An outline of a maximum matching algorithm:
Input: An undirected graph $G = (X, E)$.

Output: A maximum matching in G.

Algorithm: The algorithm is given in Figure 2.11.
In Figure 2.11, P denotes the set of edges of the path P in G, and M denotes the set of edges of the matching in G.

Let us perform the algorithm of Figure 2.11 on a given graph.

Example 2.9: Illustration of matching algorithm

Consider the graph of Figure 2.12. We are interested in finding a maximum matching in this graph.

Initialization of matching M with an arbitrary edge 34: $M = \{34\}$. The only saturated vertices are 3 and 4.

Before beginning each iteration, we perform the following test: Is there any augmenting path relative to the matching M? If the answer is "yes," we continue the iteration. Otherwise, the algorithm terminates with a maximum matching M.

To find an augmenting path relative to the matching M, we use our "eye ball," that is we try to find an augmenting path visually. Table 2.3 illustrates the different steps.

```
Algorithm:
(* Initialization of M *)
M:=Any edge of the graph G or M := ∅;
(*iteration*)
while there is an augmenting path relative to M do
begin(*whilc*)
    Let P be an augmenting path with respect to M;
    M := (M ∪ P) \ (M ∩ P);
end;(*while*)
print the edges of M;
```

FIGURE 2.11: An outline of maximum matching algorithm.

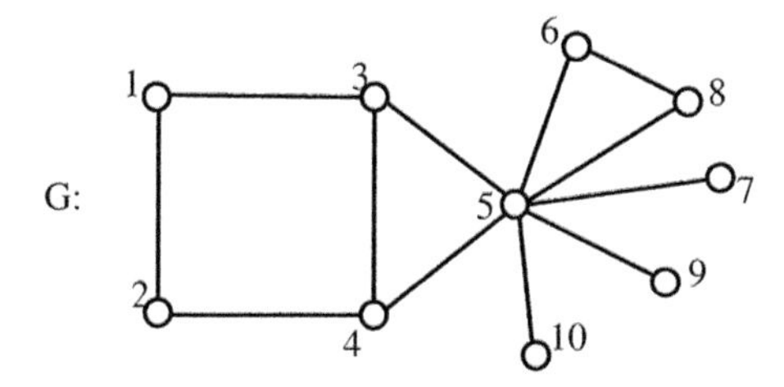

FIGURE 2.12: A graph to illustrate matching algorithm.

TABLE 2.3: Execution of matching algorithm

Iteration number	Is there an M-augmenting path P?	$M := (M \cup P) \setminus (M \cap P)$
0 (initialization)	not applicable	$\{34\}$
1	"yes," $P = (1, 3, 4, 5)$	$\{13, 45\}$
2	"yes," $P = (2, 1, 3, 4, 5, 7)$	$\{21, 34, 57\}$
3	"yes," $P = (6, 8)$	$\{21, 34, 57, 68\}$
	"no," quit the loop	

Hence a maximum matching obtained by the above algorithm is

$$\{21, 34, 57, 68\}$$

The above maximum matching algorithm is incomplete because we do not know yet how to find in a systematic and progressive manner an M-augmenting path, if one exists in the graph. This can be achieved by "a tree growing procedure" which is somewhat equivalent to the bfs. Before presenting this procedure, called "tree growing procedure," we prove the following good characterization of bipartite graphs not containing a particular type of maximum matching.

2.3.1 An application: $(k-1)$-regular subgraphs of k-regular graphs

As an application of Berge's matching algorithm, we solve the following regular subgraph problem:
Input: A k-regular simple connected graph G on *odd* number of vertices with *no* an induced star $K_{1,3}$.

Output: A $(k-1)$-regular subgraph of G.

Algorithm (Sriraman Sridharan): The vertex set is $\{1, 2, \ldots, n\}$. The graph G is represented by its adjacency matrix $M = (m_{ij})_{n \times n}$ where we set $m_{ij} = \infty$ if vertices i and j are not adjacent. In the implementation, ∞ can be replaced by a "large" integer. The algorithm uses the following results:

Step 1: Apply Floyd's algorithm to the matrix $M = (m_{ij})_{n\times n}$(see chapter 1).
The (i,j) entry of M is the length of a shortest path between vertices i and j in G
In step 2, we find a peripheral vertex of G
Step 2:
for i:=1 to n do
begin
    set $e[i] = \max(m_{ij} | 1 \leq j \leq n)$
end
The ith entry of array e gives the eccentricity of vertex i
find a vertex p such that $e[p] = \max(e[i] | 1 \leq i \leq n)$
Step 3:
Construct the subgraph $G_p = G \setminus p \setminus \Gamma(p)$
find a perfect matching M of G_p
return $G \setminus \{ p \cup M \}$

FIGURE 2.13: Algorithm to find a $(k-1)$-regular subgraph in a k-regular graph with no induced star $K_{1,3}$.

If p is a peripheral vertex, that is, a vertex p such that its eccentricity $e(p) = d(G)$ where $d(G)$ is the diameter of G, then $G\setminus p\setminus\Gamma(p)$ is *still* connected and a connected graph with no induced $K_{1,3}$ on even number of vertices has a perfect matching. For more details of the algorithm, see Sriraman Sridharan, Polynomial time algorithms for two classes of subgraph problem, RAIRO, Oper. Res. 42, 291-298, 2008.

The algorithm is given in Figure 2.13. The required subgraph is $G\setminus\{ p\cup M \}$ which is $(k-1)$-regular, where M is a perfect matching of $G\setminus p \setminus \Gamma(p)$. The reader is asked to execute the algorithm by hand on a 4-regular graph to have an understanding of the algorithm. Complexity: The complexity of the algorithm can be shown to be $O(n^3)$.

The following theorem can be viewed as a profound generalization of the following simple observation:

Observation 2.2. *Consider two non-empty sets A and B. There is no injective function $f : A \rightarrow B$ if and only if $|A| > |B|$. Stated differently, there is no injective function from the set A to the set B, if and only if for every function $f : A \rightarrow B$, there is a subset S of A satisfying the inequality $|f(S)| < |S|$. Here, $f(S) = \{ f(s) \mid s \in S \}$, the image set of S.*

Theorem 2.3 (König-Hall). *Consider a bipartite graph $G = (X_1, X_2; E)$ with bipartition X_1 and X_2. Then, the graph G has no matching saturating all the vertices of the set X_1 if and only if there is a subset S of X_1 satisfying the inequality $|N(S)| < |S|$ (in fact, $|N(S| = |S|-1)$. Here, $N(S)$, the neighbor set* *of S, denotes the set of vertices which are collectively adjacent to the vertices of S.*

Proof. We first prove the easy part of the theorem. Suppose there is a set S of X_1 satisfying the inequality $|N(S)| < |S|$. Then, we will prove that there is no matching in G saturating each vertex of the set X_1. If M is a matching saturating each vertex of X_1, then since S is a subset of X_1, there is a subset M' of M such that M' saturates each vertex of S and $|M'| = |S|$. M' is also a matching in G (since a subset of a matching is a matching) and the vertices in S are matched under M' with distinct vertices of the set $N(S)$. This implies that $|N(S) \geq |S|$, a contradiction.

Now consider the converse part. Suppose the graph G possesses no matching saturating all the vertices of X_1. We shall show that there is a subset S of X_1 satisfying the inequality $|S| < |N(S)|$. By our assumption, M does not saturate all the vertices of the set X_1. Hence there must exist a vertex $x_1 \in X_1$ unsaturated by the maximum matching M. Let R be the set of all vertices of the graph G reachable by an M-alternating path from the initial vertex x_1. Note that the path of length zero (x_1) is vacuously an M-alternating path and hence the set R is non-empty. Since the matching M is maximum, by Theorem 2.2, x_1 is the only unsaturated vertex in the set R.

Let S be the set of vertices of R belonging to the set X_1 and let T be the set of vertices of R belonging to the set X_2, that is, $S = R \cap X_1$ and $T = R \cap X_2$. Then, we shall show that

$$|T| = |S| - 1 \text{ and } |N(S)| = T.$$

Consider any M-augmenting path $P = (x_1, x_2, x_3, \ldots, x_p)$ from the initial vertex x_1. Note that an initial portion/section of this path P is also an M-augmenting path. Since G is bipartite, all the oddly subscripted vertices of the path P are in S and all the evenly subscripted vertices of P are in T. By the definition of M-augmenting path, the vertex x_2 is saturated under M with x_3, x_4 is saturated under M with x_5, etc.

Hence the vertices of the set $S \setminus x_1$ are matched under M with the vertices of the set T. This implies that

$$|T| = |S| - 1 \text{ and } N(S) \supseteq T. \tag{2.1}$$

Every vertex belonging to $N(S)$ is reachable from x_1 by an M-augmenting path. For if $s' \in N(S)$, then by the definition of $N(S)$, there is an $s \in S$ such that ss' is an edge of the graph. We shall show that there is an M-augmenting path from the vertex x_1 to the vertex s'.

Since $s \in S$, by the definition of the set S, there is an M-augmenting path Q from x_1 to s. If s' is in the path Q, then we take the initial portion of this path Q ending in s', to have an augmenting path from s to s'; otherwise, we extend the path Q by adding one more vertex s' (and the edge ss') to Q to obtain an augmenting path from x_1 to s'. Since $s' \in S$, we have by the definition of the set T, $s' \in T$. Hence we have the equality

$$N(S) = T.$$

Hence, by (2.1), we have $|N(S)| = |T| = |S| - 1 < |S|$. □

The following example illustrates the fact that Theorem 2.3 is a good characterization of a bipartite graph $G = (X_1, Y_1; E)$ possessing no matching saturating all the vertices of the set X_1.

Example 2.10: Illustration of good characterization

If a bipartite graph contains a matching saturating all the vertices of X_1, then we have to "convince or exhibit," the existence of such a matching in a "reasonable amount of time," that is, in a formal manner, in polynomial time.

Conversely, if a bipartite graph does not possess a matching saturating all the vertices of X_1, we must be able to "exhibit" the non-existence of such a matching in a "reasonable amount of time" (technically polynomial time in the size of the input length).

Consider the bipartite graph of Figure 2.14.

To "convince quickly" that the graph G of Figure 2.14 has a matching saturating all the vertices x_i's, we have to "exhibit" such a matching. One such matching is $M = \{ x_1y_1, x_2y_2, x_3y_3 \}$. Now one can "easily" verify that the set M is indeed a matching and it saturates all the vertices x_i for $i = 1, 2, 3$. For example, one can color the edges of M in the graph G and see immediately that M is a desired matching. Note that the time taken to find the matching M is irrelevant.

Conversely, consider the bipartite graph Figure 2.15.

To "convince quickly" that the graph of Figure 2.15 has no matching saturating all the vertices x_i, we must exhibit a subset S of $\{ x_1, x_2, x_3, x_4 \}$ such that $N(S) = |S| - 1$. Such a set is $S = \{ x_2, x_3, x_4 \}$. Now one can "easily" verify the equation $N(S) = |S| - 1$, because $N(S) = \{ y_2, y_3 \}$ and $|S| = 3$ and $|N(S)| = 2$. Note again here that the time to find such a set is not taken into account. We take into account only the time to find the neighbor set $N(S)$ and the verification of the equation $|N(S)| = |S| - 1$.

Definition 2.1 (Succinct certificate). *Consider a bipartite graph $G = (X_1, Y_1; E)$ having no matching saturating all the vertices of the set X_1. Then, a set $S \subset X_1$ satisfying the equality $N(S) = |S| - 1$ is called a succinct certificate to "exhibit" the non-existence of a matching saturating all the vertices of the set X_1.*

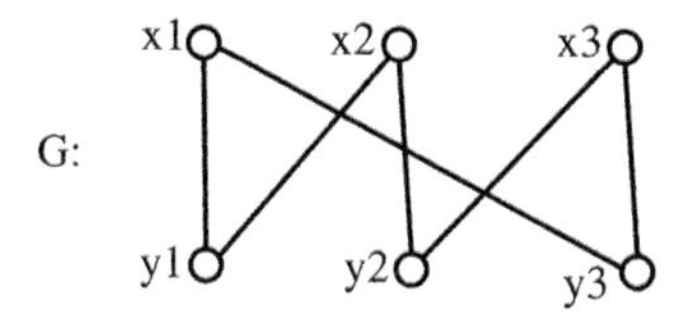

FIGURE 2.14: A perfect matching in a bipartite graph.

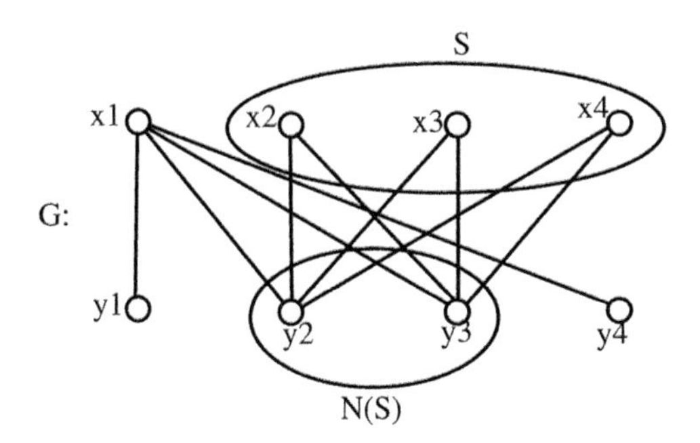

FIGURE 2.15: A graph to illustrate the idea of a succinct certificate.

We shall now prove that the dancing problem (see Example 2.7) has always an affirmative answer:

Corollary 2.1. *Consider a k-regular* ($k > 0$) *bipartite graph* $G = (X_1, X_2; E)$ *with bipartition* X_1 *and* X_2*. Then, the graph G contains a perfect matching.*

Proof. Note that the necessary condition for the existence of a perfect matching in a bipartite graph is $|X_1| = |x_2|$. We shall first prove this minimal condition.

Since G is bipartite, the number of edges of G is the same as the number of edges of the graph G incident with the vertices of the set X_1 (because there are no edges between two vertices of the set X_2). Since G is k-regular, we have $m = k|X_1|$.

By the same argument, we have $m = k|X_2|$.

Therefore, $m = k|X_1| = k|X_2|$. Since $k > 0$, $|X_1| = |X_2|$.

We shall now show that there is no subset S of X_1 satisfying the inequality $|N(S)| < |S|$.

Let S be any subset of X_1 and let E_1 and E_2 be the set of edges incident respectively with the vertices of S and with the vertices of the neighbor set $N(S)$. Since G is bipartite, by the definition of the neighbor set $N(S)$, we have the relation $E_1 \subseteq E_2$ (because there may be edges of the graph G joining a vertex of $N(S)$ and a vertex of $X_1 \setminus S$).

But $|E_1| = k|S|$ and $|E_2| = k|X_2|$. Hence,

$$k|S| \leq k|N(S)| \text{ because } |E_1| \leq |E_2|.$$

This implies that, $|S| \leq |N(S)|$ for every subset S of X_1. By the König-Hall Theorem 2.3, G contains a matching M saturating every vertex of X_1. Since $|X_1| = |X_2|$, the matching M must be a perfect matching. □

The following corollary asserts that the edge set of a regular bipartite graph containing at least one edge can be partitioned into edge disjoint union of perfect matchings.

Corollary 2.2. *The edge set of a k-regular bipartite graph ($k > 0$) can be partitioned into k edge-disjoint perfect matchings.*

Proof. Let G be a k-regular bipartite graph with $k > 0$. Then, by Corollary 2.1, G contains a perfect matching M_1. Now consider the spanning subgraph of $G - M$ of G obtained by removing the edges of the set M. Since the removal of the edges of M reduces the degree of each vertex of G by exactly one, $G - M$ is a $(k-1)$-regular bipartite graph. If $k - 1 > 0$, then we again apply the Corollary 2.1 to get a perfect matching M_2 of the graph $G - M$. We continue like this until we get a bipartite spanning subgraph with no edges. Thus, we have decomposed the edge set E of G as

$$E = M_1 \cup M_2 \cdots \cup M_k$$

with $M_i \cap M_j = \emptyset$ for all i, j with $1 \leq i < j \leq k$. □

Corollary 2.3. *Consider a $(0, 1)$ $n \times n$ square matrix A with the property that the sum of the entries in each row is equal to the sum of the entries in each column, each sum being an integer $k > 0$. Then, the matrix A can be written as the sum of k permutation matrices.*

Proof. Associate a bipartite graph $G = (I, J; E)$ where I is the set of row indices of the matrix A and J is the set of column indices of A. Two vertices i and j are joined by an edge if $i \in I$ and $j \in J$ and the (i, j) entry of A is 1. The degree of a vertex i is simply the sum of the entries of the row i is $i \in I$. Similarly, the degree of the vertex j is the sum of the j-th column where $j \in J$. Hence G is a k-regular bipartite graph. By Corollary 2.2, the edge set can be partitioned into k disjoint perfect matchings

$$E = M_1 \cup M_2 \cdots \cup M_k.$$

Each perfect matching M_i corresponds to a permutation matrix and conversely. For if $M = \{ i_1 j_1, i_2 j_2, \ldots, i_n j_n \}$ is a perfect matching in the graph G, then the corresponding $n \times n$ permutation matrix P has (i_s, j_s) entry 1 for $1 \leq s \leq n$ and all other entries of P are zeros and conversely. Hence, the given matrix

$$A = P_1 + P_2 + \cdots + P_k$$

where P_is are permutation matrices corresponding to disjoint perfect matchings. □

2.4 Matrices and Bipartite Graphs

Corollary 2.2 can be formulated in terms of $(0, 1)$ matrices, that is, matrices whose entries are either 0 or 1. A permutation matrix is a square matrix having exactly one entry 1 in each row and in each column and all other entries are *z*eros.

To each (0,1) matrix A, we can associate a bipartite graph in the following manner: The vertices of the graph G correspond to the indices of the rows and columns of the matrix A and two vertices are joined by an edge if one corresponds to a row i of A and the other to a column j of A with the (i, j) entry of the matrix A equal to 1. Conversely, we can associate a $(0, 1)$ matrix to each bipartite graph.

Example 2.11

The matrix corresponding to the bipartite 2-regular graph of Figure 2.16 is

$$\begin{array}{c} \\ x_1 \\ x_2 \\ x_3 \end{array} \begin{array}{c} \begin{array}{ccc} y_1 & y_2 & y_3 \end{array} \\ \begin{pmatrix} 0 & 1 & 1 \\ 1 & 0 & 1 \\ 1 & 1 & 0 \end{pmatrix} \end{array}$$

The perfect matching $M = \{x_1y_2, x_2y_3, x_3y_1\}$ of the graph of Figure 2.16 corresponds to the *permutation matrix* P where

$$P = \begin{bmatrix} 0 & 1 & 0 \\ 0 & 0 & 1 \\ 1 & 0 & 0 \end{bmatrix}$$

A $(0, 1)$ matrix is called a permutation matrix if its entries are either 0 or 1 and it contains exactly one entry 1 in each row and in each column. Equivalently, a permutation matrix is obtained by permuting the rows and columns of the identity matrix of order n.

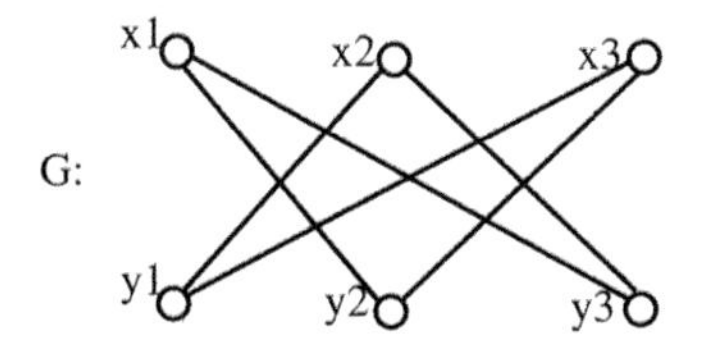

FIGURE 2.16: A bipartite graph and its corresponding (0,1) matrix.

Corollary 2.4. *Consider a* $(0,1)$ $n \times n$ *square matrix* A *with the property that the sum of the entries in each row is equal to the sum of the coefficients in each column which is equal to an integer* $k > 0$. *Then, the matrix* A *can be written as the sum of* k *permutation matrices.*

Proof. Associate a bipartite graph $G = (I, J; E)$ where I is the set of row indices of the matrix A and J is the set of column indices of A. Two vertices i and j are joined by an edge if $i \in I$ and $j \in J$ and the (i,j) entry of A is 1. The degree of a vertex i is simply the sum of the entries of the row i if $i \in I$. Similarly, the degree of the vertex j is the sum of the j-th column if $j \in J$. Hence G is a k-regular bipartite graph. By Corollary 2.2, the edge set can be partitioned into k disjoint perfect matchings

$$E = M_1 \cup M_2 \cdots \cup M_k.$$

Each perfect matching M_i corresponds to a permutation matrix and conversely. For if $M = \{ i_1 j_1, i_2 j_2, \ldots, i_n j_n \}$ is a perfect matching in the graph G, then the corresponding $n \times n$ permutation matrix P has (i_s, j_s) entry 1 for $1 \leq s \leq n$ and all other coefficients of P are zeros and conversely. Hence, the given matrix

$$A = P_1 + P_2 + \cdots + P_k$$

where P_is are permutation matrices corresponding to disjoint perfect matchings. □

Birkhoff-von Neumann theorem on bi-stochastic matrices:
As an application of the König-Hall Theorem 2.3, we shall derive the Birkhoff-von Neumann theorem on bi-stochastic matrices:

An $m \times n$ matrix $M = (m_{ij})$ is bi-stochastic if the entries are all non-negative and the sum of the coefficients of each row and each column is equal to 1.

Observation 2.3. *Any bi-stochastic matrix is necessarily a square matrix. In fact, if* M *is bi-stochastic with* m *rows and* n *columns, the sum of all entries of* M *is the sum of* n *columns of* $M = n$ *(as each column sum is 1). In symbols,*

$$\sum_{i=1}^{m} \sum_{j=1}^{n} m_{ij} = \sum_{j=1}^{n} \left(\sum_{i=1}^{m} m_{ij} \right) = \sum_{j=1}^{n} 1 = n.$$

The same sum is also equal to the sum of m *rows of* $M = m$ *(as each row sum is 1). Symbolically,*

$$\sum_{i=1}^{m} \sum_{j=1}^{n} m_{ij} = \sum_{i=1}^{m} \left(\sum_{j=1}^{n} m_{ij} \right) = \sum_{i=1}^{m} 1 = m.$$

Hence $m = n$.

Example 2.12: Bi-stochastic matrix

Clearly, every permutation matrix is bi-stochastic.

$$M = \begin{bmatrix} 0 & 1/2 & 1/4 & 1/4 \\ 1/4 & 1/4 & 0 & 1/2 \\ 1/4 & 1/4 & 1/4 & 1/4 \\ 1/2 & 0 & 1/2 & 0 \end{bmatrix}$$

M is bi-stochastic.

Determinants:

Recall that for an $n \times n$ square matrix $M = (m_{ij}$, with m_{ij} reals, the determinant of M denoted by $\det M$ is defined as the sum of the products of the entries of M taken only one entry from each row and each column of the matrix M. Symbolically,

$$\det M = \sum_{\sigma \in S_n} (-1)^{\sigma} m_{1\sigma(1)} m_{2\sigma(2)} \cdots, m_{n\sigma(n)}$$

where the sum is taken over all possible permutations S_n of the set $\{1, 2, \ldots, n\}$.

Remark 2.2. *Of course, this definition does not give an efficient way to calculate the determinants because there are $n!$ terms in the* $\det M$. *Surprisingly, determinants of order n can be calculated in polynomial time, in fact in $O(n^3)$ time by row reducing the matrix M to triangular form.*

Lemma 2.2. *Consider an $n \times n$ square matrix $M = (m_{ij})$. In the expansion of the determinant of the matrix M, each term is zero, if and only if there is a regular submatrix (a matrix which can be obtained from M by removing certain rows and columns of M) of order $k \times (n-k+1)$, $(k \leq n)$ whose entries are all zeros.*

Proof. We will apply the König-Hall Theorem 2.3 to prove the lemma.

Let us associate a bipartite graph $G = (X_1, Y_1; E)$ with $X_1 = \{x_1, x_2, \ldots, x_n\}$ and $Y_1 = \{y_1, y_2, \ldots, y_n\}$ with the edge set $E = \{x_i y_j \mid m_{ij} \neq 0 \text{ for all } i, j\}$.

Each term of the determinant is of the form $\pm m_{1i_1} m_{2i_2} \cdots m_{ni_n}$. But then this term is non-zero, if and only if each $m_{pi_p} \neq 0$ for all p with $1 \leq p \leq n$. This is true if and only if the set of edges $\{x_1 y_{i_1}, x_2 y_{i_2}, \ldots, x_n y_{i_n}\}$ form a perfect matching of the corresponding bipartite graph G (see Example 2.11), that is, if and only if G has a matching saturating each vertex of X_1.

Therefore, every term in the expansion of the $\det M$ is zero if and only if the corresponding bipartite graph contains no matching saturating each vertex of X_1. By the König-Hall theorem, this is so, if and only if there is a subset S of the set X_1 satisfying the equality $N(S) = |S| - 1$. Set $|S| = k$. Then,

$|N(S)| = k-1$ and $|X_2 \setminus N(S)| = n-k+1$. Note that by the definition of $N(S)$, there are no edges between the vertex sets S and $Y_1 - N(S)$. Hence the desired $k\times(n-k+1)$ matrix is obtained by removing the rows of the matrix M corresponding to the vertices of $X_1 - S$ and the columns of M corresponding to the vertices of the set $N(S)$. Symbolically, the matrix $M' = (m_{ij})_{S\times(Y_1\setminus N(S))}$ has all its coefficient zeros. □

Lemma 2.3. *In a bi-stochastic matrix, there is at least one term in the expansion of its determinant which is non-zero.*

Proof. Consider an $n \times n$ bi-stochastic matrix $M = (m_{ij})$. If all the terms in the expansion of the $\det M$ is zero, then by Lemma 2.2, there is a submatrix M' of dimension $k \times (n-1+k)$ all of whose terms are zero. Note that a permutation of two rows and/or two columns of a bi-stochastic matrix is again a bi-stochastic matrix. Hence, the given matrix can be restructured and partitioned into the following form:

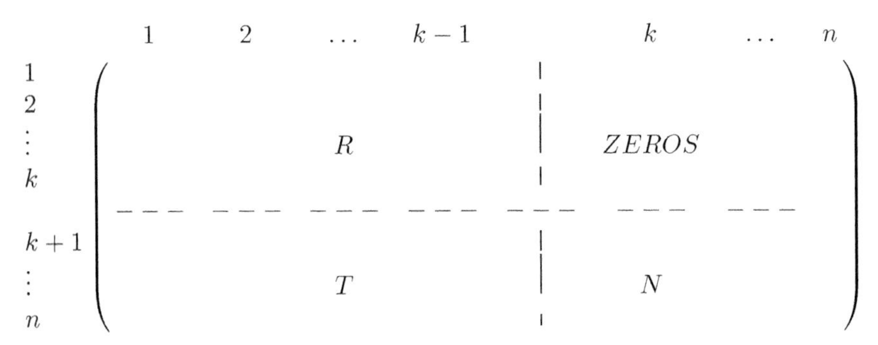

In the above partition, set $R = (r_{ij})_{k\times(k-1)}$, $T = (t_{ij})_{(n-k)\times(k-1)}$, $N = (n_{ij})_{(n-k)\times(n-k+1)}$. The above matrix is still bi-stochastic. Note that the row indices $\{1, 2, \ldots, k\}$ correspond to the set S and the column indices $\{1, 2, \ldots, k-1\}$ correspond to the set of neighbors $N(S)$ in the König-Hall Theorem 2.3.

Now the sum of all the entries of the matrix N is equal to $n-k+1$ (by adding the entries of N column-by-column).

But then the sum of the entries of the matrix T is equal to

$$\sum_{i=k+1}^{n}\left(\sum_{j=1}^{k-1} t_{ij}\right) = n-k-\text{sum of all the entries of } N = n-k-(n-k+1) = -1,$$

a contradiction because the entries of a bi-stochastic matrix are all non-negative. This contradiction establishes the result. □

We are now ready to prove the Birkhoff-von Neumann theorem on the decomposition of bi-stochastic matrices in terms of permutation matrices.

To present the theorem, we need the notion of *convex combination* of vectors. A vector of dimension n is an ordered n-tuple $(\alpha_1, \alpha_2, \ldots, \alpha_n)$ where each

α_i is a real number. Two vectors $v = (\alpha_1, \alpha_2, \ldots, \alpha_n)$ and $u = (\beta_1, \beta_2, \ldots, \beta_n)$ are declared equal, that is $u = v$ if and only if $\alpha_i = \beta_i$ for all i satisfying the inequalities $1 \leq i \leq n$. The *sum of the vectors* u and v denoted by $u + v$ is the vector $(\alpha_1 + \beta_1, \alpha_2 + \beta_2, \ldots, \alpha_n + \beta_n)$ and the scalar product αv for a real α is defined as the vector $(\alpha\alpha_1, \alpha\alpha_2, \ldots, \alpha\alpha_n)$.

Finally, the vector v is a convex combination of the vectors $v_1, v_2, \ldots, v_n$ if the vector v can be expressed as

$$v = \alpha_1 v_1 + \alpha_2 v_2 + \cdots + \alpha_n v_n$$

where each $\alpha_i \geq 0$ and $\sum_{i=1}^{n} \alpha_1 = 1$.
Before presenting the theorem, let us see an example.

Example 2.13: Convex combination

Consider the 4×4 bi-stochastic matrix

$$M = \begin{bmatrix} 1/4 & \boxed{1/4} & 1/4 & 1/4 \\ \boxed{1/2} & 1/4 & 0 & 1/4 \\ 0 & 1/2 & \boxed{1/4} & 1/4 \\ 1/4 & 0 & 1/2 & \boxed{1/4} \end{bmatrix}.$$

This matrix can be viewed as a vector (r_1, r_2, r_3, r_4) of dimension $16 = 4 \times 4$ where r_i denotes the ith row of the matrix M. We can write the matrix M as the convex combination of permutation matrices as follows:

$$M = \frac{1}{4}P_1 + \frac{1}{4}P_2 + \frac{1}{4}P_3 + \frac{1}{4}P_4$$

where

$$P_1 = \begin{bmatrix} 0 & 1 & 0 & 0 \\ 1 & 0 & 0 & 0 \\ 0 & 0 & 1 & 0 \\ 0 & 0 & 0 & 1 \end{bmatrix} \quad P_2 = \begin{bmatrix} 0 & 0 & 0 & 1 \\ 1 & 0 & 0 & 0 \\ 0 & 1 & 0 & 0 \\ 0 & 0 & 1 & 0 \end{bmatrix}$$

$$P_3 = \begin{bmatrix} 1 & 0 & 0 & 0 \\ 0 & 0 & 0 & 1 \\ 0 & 1 & 0 & 0 \\ 0 & 0 & 1 & 0 \end{bmatrix} \quad P_4 = \begin{bmatrix} 0 & 0 & 1 & 0 \\ 0 & 1 & 0 & 0 \\ 0 & 0 & 0 & 1 \\ 1 & 0 & 0 & 0 \end{bmatrix}$$

Theorem 2.4 (Birkhoff-von Neumann). *A matrix is bi-stochastic if and only if it can be expressed as the convex combination of permutation matrices.*

Algorithmic proof. By definition of scalar product, a convex combination of vectors and permutation matrices, it is easy to see that a convex combination of permutation matrices is a bi-stochastic matrix.

For the converse part, let $M = (m_{ij})$ be an $n \times n$ bi-stochastic matrix. Then, we shall prove that M can be expressed as a convex combination of permutation matrices. In fact, we shall give an algorithm to express M as a convex combination of permutation matrices.

By Lemma 2.3, in the expansion of the det M, there is a term of the form $m_{1i_1}m_{2i_2}\cdots, m_{ni_n} \neq 0$ where $(i_1, i_2, \ldots, i_n)$ is a permutation of $(1, 2, \ldots, n)$ (in Example 2.13, the product of the four en-squared terms is a term of the determinant $\neq$ zero).

Set $\alpha_1 = \min(m_{1i_1}, m_{2i_2}, \ldots, m_{ni_n})$. Note that $\alpha_1 > 0$ (in Example 2.13, $\alpha_1 = 1/4$). Now define a permutation matrix $P_1 = (p_{ij}^{(1)})$ where $p_{ki_k}^{(1)} = 1$ and $p_{ki_s}^{(1)} = 0$ if $k \neq s$ (in Example 2.13, the corresponding permutation matrix is P_1).

Set $R_1 = M - \alpha_1 P_1 = (r_{ij}^{(1)})$. The matrix R_1 possesses more zero entries than the matrix M, because at least the coefficient $r_{ki_k}^{(1)} = 0$ with $\alpha_1 = m_k i_k > 0$ for some k.

Now we apply the same argument to the matrix R_1. If every term in the expansion of the det R_1 is equal to zero, then the theorem is proved, because this implies that the matrix R_1 is the zero matrix. Otherwise, there is a term $r_1^{(1)}j_1 r_2^{(1)}j_2 \cdots r_n^{(1)}j_n$ in the expansion of det R_1 different from zero. Let $P_2 = (p_{ij}^{(2)})$ be the permutation matrix defined as

$$p_{kj_s}^{(2)} = \begin{cases} 1 & \text{if } k = s \\ 0 & \text{otherwise} \end{cases}$$

Set $\alpha_2 = \min(r_1^{(1)}j_1, r_2^{(1)}j_2, \ldots, r_n^{(1)}j_n)$. Define the matrix

$$R_2 = R_1 - \alpha_2 P_2 = M - \alpha_1 P_1 - \alpha_2 P_2.$$

As before, the matrix R_2 has more zero entries than the matrix R_1. If every term in the expansion of the det R_2 is zero, then the theorem is proved, because this implies that the matrix R_2 is the zero matrix. Otherwise, we continue the same argument.

Sooner or later we must have the equation

$$R_s = M - \alpha_1 P_1 - \alpha_2 P_2 - \cdots - \alpha_s P_s \tag{2.2}$$

with every term in the expansion of R_s equal to zero. Then, we claim that the matrix R_s is the zero matrix. Otherwise, by Equation (2.2), the sum of the entries of each row and each column of the matrix R_s is $1 - \alpha_1 - \alpha_2 - \cdots - \alpha_s > 0$. Set $\alpha = 1 - \alpha_1 - \alpha_2 - \cdots - \alpha_s$. By dividing every term of the matrix R_s we obtain a bi-stochastic matrix, that is, the matrix $(1/\alpha)R_s$ is bi-stochastic. According to Lemma 2.3, the matrix $(1/\alpha)R_s$ has a term in the expansion of $\det(1/\alpha)R_s$ different from zero. But by a property of determinants, $\det(1/\alpha)R_s = (1/\alpha^n)\det R_s$. This implies that there is a term in the expansion of $\det R_s$ different from zero, a contradiction to our assumption. Hence the matrix R_s is identically equal to zero. Therefore, by Equation (2.2)

$$M = \alpha_1 P_1 + \alpha_2 P_2 + \cdots + \alpha_s P_s$$

and the sum of each row of M = the sum of each column of $M = \alpha_1 + \alpha_2 + \cdots + \alpha_s = 1$, since M is bi-stochastic. Thus, the proof is complete. □

In the maximum matching algorithm of Figure 2.11, it is not indicated how to find an augmenting path relative to the matching M if such a path exists in the graph G. Now we shall do this for the special case of the bipartite graphs.

To present the algorithm for finding an augmenting path relative to a given matching M, we need the following definition.

Definition 2.2 (An M-alternating bfs tree). *Consider a connected bipartite graph $G = (X_1, X_2; E)$ and a matching M in the graph G with at least one unsaturated vertex of the set X_1. Perform a level order search of the graph G starting from an unsaturated vertex $x_1 \in X_1$. Let T be a bfs tree obtained by the level order search starting from the vertex x_1 (see graph G of Figure 2.2). Process the vertices in increasing order. Then, as we have already remarked (see Example 2.1), the tree T, partitions the vertex set of the graph G as $L_0(x_1) \cup L_1(x_1) \cup \cdots \cup L_k(x_1)$ where $L_i(x_1)$ consists of all the vertices at distance i from the vertex x_1 and the integer k represents the eccentricity of the vertex x_1.*

A full M-alternating bfs tree with the root vertex x_1 is obtained by the removal of all the edges of $T \setminus M$ joining a vertex of $L_{2i-1}(x_1)$ and a vertex of $L_{2i}(x_1)$ for all $i = 1, 2, \ldots$. In other words, by going down in the bfs tree T, we remove all the edges joining an oddly subscripted level vertex to an evenly subscripted level vertex not belonging to the matching M. In other words, during the level order search, we add only the edges belonging to the matching M from an odd level to an even level. A partial M-alternating tree or simply an M-alternating tree is a subtree of the full alternating tree having the same root.

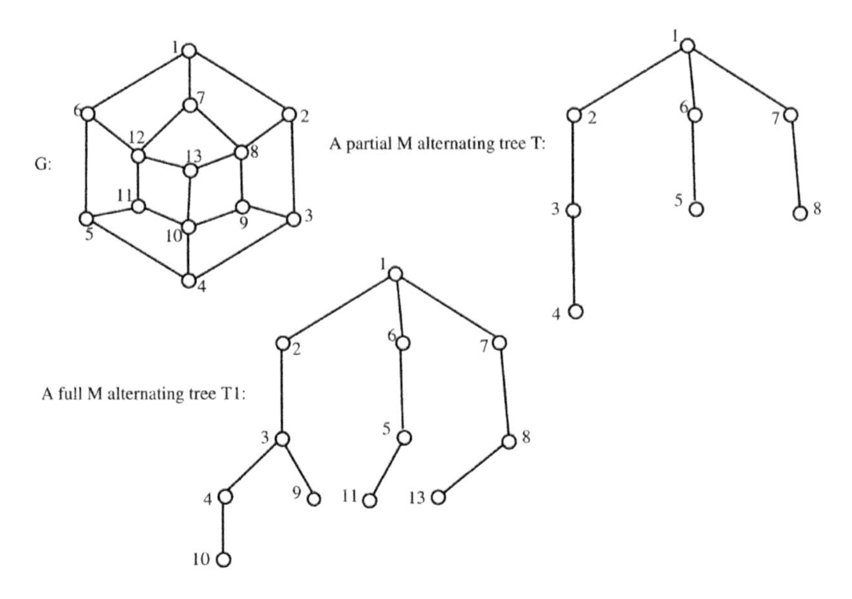

FIGURE 2.17: A bipartite graph G and a partial M-alternating tree.

The following example clarifies the definition.

Example 2.14: An M-augmenting tree

Consider the bipartite graph G of Figure 2.17 with bipartition $X_1 = \{1, 3, 8, 5, 12, 10\}$ and $X_2 = \{2, 6, 7, 4, 9, 13, 11\}$.

Consider the matching $M = \{23, 65, 78, (4, 10)\}$. Then, the tree T of Figure 2.17 is a partial M-alternating bfs tree with the root vertex 1. Note that between the level 1 and level 2 of the tree T, only the edges belonging to the matching M are added. A full alternating tree T_1 is also shown in Figure 2.17. Note that between the level 3 and level 4, only the edge $(4, 10)$ belonging to the matching M is added. The full tree T_1 cannot be grown further. The tree T is a subtree of T_1 having the same root as T.

Algorithm for finding an M-augmenting path in a bipartite graph:
Input: A bipartite graph $G = (X_1, Y_1; E)$ and an arbitrary matching M.

Output: An M-augmenting path, if such a path exists. "No," if there is no M-augmenting path.

Algorithm: see Figure 2.18.

```
Algorithm:
for each unsaturated vertex x in X1 do
begin(* while*)
    grow an M-alternating bfs tree T by definition 2.4.1;
    if we come across an unsaturated vertex y in an oddly subscripted
    level set during the growth of T
    then
   begin
       the unique path P from root x to y in T is an augmenting path;
       return;
    end
    else there is no M-augmenting path from x;
end;(*while*)
```

FIGURE 2.18: Informal algorithm for finding an M-augmenting path.

The complexity of the maximum matching algorithm in a bipartite graph:
For finding an M-augmenting path, we perform a bfs from an unsaturated vertex of a matching M. The complexity of the bfs is $O(\max(n, m))$ where n is the number of vertices and m is the number of edges. Each M augmenting path leads us to an improved matching M' where $|M'| = |M| + 1$. Since the cardinality of a maximum matching is at most $n/2$, the bfs will be called at most $n/2$ times. Hence, the complexity of the maximum matching algorithm is $O((n/2) \max(n, m)) = O(n \max(n, m))$.

What can be done once, may be repeated many times.
*D*ijkstra

Now we give an algorithm to find a succinct certificate (see Definition 2.1) in a bipartite graph $G = (X_1, Y_1; E)$, if it exists. The variable array L (whose indices are the vertices of the bipartite graph G) records the level of each vertex during the bfs from unsaturated root vertex x. More precisely, $L[v] = k$ if the vertex v is in the kth level, that is, $v \in L_k(x)$.

Algorithm to find a succinct certificate:
Input: A bipartite graph $G = (X_1, Y_1; E)$.

Output: A matching saturating all the vertices of the set X_1 or else a succinct certificate.

Algorithm: The algorithm is given in Figure 2.11.

Proof of the algorithm of Figure 2.19:
The proof that the algorithm of Figure 2.19 gives either a matching saturating all the vertices of X_1 or else a succinct certificate is the same as the proof of the necessary part of the König-Hall Theorem 2.3 and the properties (see Property 2.1 of the level sets induced by the bfs).

```
Algorithm:
Apply the maximum matching algorithm of figure 2.11 to the graph G;
Let M be the maximum matching obtained;
if M saturates all the vertices of X1
then
   begin
      print the edges of M;
      return;
   end
else
   begin
      Let A be an M alternating tree obtained from an unsaturated vertex x1 ∈ X1;
      (*S := L0(x) ∪ L2(x) ∪ L4(x) ··· *) in the tree A;
      S:=∅(* initialization of the set S*)
      for i := 1 to n do
         if xi ∈ A and L[xi] mod 2 = 0
         then S := S ∪ { xi };
      print the vertices of succinct certificate S;
   end;
```

FIGURE 2.19: Algorithm to find a matching that saturates all the vertices of the set X_1 or else a succinct certificate.

The complexity of the algorithm of Figure 2.19: The complexity is the same as that of the maximum matching algorithm which is $O(n \max(n, m))$.

As an application of the maximum matching algorithm, we shall study in the following subsection an algorithm to find a weighted perfect matching in a weighted complete bipartite graph.

2.4.1 Personnel assignment problem or weighted matching in a bipartite graph

Input: A complete bipartite simple graph $G = (X_1, Y_1; E)$ with $|X_1| = |Y_1| = n$. Let us recall that such a bipartite graph is denoted by $K_{n,n}$. The vertices of the set X_1 denote k technicians $x_1, x_2, \ldots, x_k$ and the vertices of the set X_2 stand for n jobs $y_1, y_2, \ldots, y_k$. Each edge x_iy_j for $i, j = 1, 2, \ldots, n$ of the graph $K_{n,n}$ is assigned a non-negative number $w(x_iy_j)$. This number $w(x_iy_j)$ can be considered as the efficiency (measured in terms of the profit to the company) of the technician x_i on the job y_j.

The weighted complete bipartite $K_{n,n}$ is represented as an $n \times n$ matrix $W = (w_{ij})$ where the rows of the matrix represent the n technicians $t_1, t_2, \ldots, t_n$ and the columns represent the n jobs $y_1, y_2, \ldots, y_n$ with $w_{ij} = w(x_iy_j)$.

Output: A maximum weighted perfect matching in the graph G, where the weight of a matching M is the sum of the weights of its edges.

Example 2.15: Personnel assignment problem

Consider the weighted complete bipartite graph represented by a 4×4 matrix W where

$$W = \begin{matrix} & y_1 & y_2 & y_3 & y_4 \\ x_1 \\ x_2 \\ x_3 \\ x_4 \end{matrix} \begin{pmatrix} 4 & 3 & 2 & 2 \\ 2 & 3 & 5 & 1 \\ 3 & 0 & 2 & 0 \\ 3 & 4 & 0 & 3 \end{pmatrix}$$

For example, the efficiency of the technician x_2 on the job y_3 is w_{23} which is equal to 5. A maximum weighted matching is $M = \{x_1y_4, x_2y_3, x_3y_1, x_4y_2\}$ and its weight $w(M) = w_{14} + w_{23} + w_{31} + w_{42} = 14$. This maximum matching M assigns the technician x_1 to the job y_1 and x_2 to the job y_3 and so on.

Algorithm (Brute-force): Generate all possible perfect matchings of the complete bipartite graph $K_{n,n}$ and choose one of them for which the weight is a maximum.

The most serious objection to this case-by-case method is that it works far too rarely. The reason is that there are $n!$ possible perfect matchings possible in the complete bipartite graph $K_{n,n}$ and $n!$ is an exponential function because

$$n^{n/2} \leq n! \leq \left(\frac{n+1}{2}\right)^n.$$

For example, $10! > 3.5$ million.

Example 2.16: Brute-force weighted matching algorithm

Let us execute the brute-force weighted matching algorithm on the complete weighted bipartite graph $K_{3,3}$ where the weights of the edges are given by the 3×3 matrix

$$W = \begin{matrix} & y_1 & y_2 & y_3 \\ x_1 \\ x_2 \\ x_3 \end{matrix} \begin{pmatrix} 2 & 4 & 3 \\ 1 & 2 & 4 \\ 3 & 4 & 5 \end{pmatrix}.$$

The following Table 2.4 gives all the possible $3! = 6$ perfect matchings and their corresponding weights. The last row 6 of the table gives us a maximum weighted matching whose weight is 11.

We are now ready to present the Kuhn-Munkres algorithm for the *weighted matching problem in bipartite graph.*

TABLE 2.4: Brute-force weighted matching algorithm

Number	Matchings	Weights
1	$\{x_1y_1, x_2y_2, x_3y_3\}$	9
2	$\{x_1y_1, x_2y_3, x_3y_2\}$	10
3	$\{x_1y_3, x_2y_2, x_3y_1\}$	8
4	$\{x_1y_2, x_2y_1, x_3y_3\}$	10
5	$\{x_1y_3, x_2y_1, x_3y_2\}$	8
6	$\{x_1y_2, x_2y_3, x_3y_1\}$	11

Description of the Kuhn-Munkres algorithm:

Definition 2.3 (Good labeling of the vertices of $K_{n,n}$). *Consider a weighted bipartite graph $K_{n,n}$ and the weight matrix $W = w_{ij}$ of the edges. We assign to each vertex v of the graph a non-negative number $l(v)$, called the label of the vertex v such that the weight of an edge is at most the sum of the labels of its end vertices. Symbolically,*

$$w_{ij} \leq l(x_i) + l(y_j)$$

for all $i, j = 1, 2, \ldots, n$. Such a labeling is called a good labeling of the weighted complete bipartite graph.

Example 2.17: Good labeling

Consider the weight matrix W of the complete bipartite graph $K_{4,4}$ where

$$W = \begin{array}{c} \\ x_1 \\ x_2 \\ x_3 \\ x_4 \end{array} \begin{array}{c} \begin{array}{cccc} y_1 & y_2 & y_3 & y_4 \end{array} \\ \begin{pmatrix} 4 & 3 & 2 & 2 \\ 2 & 3 & 5 & 1 \\ 3 & 0 & 2 & 0 \\ 3 & 4 & 0 & 3 \end{pmatrix} \end{array}.$$

A good labeling function l is given as follows: $l(x_1) = 4$ $l(x_2) = 5$ $l(x_3) = 3$ $l(x_4) = 4$ and $l(y_1) = l(y_2) = l(y_3) = l(y_4) = 0$. In fact, $l(x_i)$ is defined as the maximum value of the entry in the ith row and $l(y_j)$ is defined as zero for all j.

More generally, we observe that for a given weighted matrix $W = (w_{ij})$ of a complete bipartite graph $K_{n,n}$, the function l defined as

$$l(v) = \begin{cases} \max_{1 \leq j \leq n} w_{ij} & \text{if } v = x_i \\ 0 & \text{if } v = y_j \end{cases}$$

is always a good labeling.

There are n^2 edges in the complete bipartite graph $K_{n,n}$. The following lemma on which the algorithm is based says "many" of the n^2 edges can be eliminated from the graph in order to find a maximum weighted perfect matching.

Lemma 2.4. *Consider a weighted matrix $W = w_{ij}$, of a balanced complete bipartite graph $G = K_{n,n} = (X_1, Y_1; E)$ and a good labeling function l of its vertices. Let G_l be a spanning subgraph of G with vertex set $X_1 \cup Y_1$ and the edge set E_l where*

$$E_l = \{\, x_i y_j \mid l(x_i) + l(y_j) = w_{ij} \,\}.$$

In other words, the spanning subgraph G_l consists of all edges of G such that the weight of the edge is equal to the sum of the labels of its end vertices. If the graph G_l contains a perfect matching M, then M is a maximum weighted perfect matching of the graph G.

Proof. Let $M \subset E_l$ be a perfect matching in the graph G_l. Then, we have to show that M is a maximum weighted perfect matching of the graph G. Since $E_l \subset E$, M is also a perfect matching of the graph G. Let M' be a perfect matching of G. We have to show that $w(M') \leq w(M)$.
Now,

$$\begin{aligned} w(M) &= \sum_{x_i y_j \in M \subset E_l} w_{ij} \\ &= \sum_{x_i y_j \in M \subset E_l} (l(x_i) + l(y_j)) \\ &= \sum_{v \in X_1 \cup Y_1} l(v) \quad \text{(since } M \text{ saturates all the vertices)} \end{aligned} \tag{2.3}$$

But then,

$$\begin{aligned} w(M') &= \sum_{x_i y_j \in M' \subset E} w_{ij} \qquad (2.4) \\ &= \sum_{x_i y_j \in M' \subset E} (l(x_i) + l(y_j)) \\ &\leq \sum_{v \in X_1 \cup Y_1} l(v) \quad \text{(since } l \text{ is a good labeling)} \end{aligned}$$

By Equations (2.3) and (2.4), we have $w(M') \leq w(M)$. □

Example 2.18: The graph G_l

Let us refer to Example 2.17. The graph G_l corresponding to the good labeling l is drawn in Figure 2.20.

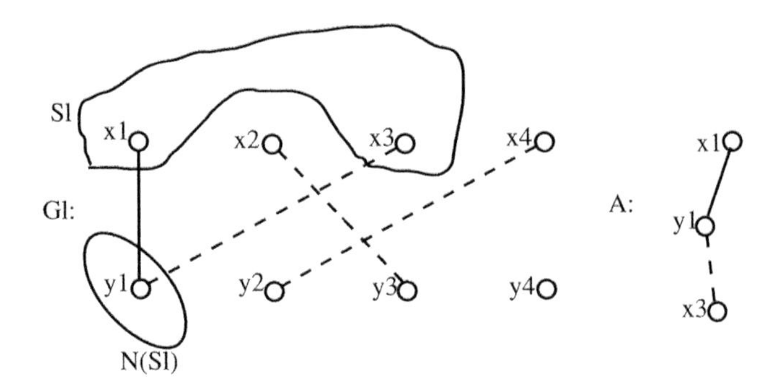

FIGURE 2.20: The graph G_l, where l is the good label.

Once the graph G_l is constructed, we apply the maximum matching algorithm of Figure 2.11 to the spanning bipartite graph G_l of the given graph G.

If the graph G_l possesses a perfect matching M, then by the Lemma 2.4, the matching M is a desired maximum weighted perfect matching in the given graph G and the algorithm "STOPS."

Otherwise, the graph G_l has a succinct certificate (see Definition 2.1) S_l. More precisely, there is a subset $S_l \in X_1$ such that the number of vertices collectively adjacent to the vertices of the set S_l in the graph G_l is exactly one less than the number of vertices in S_l. Symbolically,

$$|N_{G_l}(S_l)| = |S_l| - 1.$$

Note that this succinct certificate S_l can be obtained by an M-alternating tree of the graph G_l where M is a maximum (but not perfect) matching of the graph G_l. More precisely, if M is a maximum matching of G_l which is not perfect, then we grow a full M-alternating tree A from an unsaturated vertex $x_1 \in X_1$ as a root vertex (see Algorithm of Figure 2.18). Since M is a maximum matching, by Berge's Theorem 2.2, we obtain a tree A with only one unsaturated vertex x_1. Then, the succinct certificate S_l is given by the intersection $X_1 \cap X(A)$ (which is also the union of all oddly subscripted level sets of the tree A) where $X(A)$ is the set of vertices of the tree A.

Now, with the help of the set S_l, we shall modify the good labeling l in such a way that the tree A can still be grown further in the new graph G_l. The following definition redefines the labeling l.

Definition 2.4 (New labeling from the old one). *First of all, define the number d_l as*

$$d_l = \min_{x_i \in S_l, y_j \in Y_1 - T_l} \left(l(x_i) + l(y_j) - w(x_i y_j) \right). \tag{2.5}$$

Here, the set T_l denotes the neighbor set $N(S_l)$ in the graph G_l. Note that by the definition, the number d_l satisfies the inequality $d_l > 0$. Now the new

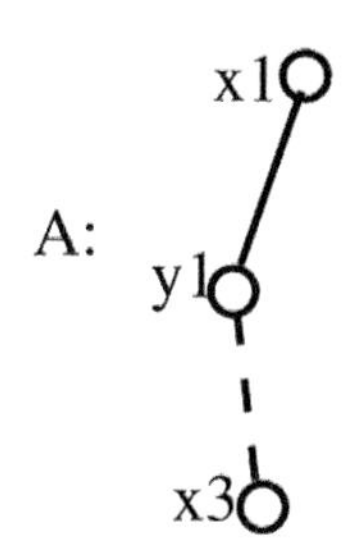

FIGURE 2.21: An M-alternating tree A of G_l.

labeling l is defined as follows:

$$l_1(v) := \begin{cases} l(v) - d_l & \textit{if } v \in S_l \\ l(v) + d_l & \textit{if } v \in T_l \\ l(v) & \textit{otherwise} \end{cases}$$

The above modification of the labeling is illustrated in the following example.

Example 2.19: New labeling from the old one

Let us refer to the graph G_l of Example 2.18. A maximum matching M in this graph G_l is $M = \{ x_2y_3, x_3y_1, x_4y_2 \}$ which is not perfect and the corresponding M-alternating tree A with the root x_1 is drawn in Figure 2.21:

The succinct certificate S_l is $S_l = X_1 \cap X(A) = \{ x_1, x_3 \}$ where $X(A)$ is the set of vertices of the tree A. The set T_l is $T_l = N_{G_l}(S_l) = \{ y_1 \}$ and the set $T_l' = Y_1 - T = \{ y_2, y_3, y_4 \}$. Then, the number $d_l = \min_{x_i \in S_l, y_j \in T_l'} (l(x_i) + l(y_j) - w(x_iy_j))$. Then, the new labeling l_1 is given as follows:

$l_1(x_1) := l(x_1) - d_l = 3$ $\quad l_(x_3) := l(x_3) - d_l = 2$ $\quad l_1(y_1) := l(y_1) + d_l = 1$ and all other vertices v keep their old l-values, that is, $l_1(v) := l(v)$ for all other vertices. The graph G_{l_1} corresponding to the labeling l_1 is drawn in Figure 2.22.

Now we construct the new spanning subgraph G_{l_1} of the graph G. Because of the definition of the labeling function l_1, in the new graph G_{l_1} all the edges between the sets S_l and T_l in the old graph G_l will also belong to the new graph G_{l_1}. In addition, at least one new edge (e.g., the edge x_1y_2 in the graph G_{l_1} of Example 2.19) between the set S_l and the set $Y_1 - T$ will appear in the new graph G_{l_1}. This new edge enables us to grow the M-alternating tree still further in the graph G_{l_1}. An edge of the old graph G_l having one end in

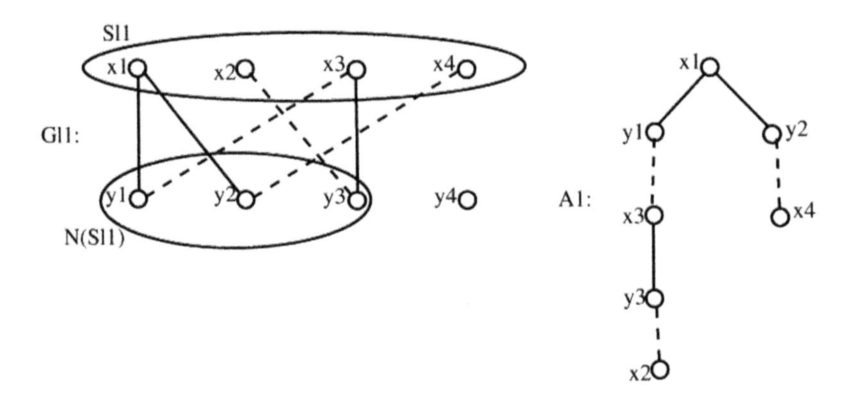

FIGURE 2.22: An M-alternating tree A_1 of G_{l_1}.

$X_1 - S_l$ and the other end in the set T_l may disappear in the new graph G_{l_1}. All other edges between the sets $X_1 - S_l$ and $Y_1 - T_l$ of the graph G_l will be conserved in the new graph G_{l_1}.

If the new graph G_{l_1} contains a perfect matching M, then by virtue of Lemma 2.4, M is a maximum weighted perfect matching in the original graph G and the algorithm STOPS.

If not, we find a maximum matching (but not perfect) and the corresponding succinct certificate in the graph G_{l_1} and we continue the algorithm as before. This continuation is explained in the example that follows:

Example 2.20: Continuation of Example 2.19

Let us refer to graph G_{l_1} of Example 2.19. A maximum matching $M = \{ x_2y_3, x_3y_1, x_4y_2 \}$ in the graph G_{l_1} which is the same as the maximum matching in the graph G_l.

The succinct certificate in the graph G_{l_1} is $S_{l_1} = X_1 \cap X(A_1) = X_1 = \{ x_1, x_2, x_3, x_4 \}$ where $X(A_1)$ is the vertex set of the alternating tree A_1 relative to the matching M and $T_{l_1} = \{ y_1, y_2, y_3 \}$ which is the set of vertices collectively adjacent to the vertices of S_{l_1} in the graph G_{l_1}. As before, we calculate the number d_{l_1} whose value is 1. We calculate the new labeling function l_2 where $l_2(x_1) = 3 - 1 = 2$, $l_2(x_2) = 5 - 1 = 4$, $l_2(x_3) = 2 - 1 = 1$, $l_2(x_4) = 4 - 1 = 3$, $l_2(y_1) = 1 + 1 = 2$, $l_2(y_2) = 0 + 1 = 1$, $l_2(y_3) = 0 + 1 = 1$ and $l_2(y_4) = l_1(y_4) = 0$.

Then, we construct the new graph G_{l_2} (see Figure 2.23) and apply the maximum matching Algorithm 2.11 to the graph G_{l_2}.

This graph possesses a perfect matching $M = \{ x_1y_2, x_2y_3, x_3y_1, x_4y_4 \}$ and by Lemma 2.4, M is also a maximum weighted perfect matching in the original graph G and the maximum weight is $w_{12} + w_{23} + w_{31} + w_{44} = 3 + 5 + 3 + 3 = 14$. Hence the algorithm STOPS.

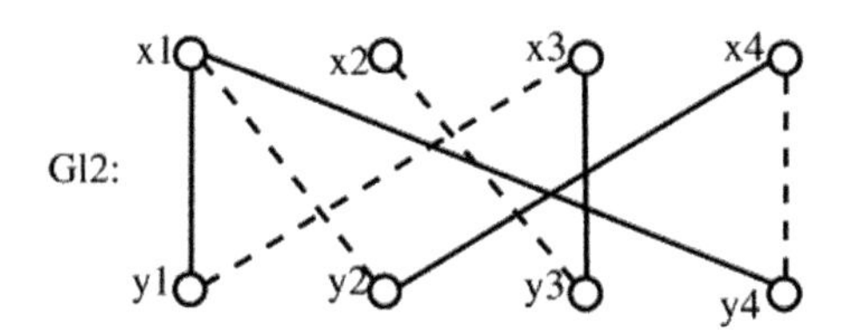

FIGURE 2.23: The graph G_{l_2}.

We now write the Kuhn-Munkres algorithm in pseudo-code: To implement this algorithm in a programming language, we may use the following data structures:

The initial weighted bipartite graph $G = K_{n,n} = (X_1, Y_1; E)$ with $X_1 = \{x_1, x_2, \ldots, x_n\}$, $Y_1 = \{y_1, y_2, \ldots, y_n\}$ is represented by an $n \times n$ matrix $W = (w_{ij})$ where w_{ij} is the weight of the edge $x_i y_j$. We shall use an array l of integer for the labeling function. The indices of the array are the vertices of the graph G. The graph G_l again is represented by an $n \times n$ matrix $B = (b_{ij})$ where

$$b_{ij} = \begin{cases} 1 & \text{if } l[x_i] + l[y_j] = w_{ij} \\ 0 & \text{otherwise} \end{cases}$$

The matching M is represented by an $n \times n$ matrix $M = m_{ij}$ where

$$m_{ij} = \begin{cases} 1 & \text{if the edge } x_i y_j \text{ is in the matching } M \\ 0 & \text{otherwise} \end{cases}$$

```
Algorithm(Kuhn-Munkres):
(* initialization of the labeling function l*)
for i := 1 to n do
begin(*for*)
    l(x_i) = max_{1<=j<=n} w_ij;
    l(y_i) = 0;
end;(*for*)
construct the graph G_l;
Apply the maximum matching algorithm of figure 2.11 to the graph G_l;
while G_l has no perfect matching do
begin(*while*)
    find a succinct certificate S_l and T_l = N(S_l) in the graph G_l;
    calculate d_l by the formula 2.5;
    for each vertex x_i in the set S_l do
        l(x_i) := l(x_i) - d_l;
    for each vertex y_i in the set T_l do
        l(y_i) := l(y_i) + d_l;
    construct the graph G_l;
end(*while*)
Let M be a perfect matching of the graph G_l;
Then M is a maximum weighted perfect matching in G;
Print the edges of M;
```

FIGURE 2.24: Kuhn-Munkres algorithm.

A subset is usually represented by its characteristic function, if the universal set is known in advance. Hence the succinct certificate S_l may be represented by a characteristic vector. More precisely, the set S_l is an array whose indices are the vertices of the set X_1 where

$$S_l[x_i] = \begin{cases} 1 & \text{if } x_i \in S_l \\ 0 & \text{otherwise} \end{cases}$$

In a similar manner, we can represent T_l, the set of neighbors of the set S_l in the graph G_l by a characteristic vector whose indices are the vertices of the set Y_1 (Figure 2.24).

2.5 Exercises

1. Write a complete program in C for the breadth-first search of a simple graph. (The graph is represented by its adjacency lists.)

2. Can we use the breadth-first search procedure to find the minimum cost in an undirected connected graph, from a given source vertex, to all other vertices, with weights associated with each edge of the graph? If the answer is "yes," write a pseudo-code to find the minimum costs of all paths from the source. The weight of each edge is > 0.

3. Write a complete program in C for the breadth-first search of a simple graph. (The graph is represented by its adjacency matrix.)

4. Describe an algorithm to find the diameter and the radius of an undirected connected graph using the breadth-first search. The weight of each edge is 1. Give the complexity of your algorithm.

5. Using the breadth-first search, write a program in C to test if a given simple graph is bipartite.

6. Can a geodetic graph be a bipartite graph? Justify your answer.

7. Using the breadth-first search, write a program in C to test if a given simple graph is geodetic.

8. Apply the maximum matching algorithm to the Petersen graph. Initially, we start with any single edge as a matching. Is the matching obtained at the end, a perfect matching?

9. Prove that the Petersen graph cannot be written as a union of three mutually disjoint perfect matchings.

10. If M_1 and M_2 are two different perfect matchings of a graph, then describe the spanning graph whose edge set is $M_1 \Delta M_2$ where Δ is the

symmetric difference of M_1 and M_2. (Here, M_1 and M_2 are considered as sets of edges.)

11. Prove that the Petersen graph is not a Hamiltonian graph; that is, it does not contain an elementary cycle passing through each vertex exactly once.

12. If G is a 3-regular simple graph on an even number of vertices containing a Hamiltonian cycle, then prove that the edge set of G can be decomposed into union of three mutually disjoint perfect matchings.

13. Apply the Kuhn-Munkres algorithm on the following matrix W corresponding to the weighted complete bipartite graph $K_{4,4}$:

$$W = \begin{array}{c} \\ x_1 \\ x_2 \\ x_3 \\ x_4 \end{array} \begin{array}{c} \begin{array}{cccc} y_1 & y_2 & y_3 & y_4 \end{array} \\ \begin{pmatrix} 4 & 4 & 3 & 2 \\ 3 & 5 & 5 & 2 \\ 3 & 2 & 2 & 4 \\ 3 & 4 & 5 & 3 \end{pmatrix} \end{array}$$

14. Write the following bi-stochastic matrix A as a convex combination of permutation matrices:

$$A = \begin{pmatrix} 1/4 & 1/2 & 0 & 1/4 \\ 1/2 & 1/4 & 1/4 & 0 \\ 0 & 1/4 & 1/2 & 1/4 \\ 1/4 & 0 & 1/4 & 1/2 \end{pmatrix}$$

15. A diagonal of a real square matrix $n \times n$ is a set of n entries of the matrix no two of which are in the same row or in the same column. The weight of a diagonal is the sum of the elements of the diagonal. Find a diagonal of maximum weight in the following 4×4 matrix A where

$$A = \begin{pmatrix} 5 & 3 & 2 & 2 \\ 2 & 3 & 5 & 3 \\ 3 & 4 & 2 & 3 \\ 3 & 4 & 2 & 5 \end{pmatrix}$$

16. Using the König-Hall theorem for perfect matching in a bipartite graph proved in this chapter, prove the following theorem due to the eminent graph theorist: Tutte.
Let G = (X,E) be a simple graph. For a subset $S \subset X$, denote by $c_o(G - S)$, the number of connected components with odd number of vertices of the induced subgraph $\langle X - S \rangle$.
Tutte's theorem: A graph G has a perfect matching if and only if

$$c_o(G - S) \leq |S| \text{ for all subsets } S \subset X$$

(For a proof, see [11].)

Chapter 3

Algebraic Structures I (Matrices, Groups, Rings, and Fields)

3.1 Introduction

In this chapter, the properties of the fundamental algebraic structures, namely, matrices, groups, rings, vector spaces, and fields are presented. In addition, the properties of finite fields which are so basic to finite geometry, coding theory, and cryptography are also discussed.

3.2 Matrices

A complex matrix A of type (m, n) or an m by n complex matrix is a rectangular arrangement of mn complex numbers in m rows and n columns in the form:

$$A = \begin{pmatrix} a_{11} & a_{12} & \dots & a_{1n} \\ a_{21} & a_{22} & \dots & a_{2n} \\ \dots & \dots & \dots & \dots \\ a_{m1} & a_{m2} & \dots & a_{mn} \end{pmatrix}.$$

A is usually written in the shortened form $A = (a_{ij})$, where $1 \le i \le m$ and $1 \le j \le n$. Here, a_{ij} is the (i, j)-th entry of A, that is, the entry common to the i-th row and j-th column of A. If $m = n$, A is a square matrix of order n. In the latter case, the vector with the entries $a_{11}, a_{22}, \dots, a_{nn}$ is known as the principal diagonal or main diagonal of A. All the matrices that we consider in this chapter are complex matrices.

3.3 Operations on Matrices: Addition, Scalar Multiplication, and Multiplication of Matrices

If $A = (a_{ij})$ and $B = (b_{ij})$ are two m by n matrices, then the sum $A + B$ is the m by n matrix $(a_{ij} + b_{ij})$, and for a scalar (that is, a complex number) α, $\alpha A = (\alpha a_{ij})$. Further, if $A = (a_{ij})$ is an m by n matrix, and $B = (b_{ij})$ is an n by p matrix, then the product AB is defined to be the m by p matrix (c_{ij}), where,

$$\begin{aligned} c_{ij} &= a_{i1}b_{1j} + a_{i2}b_{2j} + \cdots + a_{in}b_{nj} \\ &= \text{the scalar product of the } i\text{-th row vector } R_i \text{ of } A \\ &\quad\ \text{and the } j\text{-th column vector } C_j \text{ of } B. \end{aligned}$$

Thus, $c_{ij} = R_i \cdot C_j$. Both R_i and C_j are vectors of length n. It is easy to see that the matrix product satisfies both the distributive laws and the associative law, namely, for matrices A, B and C,

$$\begin{aligned} A(B+C) &= AB + AC, \\ (A+B)C &= AC + BC, \quad \text{and} \\ (AB)C &= A(BC) \end{aligned}$$

whenever these sums and products are defined.

3.3.1 Block multiplication of matrices

Let A be an m by n matrix, and B an n by p matrix so that the matrix product AB is defined, and it is an m by p matrix. Now partition the rows of A into $m_1, m_2, \ldots, m_k$ rows in order where $m_1 + m_2 \cdots + m_k = m$. Further, partition the columns of A into $c_1 + c_2 + \cdots + c_l = n$ columns in order. We partition the rows and columns of B suitably, so that the product AB can be obtained by multiplying the partitioned matrices.

We explain it by means of an example. Let $A = (a_{ij})$ be a 5×4 matrix, and $B = (b_{ij})$ be a 4×3 matrix so that AB is a 5×3 matrix. Partition the rows of A into two parts, the first part containing the first three rows of A and the second part containing the remaining two rows of A, keeping the order of the rows unchanged all the time. Now partition the columns of A into two parts with the first part having the first two columns in order, while the second part containing the remaining two columns of A in order. Thus, we can write:

$$A = \begin{pmatrix} A_{11} & A_{12} \\ A_{21} & A_{22} \end{pmatrix},$$

where A_{11} is 3×2, A_{12} is 3×2, A_{21} is 2×2 and A_{22} is 2×2.

We now partition B as

$$B = \begin{pmatrix} B_{11} & B_{12} \\ B_{21} & B_{22} \end{pmatrix}$$

where B_{11} and B_{21} have two rows each. Hence B_{12} and B_{22} have also two rows each. This forces B_{11} and B_{21} to have two columns each while B_{12} and

B_{22} have each one column. This gives

$$AB = \begin{pmatrix} A_{11} & A_{12} \\ A_{21} & A_{22} \end{pmatrix} \begin{pmatrix} B_{11} & B_{12} \\ B_{21} & B_{22} \end{pmatrix} = \begin{pmatrix} A_{11}B_{11} + A_{12}B_{21} & A_{11}B_{12} + A_{12}B_{22} \\ A_{21}B_{11} + A_{22}B_{21} & A_{21}B_{12} + A_{22}B_{22} \end{pmatrix},$$

and we find that the matrix on the right, on simplification, does indeed yield the product AB.

Note that it is not enough if we simply partition the matrices A and B. What is important is that the partitioned matrices should be conformable for multiplication. This means, in the above example, that the 8 products $A_{11}B_{11}, A_{12}B_{21}, \ldots A_{22}B_{22}$ are all defined.

The general case is similar.

3.3.2 Transpose of a matrix

If $A = (a_{ij})$ is an m by n matrix, then the n by m matrix (b_{ij}), where $b_{ij} = a_{ji}$ is called the transpose of A. It is denoted by A^t. Thus, A^t is obtained from A by interchanging the row and column vectors of A. For instance, if

$$A = \begin{pmatrix} 1 & 2 & 3 \\ 4 & 5 & 6 \\ 7 & 8 & 9 \end{pmatrix}, \quad \text{then} \quad A^t = \begin{pmatrix} 1 & 4 & 7 \\ 2 & 5 & 8 \\ 3 & 6 & 9 \end{pmatrix}.$$

It is easy to check that

i. $(A^t)^t = A$, and

ii. $(AB)^t = B^t A^t$, whenever the product AB is defined

(Note: If A is an m by p matrix, and B is a p by n matrix, then $(AB)^t$ is an n by m matrix, and $B^t A^t$ is also an n by m matrix.)

3.3.3 Inverse of a matrix

Let $A = (a_{ij})$ be an n by n matrix, that is, a matrix of order n. Let B_{ij} be the determinant minor of a_{ij} in A. Then, B_{ij} is the determinant of order $n-1$ got from A by deleting the i-th row and j-th column of A. Let A_{ij} be the cofactor of a_{ij} in $\det A$(= determinant of A) defined by $A_{ij} = (-1)^{i+j}B_{ij}$. Then, the matrix $(A_{ij})^t$ of order n is called the adjoint (or adjugate) of A, and denoted by adj A.

Theorem 3.1. *For any square matrix A of order n,*

$$A(\operatorname{adj} A) = (\operatorname{adj} A)A = (\det A)I_n,$$

where I_n is the identity matrix of order n. (I_n is the matrix of order n in which the n entries of the principal diagonal are 1 and the remaining entries are 0).

Proof. By a property of determinants, we have

$$a_{i1}A_{j1} + a_{i2}A_{j2} + \cdots + a_{in}A_{jn} = a_{1j}A_{1i} + a_{2j}A_{2i} + \cdots + a_{nj}A_{ni} = \det A \text{ or } 0$$

according to whether $i = j$ or $i \neq j$. We note that in adj A, $(A_{j1}, \ldots, A_{jn})$ is the j-th column vector, and $(A_{1j}, \ldots, A_{nj})$ is the j-th row vector. Hence, actual multiplication yields

$$A(\operatorname{adj} A) = (\operatorname{adj} A)A = \begin{pmatrix} \det A & 0 \ldots & 0 \\ 0 & \det A & \ldots & 0 \\ \ldots & \ldots & \ldots & \ldots \\ 0 & 0 & \ldots & \det A \end{pmatrix} = (\det A)I_n.$$

□

Corollary 3.1. *Let A be a non-singular matrix, that is,* $\det A \neq 0$. *Set* $A^{-1} = (1/\det A)(\operatorname{adj} A)$. *Then,* $AA^{-1} = A^{-1}A = I_n$, *where n is the order of A.*

The matrix A^{-1}, as defined in Corollary 3.1, is called the inverse of the (non-singular) matrix A. If A, B are square matrices of the same order with $AB = I$, then $B = A^{-1}$ and $A = B^{-1}$. These are easily seen by premultiplying the equation $AB = I$ by A^{-1} and postmultiplying it by B^{-1}. Note that A^{-1} and B^{-1} exist since taking determinants of both sides of $AB = I$, we get

$$\det(AB) = \det A \cdot \det B = \det I = 1,$$

and hence $\det A \neq 0$ as well as $\det B \neq 0$.

3.3.4 Symmetric and skew-symmetric matrices

A matrix A is said to be *symmetric* if $A = A^t$. A is *skew-symmetric* if $A = -A^t$. Hence if $A = (a_{ij})$, then A is symmetric if $a_{ij} = a_{ji}$ for all i and j; it is skew-symmetric if $a_{ij} = -a_{ji}$ for all i and j. Clearly, symmetric and skew-symmetric matrices are square matrices. If $A = (a_{ij})$ is skew-symmetric, then $a_{ii} = -a_{ii}$, and hence $a_{ii} = 0$ for each i. Thus, in a skew-symmetric matrix, all the entries of the principal diagonal are zero.

3.3.5 Hermitian and skew-Hermitian matrices

Let $H = (h_{ij})$ denote a complex matrix. The *conjugate* $\overline{H}$ of H is the matrix $(\overline{h}_{ij})$. The *conjugate-transpose* of H is the matrix $H^\star = (\overline{H})^t = \overline{(H^t)} = (\overline{h}_{ji}) = (h^\star_{ij})$. H is Hermitian if and only if (iff) $H^\star = H$; H is skew-Hermitian iff $H^\star = -H$. For example, the matrix $H = \begin{pmatrix} 1 & 2+3i \\ 2-3i & \sqrt{5} \end{pmatrix}$ is Hermitian, while the matrix $S = \begin{pmatrix} -i & 1+2i \\ -1+2i & 5i \end{pmatrix}$ is skew-Hermitian. Note that the diagonal entries of a skew-Hermitian matrix are all of the form ir, where r is a real number (and $i = \sqrt{-1}$).

3.3.6 Orthogonal and unitary matrices

A real matrix (that is, a matrix whose entries are real numbers) P of order n is called orthogonal if $PP^t = I_n$. If $PP^t = I_n$, then $P^t = P^{-1}$. Thus, the inverse of an orthogonal matrix is its transpose. Further, as $P^{-1}P = I_n$, we also have $P^tP = I_n$. If $R_1, \ldots, R_n$ are the row vectors of P, the relation $PP^t = I_n$ implies that $R_i \cdot R_j = \delta_{ij}$, where $\delta_{ij} = 1$ if $i = j$, and $\delta_{ij} = 0$ if $i \neq j$. A similar statement also applies to the column vectors of P. As an example, the matrix $\begin{pmatrix} \cos\alpha & \sin\alpha \\ -\sin\alpha & \cos\alpha \end{pmatrix}$ is orthogonal where α is a real number. Indeed, if (x, y) are the Cartesian coordinates of a point P referred to a pair of rectangular axes, and if (x', y') are the coordinates of the same point P with reference to a new set of rectangular axes got by rotating the original axes through an angle α about the origin (in the counter-clockwise direction), then

$$x' = x\cos\alpha + y\sin\alpha$$
$$y' = -x\sin\alpha + y\cos\alpha,$$

$$\text{that is,} \quad \begin{pmatrix} x' \\ y' \end{pmatrix} = \begin{pmatrix} \cos\alpha & \sin\alpha \\ -\sin\alpha & \cos\alpha \end{pmatrix} \begin{pmatrix} x \\ y \end{pmatrix}$$

so that rotation is effected by an orthogonal matrix.

Again, if (l_1, m_1, n_1), (l_2, m_2, n_2) and (l_3, m_3, n_3) are the direction cosines of three mutually orthogonal directions referred to an orthogonal coordinate system in the Euclidean 3-space, then the matrix $\begin{pmatrix} l_1 & m_1 & n_1 \\ l_2 & m_2 & n_2 \\ l_3 & m_3 & n_3 \end{pmatrix}$ is orthogonal. In passing, we mention that rotation in higher-dimensional Euclidean spaces is defined by means of an orthogonal matrix.

A complex matrix U of order n is called *unitary* if $UU^\star = I_n$ (Recall that $U^\star$ is the conjugate transpose of U). Again, this means that $U^\star U = I_n$. Also a real unitary matrix is simply an orthogonal matrix. The unit matrix is both orthogonal as well as unitary. For example, the matrix $U = (1/5)\begin{pmatrix} -1+2i & -4-2i \\ 2-4i & -2-i \end{pmatrix}$ is unitary.

3.3.7 Exercises

1. If $A = \begin{pmatrix} 3 & -4 \\ 1 & -1 \end{pmatrix}$, prove by induction that $A^k = \begin{pmatrix} 1+2k & -4k \\ k & 1-2k \end{pmatrix}$ for any positive integer k.

2. If $M = \begin{pmatrix} \cos\alpha & \sin\alpha \\ -\sin\alpha & \cos\alpha \end{pmatrix}$, prove that $M^n = \begin{pmatrix} \cos(n\alpha) & \sin(n\alpha) \\ -\sin(n\alpha) & \cos(n\alpha) \end{pmatrix}$, $n \in N$.

3. Compute the transpose, adjoint and inverse of the matrix $\begin{pmatrix} 1 & -1 & 0 \\ 0 & 1 & -1 \\ 1 & 0 & 1 \end{pmatrix}$.

4. If $A = \begin{pmatrix} 1 & 3 \\ -2 & 2 \end{pmatrix}$, show that $A^2 - 3A + 8I = 0$. Hence compute A^{-1}.

5. Give two matrices A and B of order 2, so that

   i. $AB \neq BA$

   ii. $(AB)^t \neq AB$

6. Prove: (i) $(AB)^t = B^tA^t$; (ii) If A and B are non-singular, $(AB)^{-1} = B^{-1}A^{-1}$.
7. Prove that the product of two symmetric matrices is symmetric iff the two matrices commute.
8. Prove: (i) $(iA)^\star = -iA^\star$; (ii) H is Hermitian iff iH is skew-Hermitian.
9. Show that every real matrix is the unique sum of a symmetric matrix and a skew-symmetric matrix.
10. Show that every complex square matrix is the unique sum of a Hermitian and a skew-Hermitian matrix.

3.4 Groups

Groups constitute an important basic algebraic structure that occurs very naturally not only in mathematics, but also in many other fields such as physics and chemistry. In this section, we present the basic properties of groups. In particular, we discuss Abelian and non-Abelian groups, cyclic groups, permutation groups and homomorphisms and isomorphisms of groups. We establish Lagrange's theorem for finite groups and the basic isomorphism theorem for groups.

3.4.1 Abelian and non-Abelian groups

Definition 3.1. *A* binary operation *on a non-empty set S is a map* $: S \times S \to S$, *that is, for every ordered pair (a, b) of elements of S, there is associated a unique element $a \cdot b$ of S. A* binary system *is a pair $(S, \cdot)$, where S is a non-empty set and $\cdot$ is a binary operation on S. The binary system $(S, \cdot)$ is* associative *if $\cdot$ is an associative operation on S, that is, for all a, b, c in S, $(a \cdot b) \cdot c = a \cdot (b \cdot c)$*

Definition 3.2. *A semigroup is an associative binary system. An element e of a binary system $(S, \cdot)$ is an identity element of S if $a \cdot e = e \cdot a = a$ for all $a \in S$.*

We use the following standard notations:

$\mathbb{Z}$ = the set of integers (positive integers, negative integers and zero),
$\mathbb{Z}^+$ = the set of positive integers,
$\mathbb{N}$ = the set of natural numbers $\{1, 2, 3, \ldots\} = \mathbb{Z}^+$,
$\mathbb{Q}$ = the set of rational numbers,
$\mathbb{Q}^+$ = the set of positive rational numbers,
$\mathbb{Q}^*$ = the set of non-zero rational numbers,
$\mathbb{R}$ = the set of real numbers,

$\mathbb{R}^*$= the set of non-zero real numbers,
$\mathbb{C}$ = the set of complex numbers, and
$\mathbb{C}^*$= the set of non-zero complex numbers.

Examples

1. $(\mathbb{N}, \cdot)$ is a semigroup, where $\cdot$ denotes the usual multiplication.
2. The operation subtraction is *not* a binary operation on $\mathbb{N}$ (for example, $3 - 5 \notin \mathbb{N}$).
3. $(\mathbb{Z}, -)$ is a binary system which is not a semigroup since the associative law is not valid in $(\mathbb{Z}, -)$; for instance, $10 - (5 - 8) \neq (10 - 5) - 8$.

We now give the definition of a group.

Definition 3.3. *A* group *is a binary system* $(\mathsf{G}, \cdot)$ *such that the following axioms are satisfied:*

(G_1): The operation $\cdot$ is associative on G, that is, for all $a, b, c \in \mathsf{G}$, $(a \cdot b) \cdot c = a \cdot (b \cdot c)$.

(G_2): (Existence of identity) There exists an element $e \in \mathsf{G}$ (called an identity element of G with respect to the operation $\cdot$) such that $a \cdot e = e \cdot a = a$ for all $a \in \mathsf{G}$.

(G_3): (Existence of inverse) To each element $a \in \mathsf{G}$, there exists an element $a^{-1} \in \mathsf{G}$ (called an inverse of a with respect to the operation $\cdot$) such that $a \cdot a^{-1} = a^{-1} \cdot a = e$.

Before proceeding to examples of groups, we show that identity element e, and inverse element a^{-1} of a, given in Definition 3.3 are unique.

Suppose G has two identities e and f with respect to the operation $\cdot$. Then,

$$\begin{aligned} e &= e \cdot f \quad (\text{as } f \text{ is an identity of } (\mathsf{G}, \cdot)) \\ &= f \quad (\text{as } e \text{ is an identity of } (\mathsf{G}, \cdot)). \end{aligned}$$

Next, let b and c be two inverses of a in $(\mathsf{G}, \cdot)$. Then,

$$\begin{aligned} b &= b \cdot e \\ &= b \cdot (a \cdot c) \text{ (as } c \text{ is an inverse of } a) \\ &= (b \cdot a) \cdot c \text{ by the associativity of } \cdot \text{ in } G \\ &= e \cdot c \text{ (as } b \text{ is an inverse of } a) \\ &= c. \end{aligned}$$

Thus, henceforth, we can talk of "The identity element e" of the group $(\mathsf{G}, \cdot)$, and "The inverse element a^{-1} of a" in $(\mathsf{G}, \cdot)$.

If $a \in \mathsf{G}$, then $a \cdot a \in \mathsf{G}$; also, $a \cdot a \cdots$ (n times) $\in \mathsf{G}$. We denote $a \cdot a \cdots$ (n times) by a^n. Further, if $a, b \in \mathsf{G}$, $a \cdot b \in \mathsf{G}$, and $(a \cdot b)^{-1} = b^{-1} \cdot a^{-1}$. (Check that $(a \cdot b)(a \cdot b)^{-1} = (a \cdot b)^{-1}(a \cdot b) = e$). More generally, if $a_1, a_2, \ldots, a_n \in \mathsf{G}$, then $(a_1 \cdot a_2 \cdots a_n)^{-1} = a_n^{-1} \cdot a_{n-1}^{-1} \cdots a_1^{-1}$, and hence $(a^n)^{-1} = (a^{-1})^n =$ (written as) a^{-n}. Then, the relation $a^{m+n} = a^m \cdot a^n$ holds for all integers m and n, with $a^0 = e$. In what follows, we drop the group operation $\cdot$ in $(\mathsf{G}, \cdot)$, and simply write group G, unless the operation is explicitly needed.

Lemma 3.1. *In a group, both the cancellation laws are valid, that is, if a, b, c are elements of a group G with $ab = ac$, then $b = c$ (left cancellation law), and if $ba = ca$, then $b = c$ (right cancellation law).*

Proof. If $ab = ac$, premultiplication by a^{-1} gives $a^{-1}(ab) = a^{-1}(ac)$. So by the associative law in G, $(a^{-1}a)b = (a^{-1}a)c$, and hence $eb = ec$. This implies that $b = c$. The other cancellation is proved similarly. □

Definition 3.4. *The* order *of a group G is the cardinality of G. The* order of an element a *of a group G is the least positive integer n such that $a^n = e$, the identity element of G. If no such n exists, the order of a is taken to be infinity.*

Definition 3.5. Abelian Group *[After the Norwegian mathematician Abel] A group G is called Abelian if the group operation of G is commutative, that is, $ab = ba$ for all $a, b \in \mathsf{G}$.*

A group G is non-Abelian if it is not Abelian, that is, there exists a pair of elements x, y in G with $xy \neq yx$.

3.4.2 Examples of Abelian groups

1. $(\mathbb{Z}, +)$ is an Abelian group, that is, the set $\mathbb{Z}$ of integers is an Abelian group under the usual addition operation. The identity element of this group is O, and the inverse of a is $-a$. $(\mathbb{Z}, +)$ is often referred to as the additive group of integers. Similarly, $(\mathbb{Q}, +)$, $(\mathbb{R}, +)$, $(\mathbb{C}, +)$ are all additive Abelian groups.

2. The sets $\mathbb{Q}^*$, $\mathbb{R}^*$ and $\mathbb{C}^*$ are groups under the usual multiplication operation.

3. Let $\mathsf{G} = \mathbb{C}[0, 1]$, the set of complex-valued continuous functions defined on $[0, 1]$. G is an Abelian group under addition. Here, if $f, g \in \mathbb{C}[0, 1]$, then $f + g$ is defined by $(f + g)(x) = f(x) + g(x), x \in [0, 1]$. The zero function O is the identity element of the group while the inverse of f is $-f$.

4. For any positive integer n, let $\mathbb{Z}_n = \{0, 1, \ldots, n-1\}$. Define addition $+$ in $\mathbb{Z}_n$ as "congruent modulo n" addition, that is, if $a, b \in \mathbb{Z}_n$, then $a + b = c$, where $c \in \mathbb{Z}_n$ and $a + b \equiv c \pmod{n}$. Then, $(\mathbb{Z}_n, +)$ is an

Abelian group. For instance, if $n = 5$, then in $\mathbb{Z}_5$, $2+2=4$, $2+3=0$, $3+3=1$, etc.

5. Let $\mathsf{G} = \{r_\alpha : r_\alpha =$ rotation of the plane about the origin through an angle α in the counter-clockwise sense$\}$. Then, if we set $r_\beta \cdot r_\alpha = r_{\alpha+\beta}$ (that is, rotation through α followed by rotation through $\beta =$ rotation through $\alpha+\beta$), then $(\mathsf{G}, \cdot)$ is a group. The identity element of $(\mathsf{G}, \cdot)$ is r_0, while $(r_\alpha)^{-1} = r_{-\alpha}$, the rotation of the plane about the origin through an angle α in the clockwise sense.

3.4.3 Examples of non-Abelian groups

1. Let $\mathsf{G} = GL(n, \mathbb{R})$, the set of all n by n non-singular matrices with real entries. Then, G is an infinite non-Abelian group under multiplication.

2. Let $\mathsf{G} = SL(n, \mathbb{Z})$ be the set of matrices of order n with integer entries having determinant 1. G is again an infinite non-Abelian multiplicative group. Note that if $A \in SL(n, \mathbb{Z})$, then $A^{-1} = (1/\det A)(\operatorname{adj} A) \in SL(n, \mathbb{Z})$ since $\det A = 1$ and all the cofactors of the entries of A are integers (see Section 3.3.3).

3. Let S_4 denote the set of all 1–1 maps $f : N_4 \to N_4$, where $N_4 = \{1, 2, 3, 4\}$. If $\cdot$ denotes composition of maps, then $(S_4, \cdot)$ is a non-Abelian group of order $4! = 24$. (See Section 3.8 for more about such groups.) For instance, let

$$f = \begin{pmatrix} 1 & 2 & 3 & 4 \\ 4 & 1 & 2 & 3 \end{pmatrix}.$$

Here, the parantheses notation signifies the fact that the image under f of a number in the top row is the corresponding number in the bottom row. For instance, $f(1) = 4$, $f(2) = 1$ and so on. Let

$$g = \begin{pmatrix} 1 & 2 & 3 & 4 \\ 3 & 1 & 2 & 4 \end{pmatrix} \quad \text{Then,} \quad g \cdot f = \begin{pmatrix} 1 & 2 & 3 & 4 \\ 4 & 3 & 1 & 2 \end{pmatrix}.$$

Note that $(g \cdot f)(1) = g\big(f(1)\big) = g(4) = 4$, while $(f \cdot g)(1) = f\big(g(1)\big) = f(3) = 2$, and hence $f \cdot g \neq g \cdot f$. In other words, S_4 is a non-Abelian group. The identity element of S_4 is the map

$$I = \begin{pmatrix} 1 & 2 & 3 & 4 \\ 1 & 2 & 3 & 4 \end{pmatrix} \quad \text{and} \quad f^{-1} = \begin{pmatrix} 1 & 2 & 3 & 4 \\ 2 & 3 & 4 & 1 \end{pmatrix}.$$

3.4.4 Group tables

The structure of a finite group G can be completely specified by means of its *group table* (sometimes called *multiplication table*). This is formed by listing the elements of G in some order as $\{g_1, \ldots, g_n\}$, and forming an n by n double array (g_{ij}), where $g_{ij} = g_i g_j$, $1 \le i \le j \le n$. It is customary to take $g_1 = e$, the identity element of G.

TABLE 3.1: Group table of Klein's 4-group

	e	a	b	c
e	e	a	b	c
a	a	e	c	b
b	b	c	e	a
c	c	b	a	e

Examples Continued

1. [Klein's 4-group K_4] This is a group of order 4. If its elements are e, a, b, c, the group table of K_4 is given by Table 3.1.

   It is observed from the table that

   $$ab = ba = c, \quad \text{and} \quad a(ba)b = a(ab)b$$

   This gives

   $$c^2 = (ab)(ab) = a(ba)b = a(ab)b = a^2b^2 = ee = e.$$

   Thus, every element of K_4 other than e is of order 2.

3.5 A Group of Congruent Transformations (Also called Symmetries)

We now look at the congruent transformations of an equilateral triangle ABC. Assume without loss of generality that the side BC of the triangle is horizontal so that A is in the vertical through the middle point D of BC.

Let us denote by r, the rotation of the triangle about its center through an angle of $120°$ in the counter-clockwise sense and let f denote the flipping of the triangle about the vertical through the middle point of the base. f interchanges the base vertices and leaves all the points of the vertical through the third vertex unchanged. Then, fr denotes the transformation r followed by f and so on.

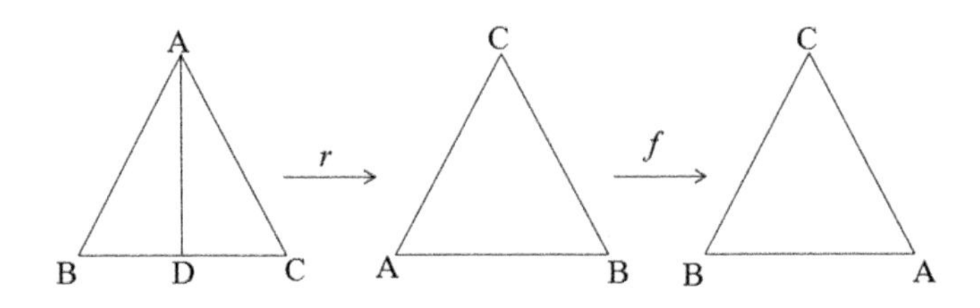

TABLE 3.2: Group table of the dihedral group D_3

	$r^3 = e$	r	r^2	f	rf	r^2f
$r^3 = e$	e	r	r^2	f	rf	r^2f
r	r	r^2	e	rf	r^2f	f
r^2	r^2	e	r	r^2f	f	rf
f	f	fr	fr^2	e	r^2	r
rf	rf	f	r^2f	r	e	r^2
r^2f	r^2f	rf	f	r^2	r	e

Thus, fr leaves B fixed and flips A and C in $\triangle ABC$. There are six congruent transformations of an equilateral triangle and they form a group as per group Table 3.2

For instance, r^2fr and rf are obtained as follows:

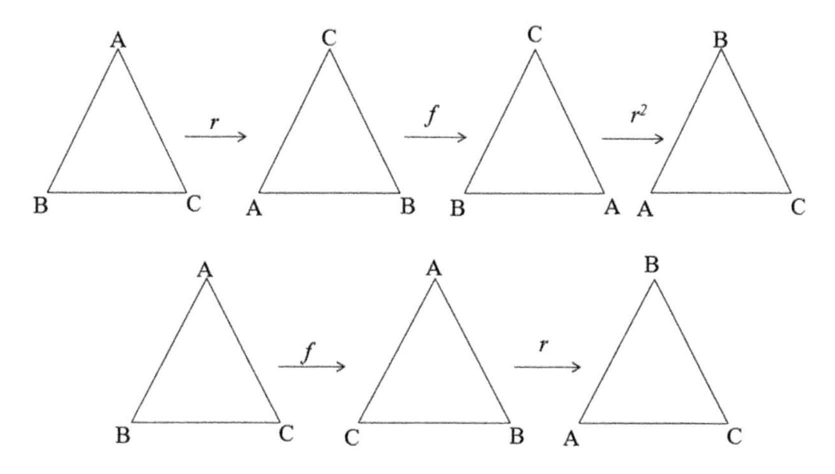

Thus, $r^2fr = rf$, and similarly the other products can be verified. The resulting group is known as the dihedral group D_3. It is of order $2 \cdot 3 = 6$.

3.6 Another Group of Congruent Transformations

Let D_4 denote the transformations that leave a square invariant. If r denotes a rotation of 90° about the center of the square in the counterclockwise sense, and f denotes the flipping of the square about one of its diagonals, then the defining relations for D_4 are given by:

$$r^4 = e = f^2 = (rf)^2.$$

D_4 is the dihedral group of order $2 \times 4 = 8$. Both D_3 and D_4 are non-Abelian groups. The dihedral group D_n of order $2n$ is defined in a similar fashion as the group of congruent transformations of a regular polygon of n sides. The groups D_n, $n \geq 3$, are all non-Abelian. D_n is of order $2n$ for each n.

3.7 Subgroups

Definition 3.6. *A subset H of a group $(\mathsf{G}, \cdot)$ is a subgroup of $(\mathsf{G}, \cdot)$ if $(H, \cdot)$ is a group.*

Definition 3.6 shows that the group operation of a subgroup H of G is the same as that of G.

3.7.1 Examples of subgroups

1. $\mathbb{Z}$ is a subgroup of $(\mathbb{Q}, +)$.
2. $2\mathbb{Z}$ is a subgroup of $(\mathbb{Z}, +)$. (Here, $2\mathbb{Z}$ denotes the set of even integers).
3. $\mathbb{Q}^*$ is a subgroup of $(\mathbb{R}^*, \cdot)$. (Here, $\cdot$ denotes multiplication).
4. Let H be the subset of maps of S_4 that fix 1, that is, $H = \{f \in S_4 : f(1) = 1\}$. Then, H is a subgroup of S_4).
   Note that the set $\mathbb{N}$ of natural numbers does not form a subgroup of $(\mathbb{Z}, +)$.

3.7.2 Subgroup generated by a subset of a group

Definition 3.7. *Let S be a non-empty subset of a group G. The subgroup generated by S in G, denoted by $\langle S \rangle$, is the intersection of all subgroups of G containing S.*

Proposition 3.1. *The intersection of any family of subgroups of G is a subgroup of G.*

Proof. Consider a family $\{\mathsf{G}_\alpha\}_{\alpha \in I}$ of subgroups of G, and let $H = \cap_{\alpha \in I} \mathsf{G}_\alpha$. If $a, b \in H$, then $a, b \in \mathsf{G}_\alpha$ for each $\alpha \in I$, and since G_α is a subgroup of $\mathsf{G}, ab \in \mathsf{G}_\alpha$ for each $\alpha \in I$. Therefore, $ab \in H$, and similarly $a^{-1} \in H$ and $e \in H$. The associative law holds in H, a subset of G, as it holds in G. Thus, H is a subgroup of G. □

Corollary 3.2. *Let S be a non-empty subset of a group G. Then, $\langle S \rangle$ is the smallest subgroup of G containing S.*

Proof. By definition, $\langle S \rangle$ is contained in every subgroup H of G containing S. Since $\langle S \rangle$ is itself a subgroup of G containing S, it is the smallest subgroup of G containing S. □

3.8 Cyclic Groups

Definition 3.8. *Let* G *be a group and* a*, an element of* G*. Then, the subgroup generated by* a *in* G *is* $\langle\{a\}\rangle$*, that is, the subgroup generated by the singleton subset* $\{a\}$*. It is also denoted simply by* $\langle a\rangle$*.*

By Corollary 3.2, $\langle a\rangle$ is the smallest subgroup of G containing a. As $a \in \langle a\rangle$, all the powers of $a^n, n \in \mathbb{Z}$, also belong to $\langle a\rangle$. But then, as may be checked easily, the set $\{a^n : n \in \mathbb{Z}\}$ of powers of a is already a subgroup of G. Hence $\langle a\rangle = \{a^n : n \in \mathbb{Z}\}$. Note that $a^0 = e$, the identity element of G, and $a^{-n} = (a^{-1})^n$, the inverse of a^n. This makes $a^m \cdot a^n = a^{m+n}$ for all integers m and n. The subgroup $A = \langle a\rangle$ of G is called *the subgroup generated by* a, and a is called a generator of A.

Now since $\{a^n : n \in \mathbb{Z}\} = \{(a^{-1}) : n \in \mathbb{Z}\}$, a^{-1} is also a generator of $\langle a\rangle$. Suppose a is of finite order m in $\langle a\rangle$. This means that m is the least positive integer with the property that $a^m = e$. Then, the elements $a^1 = a, a^2, \ldots, a^{m-1}, a^m = e$ are all distinct. Moreover, for any integer n, by Euclidean algorithm for integers, there are integers q and r such that

$$n = qm + r, \quad 0 \leq r < m.$$

Then,

$$a^n = a^{(qm+r)} = (a^m)^q a^r = e^q a^r = ea^r = a^r, \quad 0 \leq r < m,$$

and hence $a^n \in \langle a\rangle$. Thus, in this case,

$$\langle a\rangle = \left\{a, a^2, \ldots, a^{m-1}, a^m = e\right\}.$$

In the contrary case, there exists no positive integer m such that $a^m = e$. Then, all the powers $a^r : r \in \mathbb{Z}$ are distinct. If not, there exist integers r and s, $r \neq s$, such that $a^r = a^s$. Suppose $r > s$. Then, $r - s > 0$ and the equation $a^r = a^s$ gives $a^{-s}a^r = a^{-s}a^s$, and therefore $a^{r-s} = a^0 = e$, a contradiction. In the first case, the cyclic group $\langle a\rangle$ is of order m while in the latter case, it is of infinite order.

3.8.1 Examples of cyclic groups

1. The additive group of integers $(\mathbb{Z}, +)$ is an infinite cyclic group. It is generated by 1 as well as -1.

2. The group of n-th roots of unity, $n \geq 1$. Let G be the set of n-th roots of unity so that

$$\mathsf{G} = \left\{\omega, \omega^2, \ldots, \omega^n = 1; \quad \omega = \cos\frac{2\pi}{n} + i\sin\frac{2\pi}{n}\right\}.$$

Then, G is a cyclic group of order n generated by ω, that is, $\mathsf{G} =< \omega >$. In fact $\omega^k, 1 \leq k \leq n$, also generates k iff $(k, n) = 1$. Hence, the number of generators of G is $\phi(n)$, where ϕ is the Euler's totient function.

If $\mathsf{G} = \langle a \rangle = \{a^n : n \in \mathbb{Z}\}$, then since for any two integers n and m, $a^n a^m = a^{n+m} = a^m a^n$, G is Abelian. In other words, every cyclic group is Abelian. However, the converse is not true. K_4, the Klein's 4-group (See Table 3.1 of Section 3.4) is Abelian but not cyclic since K_4 has no element of order 4.

Theorem 3.2. *Any subgroup of a cyclic group is cyclic.*

Proof. Let $\mathsf{G} = \langle a \rangle$ be a cyclic group, and H, a subgroup of G. If $H = \{e\}$, then H is trivially cyclic. So assume that $H \neq \{e\}$. As the elements of G are powers of a, $a^n \in H$ for some non-zero integer n. Then, its inverse a^{-1} also belongs to H, and of n and $-n$ at least one of them is a positive integer. Let s be the least positive integer such that $a^s \in H$. (recall that $H \neq \{e\}$ as per our assumption). We claim that $H = \langle a^s \rangle$, the cyclic subgroup of G generated by a^s. To prove this, we have to show that each element of H is a power of a^s. Let g be any element of H. As $g \in \mathsf{G}$, $g = a^m$ for some integer m. By division algorithm,

$$m = qs + r, \quad 0 \leq r < s.$$

Hence $a^r = a^{m-qs} = a^m (a^s)^{-q} \in H$ as $a^m \in H$ and $a^s \in H$. Thus, $a^r \in H$. This however implies, by the choice of s, $r = 0$ (otherwise $a^r \in H$ with $0 < r < s$). Hence $a^r = a^0 = e = a^m (a^s)^{-q}$, and therefore, $g = a^m = (a^s)^q, \quad q \in \mathbb{Z}$. Thus, every element of H is a power of a^s and so $H \subset \langle a^s \rangle$. Now since $a^s \in H$, all powers of a^s also $\in H$, and so $\langle a^s \rangle \subset H$. Thus, $H = \langle a^s \rangle$ and therefore H is cyclic. □

Definition 3.9. *Let S be any non-empty set. A permutation on S is a bijective mapping from S to S.*

Lemma 3.2. *If σ_1 and σ_2 are permutations on S, then the map $\sigma = \sigma_1 \sigma_2$ defined on S by*

$$\sigma(s) = \sigma_1 \sigma_2(s) = \sigma_1 \left(\sigma_2(s) \right), \quad s \in S$$

is also a permutation on S.

Proof. Indeed, we have, for s_1, s_2 in S, $\sigma(s_1) = \sigma(s_2)$ gives that $\sigma_1(\sigma_2 s_1) = \sigma_1(\sigma_2 s_2)$. This implies, as σ_1 is $1-1$, $\sigma_2 s_1 = \sigma_2 s_2$. Again, as σ_2 is $1-1$, this gives that $s_1 = s_2$. Thus, σ is $1-1$. For a similar reason, σ is onto. □

Let $\mathbb{B}$ denote the set of all bijections on S. Then, it is easy to verify that $(\mathbb{B}, \cdot)$, where $\cdot$ is the composition map, is a group. The identity element of this group is the identity function e on S.

The case when S is a finite set is of special significance. So let $S = \{1, 2, \ldots, n\}$. The set $\mathbb{P}$ of permutations of S forms a group under the composition operation. Any $\sigma \in \mathbb{P}$ can be conveniently represented as:

$$\sigma : \quad \begin{pmatrix} 1 & 2 & \ldots & n \\ \sigma(1) & \sigma(2) & \ldots & \sigma(n) \end{pmatrix}.$$

Then, σ^{-1} is just the permutation

$$\sigma^{-1} : \quad \begin{pmatrix} \sigma(1) & \sigma(2) & \ldots & \sigma(n) \\ 1 & 2 & \ldots & n \end{pmatrix}$$

What is the order of the group $\mathbb{P}$? Clearly $\sigma(1)$ has n choices, namely, any one of $1, 2, \ldots, n$. Having chosen $\sigma(1), \sigma(2)$ has $n-1$ choices (as σ is $1-1$, $\sigma(1) \neq \sigma(2)$). For a similar reason, $\sigma(3)$ has $n-2$ choices and so on, and finally $\sigma(n)$ has just one left out choice. Thus, the total number of permutations on S is

$$n \cdot (n-1) \cdot (n-2) \cdots 2 \cdot 1 = n!.$$

In other words, the group $\mathbb{P}$ of permutations on a set of n elements is of order $n!$ It is denoted by S_n and is called the symmetric group of degree n. Any subgroup of S_n is called a permutation group of degree n.

Example

Let $S = \{1, 2, 3, 4, 5\}$, and let σ and $\tau \in S_5$ be given by

$$\sigma = \begin{pmatrix} 1 & 2 & 3 & 4 & 5 \\ 2 & 3 & 5 & 4 & 1 \end{pmatrix}, \quad \tau = \begin{pmatrix} 1 & 2 & 3 & 4 & 5 \\ 5 & 2 & 4 & 1 & 3 \end{pmatrix}.$$

Then,

$$\sigma \cdot \tau = \sigma\tau = \begin{pmatrix} 1 & 2 & 3 & 4 & 5 \\ 1 & 3 & 4 & 2 & 5 \end{pmatrix}, \quad \text{and} \quad \sigma^2 = \sigma \cdot \sigma = \begin{pmatrix} 1 & 2 & 3 & 4 & 5 \\ 3 & 5 & 1 & 4 & 2 \end{pmatrix}.$$

Definition 3.10. *A cycle in S_n is a permutation $\sigma \in S_n$ that can be represented in the form $(a_1, a_2, \ldots, a_r)$, where the a_i, $1 \leq i \leq r$, $r \leq n$, are all in S, and $\sigma(a_i) = a_{i+1}$, $1 \leq i \leq r-1$, and $\sigma(a_r) = a_1$, that is, each a_i is mapped cyclically to the next element (or) number a_{i+1} and σ fixes the remaining a_is. For example, if*

$$\sigma = \begin{pmatrix} 1 & 3 & 2 & 4 \\ 3 & 2 & 1 & 4 \end{pmatrix} \in S_4,$$

then σ can be represented by (132). Here, σ leaves 4 fixed.

Now consider the permutation

$$p = \begin{pmatrix} 1 & 2 & 3 & 4 & 5 & 6 & 7 \\ 3 & 4 & 2 & 1 & 6 & 5 & 7 \end{pmatrix}.$$

Clearly, p is the product of the cycles

$$(1324)(56)(7) = (1324)(56).$$

Since 7 is left fixed by p, 7 is not written explicitly. In this way, every permutation on n symbols is a product of disjoint cycles. The number of symbols in a cycle is called the length of the cycle. For example, the cycle (1 3 2 4) is a cycle of length 4. A cycle of length 2 is called a transposition. For example, the permutation (1 2) is a transposition. It maps 1 to 2, and 2 to 1. Now consider the product of transpositions $t_1 = (13)$, $t_2 = (12)$ and $t_3 = (14)$. We have

$$t_1t_2t_3 = (13)(12)(14) = \begin{pmatrix} 1 & 3 \\ 3 & 1 \end{pmatrix} \begin{pmatrix} 1 & 2 \\ 2 & 1 \end{pmatrix} \begin{pmatrix} 1 & 4 \\ 4 & 1 \end{pmatrix} = \begin{pmatrix} 1 & 2 & 3 & 4 \\ 4 & 3 & 1 & 2 \end{pmatrix} = (1423)\,.$$

To see this, note that

$$(t_1t_2t_3)(4) = (t_1t_2)(t_3(4)) = (t_1t_2)(1) = t_1(t_2(1)) = t_1(2) = 2, \quad \text{and so on.}$$

In the same way, any cycle $(a_1a_2 \dots a_n) = (a_1a_n)(a_1a_{n-1}) \cdots (a_1a_2)$ is a product of transpositions. Since any permutation is a product of disjoint cycles and any cycle is a product of transpositions, it is clear that any permutation is a product of transpositions. Now in the expression of a cycle as a product of transpositions, the number of transpositions need not be unique. For instance, $(12)(12) =$ identity permutation, and

$$(1324) = (14)(12)(13) = (12)(12)(14)(12)(13).$$

However, this number is always odd or always even.

Theorem 3.3. *Let σ be any permutation on n symbols. Then, in whatever way σ is expressed as a product of transpositions, the number of transpositions is always odd or always even.*

Proof. Assume that σ is a permutation on $\{1, 2, \dots, n\}$. Let the product

$$\begin{aligned} P =&(a_1 - a_2)(a_1 - a_3) \cdots (a_1 - a_n) \\ &(a_2 - a_3) \cdots (a_2 - a_n) \\ &\ddots \ddots \\ &(a_{n-1} - a_n) \\ =& \prod_{1 \le i < j \le n} (a_i - a_j) = \det \begin{vmatrix} 1 & 1 & \cdots & 1 \\ a_1 & a_2 & \cdots & a_n \\ a_1^2 & a_2^2 & \cdots & a_n^2 \\ \cdots & \cdots & \cdots & \cdots \\ a_1^{n-1} & a_2^{n-1} & \cdots & a_n^{n-1} \end{vmatrix}. \end{aligned}$$

Any transposition $(a_i a_j)$ applied to the product P changes P to $-P$ as this amounts to the interchange of the i-th and j-th columns of the above determinant. Now σ when applied to P has a definite effect, namely, either it changes P to $-P$ or leaves P unchanged. In case σ changes P to $-P$, σ must always be a product of an odd number of transpositions; otherwise, it must be the product of an even number of transpositions. □

(The determinant (call it D) on the right is known as the Vander Monde determinant. If we set, for $i < j, a_i = a_j$ in D, then two columns become identical, and hence $D = 0$. This means that $(a_i - a_j)$ is a factor of D. Further the coefficient of $a_2 a_3^2 \ldots a_n^{n-1}$ on both sides is 1. Hence the last equality.)

Theorem 3.4. *The even permutations in S_n form a subgroup A_n of S_n, called the alternating group of degree n.*

Proof. If σ_1 and σ_2 are even permutations in S_n, then each of them is a product of an even number of transpositions and hence so is their product. Further if $\sigma = t_1 t_2 \ldots t_r$, where each t_i is a transposition, then $\sigma^{-1} = t_r^{-1} t_{r-1}^{-1} \ldots t_2^{-1} t_1^{-1}$. Hence the inverse of an even permutation is even. Further the identity permutation is even since for any transposition $t, e = t \circ t = t^2$. □

Definition 3.11. *A permutation is odd or even according to whether it is expressible as a product of an odd number or even number of transpositions.*

Example 3.1

$$\text{Let} \quad \sigma = \begin{pmatrix} 1 & 2 & 3 & 4 & 5 & 6 & 7 & 8 & 9 \\ 4 & 5 & 1 & 2 & 3 & 7 & 9 & 8 & 6 \end{pmatrix}.$$

$$\begin{aligned} \text{Then,} \quad \sigma &= (14253)(679)(8) \\ &= (13)(15)(12)(14)(69)(67) \\ &= \text{a product of an even number of transpositions.} \end{aligned}$$

Hence σ is an even permutation.

3.9 Lagrange's Theorem for Finite Groups

We now establish the most famous basic theorem on finite groups, namely, Lagrange's theorem. For this, we need the notion of left and right cosets of a subgroup.

Definition 3.12. *Let* G *be a group and* H *a subgroup of* G*. For* $a \in$ G*, the left coset* aH *of* a *in* G *is the subset* $\{ah : h \in H\}$ *of* G*. The right coset* Ha *of* a *is defined in an analogous manner.*

Lemma 3.3. *Any two left cosets of a subgroup* H *of a group* G *are equipotent (that is, they have the same cardinality). Moreover, they are equipotent to* H*.*

Proof. Let aH and bH be two left cosets of H in G. Consider the map

$$\phi : aH \to bH$$

defined by $\phi(ah) = bh$, $h \in H$. ϕ is $1-1$ since $\phi(ah_1) = \phi(ah_2)$, for $h_1, h_2 \in H$, implies that $bh_1 = bh_2$ and therefore $h_1 = h_2$. Clearly, ϕ is onto. Thus, ϕ is a bijection of aH onto bH. In other words, aH and bH are equipotent. Since $eH = H$, H itself is a left coset of H and so all left cosets of H are equipotent to H. (Recall that two sets are equipotent if there is a bijection between them.) □

Lemma 3.4. *The left coset* aH *is equal to* H *iff* $a \in H$*.*

Proof. If $a \in H$, then $aH = \{ah : \ h \in H\} \subset H$ as $ah \in H$. Further if $b \in H$, then $a^{-1}b \in H$, and so $a(a^{-1}b) \in aH$. But $a(a^{-1}b) = b$. Hence $b \in aH$, and therefore $aH = H$. (In particular if H is a group, then multiplication of the elements of H by any element $a \in H$ just gives a permutation of H.)

Conversely, if $aH = H$, then $a = ae \in aH = H$, as $e \in H$. □

Example 3.2

It is not necessary that $aH = Ha$ for all $a \in$ G. For example, consider S_3, the symmetric group of degree 3. The $3! = 6$ permutations of S_3 are given by

$$S_3 = \begin{cases} e = \begin{pmatrix} 1 & 2 & 3 \\ 1 & 2 & 3 \end{pmatrix}, & \begin{pmatrix} 1 & 2 & 3 \\ 1 & 3 & 2 \end{pmatrix} = (23), \quad \begin{pmatrix} 1 & 2 & 3 \\ 3 & 2 & 1 \end{pmatrix} = (13) \\ \begin{pmatrix} 1 & 2 & 3 \\ 2 & 1 & 3 \end{pmatrix} = (12), & \begin{pmatrix} 1 & 2 & 3 \\ 2 & 3 & 1 \end{pmatrix} = (123), \quad \begin{pmatrix} 1 & 2 & 3 \\ 3 & 1 & 2 \end{pmatrix} = (132). \end{cases}$$

Let H be the subgroup $\{e, (12)\}$. For $a = (123)$, we have

$$\begin{aligned} aH &= \{(123)e = (123), \quad (123)(12) = (13)\}, \quad \text{and} \\ Ha &= \{e(123) = (123), \quad (12)(123) = (23)\}, \end{aligned}$$

so that $aH \neq Ha$.

Proposition 3.2. *The left cosets* aH *and* bH *are equal iff* $a^{-1}b \in H$*.*

Proof. $aH = bH \Leftrightarrow a^{-1}(aH) = a^{-1}(bH) \Leftrightarrow H = (a^{-1}b)H \Leftrightarrow a^{-1}b \in H$ by Lemma 3.4. □

Lemma 3.5. *Any two left cosets of the same subgroup of a group are either identical or disjoint.*

Proof. Suppose aH and bH are two left cosets of the subgroup H of a group G, where $a, b \in \mathsf{G}$. If aH and bH are disjoint, there is nothing to prove. Otherwise, $aH \cap bH \neq \phi$, and therefore, there exist $h_1, h_2 \in H$ with $ah_1 = bh_2$. This however means that $a^{-1}b = h_1h_2^{-1} \in H$. So by Proposition 3.2, $aH = bH$. □

Example 3.3

For the subgroup H of Example 3.2, we have seen that $(123)H = \{(123), (13)\}$. Now $(12)H = \{(12)e, (12)(12)\} = \{(12), e\} = H$, and hence $(123)H \cap (12)H = \phi$. Also $(23)H = \{(23)e, (23)(12)\} = \{(23), (132)\}$, and $(13)H = (13)\{e, (12)\} = \{(13), (123)\} = (123)H$. Note that $(13)^{-1}(123) = (13)(123) = (12) \in H$ (refer to Proposition 3.2).

Theorem 3.5. Lagrange's Theorem *[After the French mathematician J. L. Lagrange] The order of any subgroup of a finite group G divides the order of G.*

Proof. Let H be a subgroup of the finite group G. We want to show that $o(H)|o(\mathsf{G})$. We show this by proving that the left cosets of H in G form a partition of G. First of all, if g is any element of G, $g = ge \in gH$. Hence every element of G is in some left coset of H. Now by Lemma 3.5, the distinct cosets of H are pairwise disjoint and hence form a partition of G. Again, by Lemma 3.3, all the left cosets of H have the same cardinality as H, namely, $o(H)$. Thus, if there are l left cosets of H in G, we have

$$l \cdot o(H) = o(\mathsf{G}) \tag{3.1}$$

Consequently, $o(H)$ divides $o(\mathsf{G})$. □

Definition 3.13. *Let H be a subgroup of a group G. Then, the number (may be infinite) of left cosets of H in G is called the index of H in G and denoted by $i_{\mathsf{G}}(H)$*

If G is a finite group, then Equation (3.1) in the proof of Theorem 3.5 shows that

$$o(\mathsf{G}) = o(H)i_{\mathsf{G}}(H)$$

Theorem 3.6 (An application of Lagrange's theorem). *If p is a prime, and n any positive integer, then*

$$n|\phi(p^n - 1),$$

where ϕ is Euler's totient function. First we prove a lemma.

Lemma 3.6. *If $m \geq 2$ is a positive integer, and S, the set of all positive integers less than m and prime to it, then S is a multiplicative group modulo m.*

Proof. If $(a, m) = 1$ and $(b, m) = 1$, then $(ab, m) = 1$. For if p is a prime factor of ab and m, then as p divides ab, p must divide either a or b, say, $p|a$. Then, $(a, m) \geq p$, a contradiction. Moreover, if $ab \pmod{m} \equiv c$, $1 \leq c < m$, then $(c, m) = 1$. Thus, $ab \pmod{m} = c \in S$. Also as $(1, m) = 1$, $1 \in S$. Now for any $a \in S$, by Euclidean algorithm, there exists $b \in \mathbb{N}$ such that $ab \equiv 1 \pmod{m}$. Then, $(b, m) = 1$ (if not, there exists a prime p with $p|b$ and $p|m$, then $p|1$, a contradiction). Thus, a has an inverse $b(\bmod\, m)$ in S. Thus, S is a multiplicative group modulo m, and $o(S) = \phi(m)$. □

Proof. (Proof of Theorem 3.6) We apply Lemma 3.6 by taking $m = p^n - 1$. Let $H = \{1, p, p^2, \ldots, p^{n-1}\}$. All the numbers in H are prime to m and hence $H \subset S$ (where S is as defined in Lemma 3.6). Further $p^j \cdot p^{n-j} = p^n \equiv 1$ (mod m). Hence every element of H has an inverse modulo m. Therefore (as the other group axioms are trivially satisfied by H), H is a subgroup of order n of S. By Lagrange's theorem, $o(H)|o(S)$, and so $n|\phi(p^n - 1)$. □

As another application of Lagrange's theorem, we have the following result.

Theorem 3.7. *Any group of prime order is cyclic.*

Proof. Let G be a group of prime order p. Suppose H is a subgroup of G with order q. Then, by Lagrange's theorem $q|p$. Hence $q = 1$ or $q = p$, as p is prime. This means that $G = \{e\}$ or $G = H$ and any element $a(\neq e)$ generates G since the cyclic subgroup $\langle a \rangle$ of G is equal to G. □

3.10 Homomorphisms and Isomorphisms of Groups

Consider the two groups:

$$\begin{aligned} \mathsf{G}_1 &= \text{the multiplicative group of the sixth root of unity} \\ &= \{\omega^6 = 1, \omega, \omega^2, \ldots, \omega^5 : \omega = \text{a primitive sixth root of unity}\} \\ \text{and} \quad \mathsf{G}_2 &= \text{the additive group } \mathbb{Z}_6 = \{0, 1, 2, \ldots, 5\}. \end{aligned}$$

G_1 is a multiplicative group while G_2 is an additive group. However, structure-wise, they are just the same. By this we mean that if we can make a suitable identification of the elements of the two groups, then they behave in the same manner. If we make the correspondence

$$\omega^i \longleftrightarrow i, 0 \leq i \leq 5,$$

we see that $\omega^i\omega^j \longleftrightarrow i + j$ as $\omega^i\omega^j = \omega^{i+j}$, when $i + j$ is taken modulo 6. For instance, in G_1, $\omega^3\omega^4 = \omega^7 = \omega^1$, while in G_2, $3 + 4 = 1$ as $7 \equiv 1 \pmod 6$. The order of ω in $\mathsf{G}_1 = 6 =$ the (additive) order of 1 in G_2. G_1 has $\{1, \omega^2, \omega^4\}$ as a subgroup while G_2 has $\{0, 2, 4\}$ as a subgroup and so on. It is clear that we can replace 6 by any positive integer n and a similar result holds good. In the above situation, we say G_1 and G_2 are isomorphic groups. We now formalize the above concept.

Definition 3.14. *Let* G *and* G' *be groups (distinct or not). A homomorphism from* G *to* G' *is a map* $f : G \to G'$ *such that*

$$f(ab) = f(a)f(b). \tag{3.2}$$

In Definition 3.14, the multiplication operation has been used to denote the group operations in G_1 and G_2. If, for instance, G_1 is an additive group and G_2 is a multiplicative group, Equation 3.2 should be changed to

$$f(a + b) = f(a)f(b)$$

and so on.

Definition 3.15. *An isomorphism from a group* G *to a group* G' *is a bijective homomorphism from* G *to* G'*, that is, it is a map* $f : G \to G'$ *which is both a bijection and a group homomorphism.*

It is clear that if $f : G \to G'$ is an isomorphism from G to G', then $f^{-1} : G' \to G$ is an isomorphism from $G' \to G$. Hence if there exists a group isomorphism from G to G', we can say without any ambiguity that G and G' are isomorphic groups. A similar statement cannot be made for group homomorphism. If G is isomorphic to G', we write: $G \simeq G'$.

Examples of Groups, Homomorphisms and Isomorphisms

1. Let $G = (\mathbb{Z}, +)$, and $G' = (n\mathbb{Z}, +), n \neq 0$ ($n\mathbb{Z}$ is the set gotten by multiplying all integers by n). The map $f : G \to G'$ defined by $f(m) = mn, m \in G$, is a group homomorphism from G onto G'.

2. Let $G = (\mathbb{R}, +)$ and $G' = (\mathbb{R}^+, \cdot)$. The map $f : G \to G'$ defined by $f(x) = e^x, x \in G$, is a group homomorphism from G onto G'.

3. Let $G = (\mathbb{Z}, +)$, and $G' = (\mathbb{Z} \times \mathbb{Z}, +)$. The map $f : G \to G'$ defined by $f(n) = (0, n), n \in \mathbb{Z}$ is a homomorphism from G to G'.

4. Let $G = \mathbb{R}^2 = \mathbb{R} \times \mathbb{R}$, the real plane with addition $+$ as group operation, that is $(x, y) + (x', y') = (x + x', y + y')$, and $P_X : \mathbb{R}^2 \to \mathbb{R}$ be defined by $P_X(x, y) = x$, the projection of $\mathbb{R}^2$ on the X-axis. P_X is a homomorphism from $(\mathbb{R}^2, +)$ to $(\mathbb{R}, +)$.

We remark that the homomorphism in the last list of examples is not onto while those in Examples 1, 2 and 4 are onto. The homomorphisms in Examples 1 and 2 are isomorphisms. The isomorphism in Example 1 is an isomorphism of G onto a proper subgroup of G. We now check that the map f of Example 2 is an isomorphism. First, it is a group homomorphism since

$$f(x+y) = e^{x+y} = e^x e^y = f(x)f(y).$$

(Note that the group operation in G is addition while in G′, it is multiplication). Next we check that f is $1-1$. In fact, $f(x) = f(y)$ gives $e^x = e^y$, and therefore $e^{x-y} = 1$. This means, as the domain of f is $\mathbb{R}$, $x - y = 0$, and hence $x = y$. Finally, f is onto. If $y \in \mathbb{R}^+$, then there exists x such that $e^x = y$; in fact, $x = log_e y$, is the unique preimage of y. Thus, f is a $1-1$, onto group homomorphism and hence it is a group isomorphism.

3.11 Properties of Homomorphisms of Groups

Let $f : \mathsf{G} \to \mathsf{G}'$ be a group homomorphism. Then, f satisfies the following properties:

Property 3.1. *$f(e) = e'$, that is, the image of the identity element e of G under f is the identity element e' of G′.*

Proof. For $x \in \mathsf{G}$, the equation $xe = x$ in G gives, as f is a group homomorphism, $f(x)f(e) = f(xe) = f(x) = f(x)e'$ in G′. As G′ is a group, both the cancellation laws are valid in G′. Hence cancellation of $f(x)$ gives $f(e) = e'$. □

Property 3.2. *The image $f(a^{-1})$ of the inverse of an element a of G is the inverse of $f(a)$ in G′, that is, $f(a^{-1}) = (f(a))^{-1}$.*

Proof. The relation $aa^{-1} = e$ in G gives $f(aa^{-1}) = f(e)$. But by Property 3.1, $f(e) = e'$, and as f is a homomorphism, $f(aa^{-1}) = f(a)f(a^{-1})$. Thus, $f(a)f(a^{-1}) = e'$ in G′. This implies that $f(a^{-1}) = (f(a))^{-1}$. □

Property 3.3. *The image $f(\mathsf{G}) \subset \mathsf{G}'$ is a subgroup of G′. In other words, the homomorphic image of a group is a group.*

Proof.

i. Let $f(a), f(b) \in f(\mathsf{G})$, where $a, b \in \mathsf{G}$. Then, $f(a)f(b) = f(ab) \in f(\mathsf{G})$, as $ab \in \mathsf{G}$.

ii. The associative law is valid in $f(\mathsf{G})$. As $f(\mathsf{G}) \subset \mathsf{G}'$ and G′ being a group, $f(\mathsf{G})$ satisfies the associative law.

iii. By Property 3.1, the element $f(e) \in f(\mathsf{G})$ acts as the identity element of $f(\mathsf{G})$.

iv. Let $f(a) \in f(\mathsf{G})$, $a \in \mathsf{G}$. By Property 3.2, $(f(a))^{-1} = f(a^{-1}) \in f(\mathsf{G})$, as $a^{-1} \in \mathsf{G}$.

Thus, $f(\mathsf{G})$ is a subgroup of G'. □

Theorem 3.8. *Let $f : \mathsf{G} \to \mathsf{G}'$ be a group homomorphism and $K = \{a \in \mathsf{G} : f(a) = e'\}$, that is, K is the set of all those elements of* G *that are mapped by f to the identity element e' of* G. *Then, K is a subgroup of* G.

Proof. It is enough to check that if $a, b \in K$, then $ab^{-1} \in K$. (Then, $aa^{-1} = e \in K$, and $ea^{-1} = a^{-1} \in K$. Further, as $b^{-1} \in K, a(b^{-1})^{-1} = ab \in K$). Now $f(ab^{-1}) =$ (as f is a group homomorphism)

$$f(a)f(b^{-1}) = f(a)\left(f(b)\right)^{-1} = e'(e')^{-1} = e'e' = e'.$$

and hence $ab^{-1} \in K$. Thus, K is a subgroup of G. □

Definition 3.16. *The subgroup K defined in the statement of Theorem 3.8 is called the kernel of the group homomorphism f.*

As before, let $f : \mathsf{G} \to \mathsf{G}'$ be a group homomorphism.

Property 3.4. *For $a, b \in \mathsf{G}$, $f(a) = f(b)$ iff $ab^{-1} \in K$, the kernel of f.*

Proof.

$$\begin{aligned} f(a) = f(b) &\Leftrightarrow f(a)\left(f(b)\right)^{-1} = e', \text{ the identity element of } G' \\ &\Leftrightarrow f(a)f(b^{-1}) = e' \quad \text{(By Property 3.2)} \\ &\Leftrightarrow f(ab^{-1}) = e' \\ &\Leftrightarrow ab^{-1} \in K. \end{aligned}$$

□

Property 3.5. *f is a $1-1$ map iff $K = \{e\}$.*

Proof. Let f be $1-1, a \in K$. Then, by the definition of the kernel, $f(a) = e'$. But $e' = f(e)$, by Property 3.1. Thus, $f(a) = f(e)$, and this implies, as f is $1-1$, that $a = e$.

Conversely, assume that $K = \{e\}$, and let $f(a) = f(b)$. Then, by Property 3.4, $ab^{-1} \in K$. Thus, $ab^{-1} = e$ and so $a = b$. Hence f is $1-1$. □

Property 3.6. *A group homomorphism $f : \mathsf{G} \to \mathsf{G}'$ is an isomorphism iff $f(\mathsf{G}) = \mathsf{G}'$ and K(= the kernel of f) $= \{e\}$.*

Proof. f is an isomorphism iff f is a $1-1$, onto homomorphism. Now f is $1-1$ iff $K = \{e\}$, by Property 3.5. Further, f is onto iff $f(\mathsf{G}) = \mathsf{G}'$. □

Property 3.7 (Composition of homomorphisms). *Let $f : \mathsf{G} \to \mathsf{G}'$ and $g : \mathsf{G}' \to \mathsf{G}''$ be group homomorphisms. Then, the composition map $h = gof : \mathsf{G} \to \mathsf{G}''$ is also a group homomorphism.*

Proof. h is a group homomorphism iff $h(ab) = h(a)h(b)$ for all $a, b \in \mathsf{G}$.

$$\begin{aligned} \text{Now} \quad h(ab) &= (gof)(ab) = g\left(f(ab)\right) \\ &= g\left(f(a)f(b)\right), \quad \text{as } f \text{ is a group homomorphism} \\ &= g\left(f(a)gf(b)\right), \quad \text{as } g \text{ is a group homomorphism} \\ &= (g \cdot f)(a) \cdot (g \cdot f)(b) \\ &= h(a)h(b). \end{aligned}$$

□

3.12 Automorphism of Groups

Definition 3.17. *An automorphism of a group* G *is an isomorphism of* G *onto itself.*

Example 3.4

Let $\mathsf{G} = \{\omega^0 = 1, \omega, \omega^2\}$ be the group of cube roots of unity, where $\omega = \cos(2\pi/3) + i\sin(2\pi/3)$. Let $f : \mathsf{G} \to \mathsf{G}'$ be defined by $f(\omega) = \omega^2$. To make f a group homomorphism, we have to set $f(\omega^2) = f(\omega \cdot \omega) = f(\omega)f(\omega) = \omega^2 \cdot \omega^2 = \omega$, and $f(1) = f(\omega^3) = (f(\omega))^3 = (\omega^2)^3 = (\omega^3)^2 = 1^3 = 1$. In other words, the homomorphism $f : \mathsf{G} \to \mathsf{G}'$ is uniquely defined on G once we set $f(\omega) = \omega^2$. Clearly, f is onto. Further, only 1 is mapped to 1 by f, while the other two elements ω and ω^2 are interchanged by f. Thus, Ker $f = \{1\}$. So by Property 3.7, f is an isomorphism of G onto G, that is, an automorphism of G.

Our next theorem shows that there is a natural way of generating at least one set of automorphisms of a group.

Theorem 3.9. *Let* G *be a group and* $a \in \mathsf{G}$. *The map* $f_a : \mathsf{G} \to \mathsf{G}$ *defined by* $f_a(x) = axa^{-1}$ *is an automorphism of* G.

Proof. First we show that f_a is a homomorphism. In fact, for $x, y \in \mathsf{G}$,

$$\begin{aligned} f_a(xy) &= a(xy)a^{-1} \quad \text{by the definition of } f_a \\ &= a(xa^{-1}ay)a^{-1} \\ &= (axa^{-1})(aya^{-1}) \\ &= f_a(x)f_a(y). \end{aligned}$$

Thus, f_a is a group homomorphism. Next we show that f_a is $1-1$. Suppose for $x, y \in \mathsf{G}$, $f_a(x) = f_a(y)$. This gives $axa^{-1} = aya^{-1}$, and so by the two cancellation laws that are valid in a group, $x = y$. Finally, if $y \in \mathsf{G}$, then $a^{-1}ya \in \mathsf{G}$, and $f_a(a^{-1}ya) = a(a^{-1}ya)a^{-1} = (aa^{-1})y(aa^{-1}) = eye = y$, and so f is onto. Thus, f is an automorphism of the group G. □

Definition 3.18. *A map of the form f_a for some $a \in \mathsf{G}$ defined by $f_a(x) = axa^{-1}, x \in G$, is called an inner automorphism of* G.

3.13 Normal Subgroups

Definition 3.19. *A subgroup N of a group G is called a normal subgroup of G (equivalently, N is normal in G) if*

$$aNa^{-1} \subseteq N \quad \text{for each } a \in \mathsf{G}. \tag{3.3}$$

In other words, N is normal in G if N is left invariant by the inner automorphisms f_a for each $a \in \mathsf{G}$. We state this observation as a proposition.

Proposition 3.3. *The normal subgroups of a groups G are those subgroups of G that are left invariant by all the inner automorphisms of G.*

Now the condition $aNa^{-1} \subseteq N$ for each $a \in \mathsf{G}$ shows, by replacing a by a^{-1}, $a^{-1}N(a^{-1})^{-1} = a^{-1}Na \subseteq N$. The latter condition is equivalent to

$$N \subseteq aNa^{-1} \quad \text{for each } a \in \mathsf{G}. \tag{3.4}$$

The conditions (3.3) and (3.4) give the following equivalent definition of a normal subgroup.

Definition 3.20. *A subgroup N of a group G is normal in G iff $aNa^{-1} = N$ (equivalently, $aN = Na$) for every $a \in \mathsf{G}$.*

Examples of normal subgroups

1. Let $\mathsf{G} = S_3$, the group of $3! = 6$ permutations on $\{1, 2, 3\}$. Let $N = \{e, (123), (132)\}$. Then, N is a normal subgroup of S_3. First of all note that N is a subgroup of G. In fact, we have $(123)^2 = (132)$, $(132)^2 = (123)$, and $(123)(132) = e$. Hence $(123)^{-1} = (132)$ and $(132)^{-1} = (123)$. Let $a \in S_3$. If $a \in N$, then $aN = N = Na$ (See Lemma 3.4). So let $a \in S \setminus N$. Hence $a = (12), (23)$ or (13). If $a = (12)$, then $aNa^{-1} = \{(12)e(12),\ (12)(123)(12),\ (12)(132)(12)\} = \{e,\ (132),\ (123)\}$. In a similar manner, we have $(23)N(23) = N$ and $(13)N(13) = N$. Thus, N is a normal subgroup of S_3.

2. Let $H = \{e, (12)\} \subset S_3$. Then, H is a subgroup of G that is not normal in S_3. In fact, if $a = (23)$, we have

$$\begin{aligned} aHa^{-1} &= (23)\,\{e,\ (12)\}\,(23) \\ &= \{(23)e(23),\ (23)(12)(23)\} \\ &= \{e,\ (13)\} \neq H \end{aligned}$$

Hence H is not a normal subgroup of S_3.

Definition 3.21. *The center of a group* G *consists of those elements of* G *each of which commutes with all the elements of* G. *It is denoted by* $C(\mathsf{G})$. *Thus,*

$$C(\mathsf{G}) = \{x \in \mathsf{G} : xa = ax \text{ for each } a \in \mathsf{G}\}$$

For example, $C(S_3) = \{e\}$, that is, the center of S_3 is trivial. Also, it is easy to see that the center of an Abelian group G is G itself.

Clearly the trivial subgroup $\{e\}$ is normal in G and G is normal in G. (Recall that $a\mathsf{G} = \mathsf{G}$ for each $a \in \mathsf{G}$.)

Proposition 3.4. *The center* $C(\mathsf{G})$ *of a group* G *is a normal subgroup of* G.

Proof. We have for $a \in \mathsf{G}$,

$$\begin{aligned} aC(\mathsf{G})a^{-1} &= \{aga^{-1} : g \in C(\mathsf{G})\} \\ &= \{(ag)a^{-1} : g \in C(\mathsf{G})\} \\ &= \{(ga)a^{-1} : g \in C(\mathsf{G})\} \\ &= \{g(aa^{-1}) : g \in C(\mathsf{G})\} \\ &= \{g : g \in C(\mathsf{G})\} \\ &= C(\mathsf{G}) \end{aligned}$$

□

Theorem 3.10. *Let* $f : \mathsf{G} \to \mathsf{G}'$ *be a group homomorphism. Then,* $K = \mathrm{Ker} f$ *is a normal subgroup of* G.

Proof. We have, for $a \in \mathsf{G}$, $aKa^{-1} = \{aka^{-1} : k \in K\}$. Now $f(aka^{-1}) = f(a)f(k)f(a^{-1}) = f(a)e'f(a^{-1}) = f(a)f(a^{-1}) = f(a)(f(a))^{-1} = e'$. Hence $aka^{-1} \in K$ for each $k \in K$ and so $aKa^{-1} \subset \mathrm{Ker}\, f = K$ for every $a \in \mathsf{G}$. This implies that K is a normal subgroup of G. □

3.14 Quotient Groups (or Factor Groups)

Let G be a group, and H a normal subgroup of G. Let G/H (read G modulo H) be the set of all left cosets of H. Recall that when H is a normal subgroup of G, there is no distinction between the left coset aH and the right coset Ha of H. The fact that H is a normal subgroup of G enables us to define a group operation in G/H. We set, for any two cosets aH and bH of H in G, $aH \cdot bH = (ab)H$. This definition is well defined. By this we mean that if we take different representative elements instead of a and b to define the cosets aH and bH, still we end up with the same product. To be precise, let

$$aH = a_1H, \quad \text{and} \quad bH = b_1H \tag{3.5}$$

$$\text{Then, } a^{-1}a_1 \in H \quad \text{and} \quad b^{-1}b_1 \in H.$$

Let $a^{-1}a_1 = h \in H$. This gives: $(ab)^{-1}(a_1b_1) = b^{-1}(a^{-1}a_1)b_1 = b^{-1}hb_1 = (b^{-1}hb)(b^{-1}b_1) \in H$, since $b^{-1}hb \in H$ (H being a normal subgroup of G), and $b^{-1}b_1 \in H$. Consequently, $(aH) \cdot (bH) = (a_1H)(b_1H)$ as the product on the left is $(ab)H$ while the poduct on the right is $(a_1b_1)H$. Hence the binary operation on the set of cosets of H in G is well defined in the sense that it is independent of the representative elements chosen for the cosets of H in G.

Further $eH = H$ acts as the identity element of G/H as

$$(aH)(eH) = (ae)H = aH = (ea)H = eH \cdot aH$$

Finally, the inverse of aH is $a^{-1}H$ since

$$(aH)(a^{-1}H) = (aa^{-1})H = eH = H,$$

and for a similar reason $(a^{-1}H)(aH) = H$.
Thus, G/H is a group under this binary operation. G/H is called the quotient group or factor group of G modulo H.

Example 3.5

We now present an example of a quotient group. Let $\mathsf{G} = (\mathbb{R}^2, +)$, the additive group of points of the plane $\mathbb{R}^2$. (If (x_1, y_1) and (x_2, y_2) are two points of $\mathbb{R}^2$, their sum $(x_1, y_1) + (x_2, y_2)$ is defined as $(x_1 + x_2, y_1 + y_2)$. The identity element of this group $(0, 0)$ and the inverse of (x, y) is $(-x, -y)$). Let H be the subgroup: $\{(x, 0) : x \in \mathbb{R}\}$ = X-axis. If (a, b) is any point of $\mathbb{R}^2$, then

$$(a, b) + H = \{(a, b) + (x, 0) = (a + x, b) : x \in \mathbb{R}\}$$

=line through (a, b) parallel to X-axis. Clearly if $(a, b) + H = (a', b') + H$, then $((a' - a), (b' - b)) \in H)$ = X-axis and therefore the Y-coordinate $b' - b = 0$ and so $b' = b$. In other words, the line through (a, b) and the line through (a', b'), both parallel to the X-axis, are the same iff $b = b'$, as is expected (See Figure 3.1). For this reason, this line may be taken as $(0, b) + H$. Thus, the cosets of H in $\mathbb{R}$ are the lines parallel to the X-axis and therefore the elements of the quotient group $\mathbb{R}/H$ are the lines parallel to the X-axis. If $(a, b) + H$ and $(a', b') + H$ are two elements of $\mathbb{R}/H$, we define their sum to be $(a + a', b + b') + H = (0,\, b + b') + H$, the line through $(0,\, b + b')$ parallel to the X-axis. Note that $(\mathbb{R}^2, +)$ is an Abelian group and so H is a normal subgroup of $\mathbb{R}^2$. Hence the above sum is well defined. The above addition defines a group structure on the set of lines parallel to the X-axis, that is, the elements of $\mathbb{R}/H$. The identity element of the quotient group is the X-axis $= H$, and the inverse of $(0, b) + H$ is $(0, -b) + H$.

Our next result exhibits the importance of factor groups.

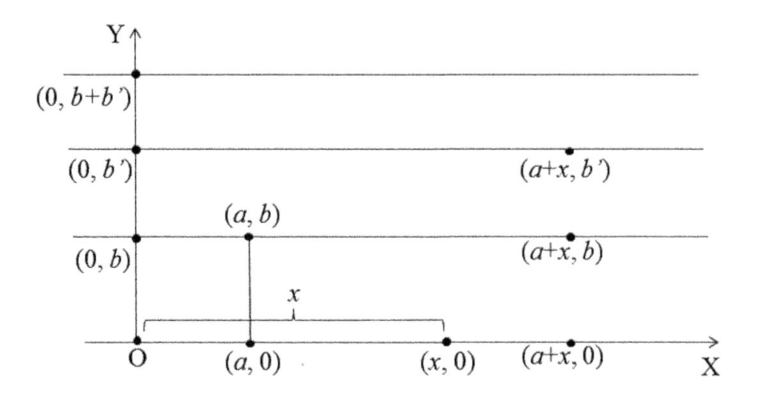

FIGURE 3.1: An example for quotient group.

3.15 Basic Isomorphism Theorem for Groups

Theorem 3.11. *If there exists a homomorphism f from a group G onto a group G' with kernel K, then $\mathsf{G}/K \simeq \mathsf{G}'$.*

Proof. We have to prove that there exists an isomorphism ϕ from the factor group G/K onto G' (observe that the factor group G/K is defined, as K is a normal subgroup of G (see Theorem 3.10)). Define $\phi : \mathsf{G}/K \rightarrow \mathsf{G}'$ by $\phi(gK) = f(\mathrm{g})$ (the mapping ϕ is pictorially depicted in Figure 3.1. See Example 1 below).

It is possible that $gK = g'K$ with $g \neq g'$ in G. Hence we need to establish that ϕ is a well-defined map. As per our definition of ϕ, $\phi(gK) = f(\mathrm{g})$, while $\phi(g'K) = f(g')$. Hence our definition of ϕ will be valid only if $f(\mathrm{g}) = f(g')$. Now $gK = g'K$ implies that ${g'}^{-1}g \in K$. Let ${g'}^{-1}g = k \in K$. Then, $f(k) = e'$, the identity element of G'. Moreover, $e' = f({g'}^{-1}g) = f({g'}^{-1})f(g) = f(g')^{-1}f(g)$ (as f is a group homomorphism, Property 3.2 holds). Thus, $f(g') = f(g)$, and f is well defined. We next show that ϕ is a group isomorphism.

i. ϕ is a group homomorphism: We have for g_1K, g_2K in G/K,

$$\begin{aligned}\phi((g_1K)(g_2K)) &= \phi((g_1g_2)K)\\ &= f(g_1g_2), \quad \text{by the definition of } \phi\\ &= f(g_1)f(g_2), \quad \text{as } f \text{ is a group homomorphism}\\ &= \phi(g_1K)\phi(g_2K)\end{aligned}$$

Thus, ϕ is a group homomorphism.

ii. ϕ is $1-1$: Suppose $\phi(g_1K) = \phi(g_2K)$, where $g_1K, g_2K \in \mathsf{G}/K$. This gives that $f(g_1) = f(g_2)$, and therefore $f(g_1)f(g_2)^{-1} = e'$, the identity element of G'. But $f(g_1)f(g_2)^{-1} = f(g_1)f(g_2^{-1}) = f(g_1g_2^{-1})$. Hence $g_1g_2^{-1} = e'$, and so $g_1g_2^{-1} \in K$, and consequently, $g_1K = g_2K$ (by Property 3.4). Thus, ϕ is $1-1$.

iii. ϕ is onto: Let $g' \in \mathsf{G}'$. As f is onto, there exists $g \in \mathsf{G}$ with $f(g) = g'$. Now $gK \in \mathsf{G}/K$, and $\phi(gK) = f(g) = g'$. Thus, ϕ is onto and hence ϕ is a group isomorphism. □

3.15.1 Examples of factor groups

1. Let us see as to what this isomorphism means with regard to the factor group $\mathbb{R}^2/K$ given in Example 3.5. Define $f : \mathbb{R}^2 \to \mathbb{R}$ by $f(a,b) = (0,b)$, the projection of the point $(a,b) \in \mathbb{R}^2$ on the Y-axis $= \mathbb{R}$. The identity element of the image group is the origin $(0,0)$. Clearly K is the set of all points $(a,b) \in \mathbb{R}^2$ that are mapped to $(0,0)$, that is, the set of those points of $\mathbb{R}^2$ whose projections on the Y-axis coincide with the origin. Thus, K is the X-axis($= \mathbb{R}$). Now $\phi : \mathsf{G}/K = \mathbb{R}^2/\mathbb{R} \to \mathsf{G}'$ is defined by $\phi((a,b)+K) = f(a,b) = (0,b)$. This means that all points of the line through (a,b) parallel to the X-axis are mapped to their common projection on the Y-axis, namely, the point $(0,b)$. Thus, the isomorphism between G/K and G' is obtained by mapping each line parallel to the X-axis to the point where the line meets the Y-axis. (Note: $G/K = \mathbb{R}^2/(X\text{-}axis)$, and $G' = Y$-axis.)

2. We now consider another example. Let $\mathsf{G} = S_n$, the symmetric group of degree n, and $\mathsf{G}' = \{-1, 1\}$, the multiplicative group of two elements (with multiplication defined in the usual way). 1 is the identity element of G' and the inverse of -1 is -1. Define $f : \mathsf{G} \to \mathsf{G}'$ by setting

$$f(\sigma) = \begin{cases} 1 & \text{if } \sigma \in S_n \text{ is an even permutation, that is, if } \sigma \in A_n \\ -1 & \text{if } \sigma \in S_n \text{ is an odd permutation.} \end{cases}$$

Recall that A_n is a subgroup of S_n. Now $|S_n| = n!$ and $|A_n| = n!/2$. Further, if α is an odd permutation in S_n, and $\sigma \in A_n$, then $\alpha\sigma$ is an odd permutation, and hence $|\alpha A_n| = n!/2$. Let B_n denote the set of odd permutations in S_n. Then, $S_n = A_n \cup B_n$, $A_n \cap B_n = \phi$, and $\alpha A_n = B_n$ for each $\alpha \in B_n$. Thus, S_n/A_n has exactly two distinct cosets of A_n, namely, A_n and $S_n \setminus A_n = B_n$.

The mapping $f : S_n \to \{1,-1\}$ defined by $f(\alpha) = 1$ or -1 according to whether the permutation α is even or odd clearly defines a group homomorphism. The kernel K of this homomorphism is A_n, and we have $S_n \setminus A_n \simeq \{1,-1\} = G'$. The isomorphism is obtained by mapping the coset σA_n to 1 or -1 according to whether σ is an even or an odd permutation.

3.16 Exercises

1. Let G $= SL(n, \mathbb{C})$ be the set of all invertible complex matrices A of order n. If the operation $\cdot$ denotes matrix multiplication, show that G is a group under $\cdot$.

2. Let G denote the set of all real matrices of the form $\left(\begin{smallmatrix} a & 0 \\ b & 1 \end{smallmatrix}\right)$ with $a \neq 0$. Show that G is a group under matrix multiplication.

3. Which of the following semigroups are groups?

   i. $(\mathbb{Q}, \cdot)$

   ii. $(\mathbb{R}^*, \cdot)$

   iii. $(\mathbb{Q}, +)$

   iv. $(\mathbb{R}^*, \cdot)$

   v. The set of all 2 by 2 real matrices under matrix multiplication.

   vi. The set of all 2 by 2 real matrices of the form $\left(\begin{smallmatrix} a & 0 \\ b & 1 \end{smallmatrix}\right)$.

4. Prove that a finite semigroup in which the right and left cancellation laws are valid is a group, that is, if H is a finite semigroup in which both the cancellation laws are valid (that is, $ax = ay$ implies that $x = y$, and $xa = ya$ implies that $x = y$, where $a, x, y \in H$), is a group.

5. Prove that in semigroup G in which the equations $ax = b$ and $yc = d$, are solvable in G, where $a, b, c, d \in$ G, is a group.

6. In the group $GL(2, \mathbb{C})$ of 2×2 complex non-singular matrices, find the order of the following elements:

   (i) $\left(\begin{smallmatrix} 1 & 0 \\ 0 & -i \end{smallmatrix}\right)$ (ii) $\left(\begin{smallmatrix} 1 & 1 \\ 0 & 1 \end{smallmatrix}\right)$ (iii) $\left(\begin{smallmatrix} i & 0 \\ 0 & -i \end{smallmatrix}\right)$ (iv) $\left(\begin{smallmatrix} -2+3i & -1+2i \\ 1-i & 3-2i \end{smallmatrix}\right)$

7. Let G be a group, and $\phi :$ G $\to$ G be defined by $\phi(g) = g^{-1}, g \in$ G. Show that ϕ is an automorphism of G iff G is Abelian.

8. Prove that any group of even order has an element of order 2. (Hint: $a \neq e$ and $o(a) \neq 2$ iff $a \neq a^{-1}$.) Pair off such elements (a, a^{-1}).

9. Give an example of a non-cyclic group where each of its proper subgroup is cyclic.

10. Show that no group can be the set union of two of its proper subgroups.

11. Show that $S = \{3, 5\}$ generates the group $(\mathbb{Z}, +)$.

12. Give an example of an infinite non-Abelian group.

13. Show that the permutations $T = \{12345\}$ and $S = (25)(34)$ generate a subgroup of order 10 in S_5.

14. Find if the following permutations are odd:

    (i) $(123)(456)$ (ii) $(1546)(2)(3)$ (iii) $\begin{pmatrix} 1 & 2 & 3 & 4 & 5 & 6 & 7 & 8 & 9 \\ 2 & 5 & 4 & 3 & 1 & 7 & 6 & 9 & 8 \end{pmatrix}$

15. If $G = \{a_1, \ldots, a_n\}$ is a finite Abelian group of order n, show that $(a_1 a_2 \ldots a_n)^2 = e$.

16. Let $G = \{a \in \mathbb{R} : a \neq -1\}$. If the binary operation $*$ is defined in G by $a * b = a + b + ab$ for all $a, b \in G$, show that $(G, *)$ is a group.

17. Let $G = \{a \in \mathbb{R} : a \neq 1\}$. If the binary operation $*$ is defined on G by $a * b = a + b - ab$ for all $a, b \in G$, show that $(G, *)$ is not a group.

18. Let $\sigma = (i_1 i_2 \ldots i_r)$ be a cycle in S_n of length r. Show that the order of σ (= the order of the subgroup generated by σ) is r.

19. Prove that the center of a group is a normal subgroup G.

20. Let α, β, γ be permutations in S_4 defined by

    $$\alpha = \begin{pmatrix} 1 & 2 & 3 & 4 \\ 1 & 4 & 3 & 2 \end{pmatrix}, \quad \beta = \begin{pmatrix} 1 & 2 & 3 & 4 \\ 2 & 1 & 4 & 3 \end{pmatrix}, \quad \gamma = \begin{pmatrix} 1 & 2 & 3 & 4 \\ 3 & 1 & 2 & 4 \end{pmatrix}.$$

    Find (i) α^{-1}, (ii) $\alpha^{-1}\beta\gamma$, (iii) $\beta\gamma^{-1}$.

21. Show that any infinite cyclic group is isomorphic to $(\mathbb{Z}, +)$.

22. Show that the set $\{e^{in} : n \in \mathbb{Z}\}$ forms a multiplicative group. Show that this is isomorphic to $(\mathbb{Z}, +)$. Is this group cyclic?

23. Find a homomorphism of the additive group of integers to itself that is not onto.

24. Give an example of a group that is isomorphic to one of its proper subgroups.

25. Prove that $(\mathbb{Z}, +)$ is not isomorphic to $(\mathbb{Q}, +)$. (Hint: Suppose $\exists$ an isomorphism $\phi : \mathbb{Z} \to \mathbb{Q}$. Let $\phi(5) = a \in \mathbb{Q}$. Then, $\exists\, b \in \mathbb{Q}$ with $2b = a$. Let $x \in \mathbb{Z}$ be the preimage of b. Then, $2x = 5$ in $\mathbb{Z}$, which is not true.)

26. Prove that the multiplicative groups $\mathbb{R}^\star$ and $\mathbb{C}^\star$ are not isomorphic.

27. Give an example of an infinite group in which every element is of finite order.

28. Give the group table of the group S_3. From the table, find the center of S_3.

29. Show that if a subgroup H of a group G is generated by a subset S of G, then H is a normal subgroup iff $aSa^{-1} \subset \langle S \rangle$ for each $a \in G$.

30. Show that a group G is Abelian iff the center $C(G)$ of G is G.

31. Let G be a group. Let $[\mathsf{G},\mathsf{G}]$ denote the subgroup of G generated by all elements of G of the form $aba^{-1}b^{-1}$ (called the commutator of a and b) for all pairs of elements $a, b \in \mathsf{G}$. Show that $[\mathsf{G},\mathsf{G}]$ is a normal subgroup of G. [Hint: For $c \in \mathsf{G}$, we have $c(aba^{-1}b^{-1})c^{-1} = (cac^{-1})(cbc^{-1})(cac^{-1})^{-1}(cbc^{-1})^{-1} \in [\mathsf{G},\mathsf{G}]$. Now apply Exercise 29.]

32. Show that a group G is Abelian $\Leftrightarrow [\mathsf{G},\mathsf{G}] = \{e\}$.
Remark: $[\mathsf{G},\mathsf{G}]$ is called the *commutator subgroup* of G. In general, the commutators of a group G need not form a subgroup of G, and it is for this reason that we take the subgroup generated by the commutators of G. There is no known elementary counter-example. For a counter-example, see for instance, *Theory of Groups* [34].

33. Let G be the set of all roots of unity, that is, $\mathsf{G} = \{\omega \in \mathbb{C} : \omega^n = 1$ for some $n \in \mathbb{N}\}$. Prove that G is an Abelian group that is not cyclic.

34. If A and B are normal subgroups of a group G such that $A \cap B = \{e\}$. Then, show that for $\forall a \in A$ and $\forall b \in B$, $ab = ba$.

35. If H is the only subgroup of a given finite order in a group G, show that H is normal in G.

36. Show that any subgroup of a group G of index 2 is normal in G.

37. Prove that the subgroup $\{e, (13)\}$ of S_3 is not normal in S_3.

38. Prove that the subgroup $\{e,\ (123),\ (132)\}$ is a normal subgroup of S_3.

3.17 Rings

The study of commutative rings arose as a natural abstraction of the algebraic properties of the set of integers, while that of fields arose out of the sets of rational, real and complex numbers.

We begin with the definition of a ring and then proceed to establish some of its basic properties.

3.17.1 Rings, definitions and examples

Definition 3.22. *A ring is a non-empty set A with two binary operations, denoted by $+$ and $\cdot$ (called addition and multiplication, respectively) satisfying the following axioms:*

R_1: *$(A, +)$ is an Abelian group. (The identity element of $(A, +)$ is denoted by 0).*

R_2: *$\cdot$ is associative, that is, $a \cdot (b \cdot c) = (a \cdot b) \cdot c$ for all $a, b, c \in A$.*

R_3: *For all $a, b, c \in A$,*

$$a \cdot (b + c) = a \cdot b + a \cdot c \quad \text{(left distributive law)}$$
$$(a + b) \cdot c = a \cdot c + b \cdot c \quad \text{(right distributive law).}$$

It is customary to write ab instead of $a \cdot b$.

Examples of Rings

1. $A = \mathbb{Z}$, the set of all integers with the usual addition $+$ and the usual multiplication taken as $\cdot$.

2. $A = 2\mathbb{Z}$, the set of even integers with the usual addition and multiplication.

3. $A = \mathbb{Q}$, $\mathbb{R}$ or $\mathbb{C}$ with the usual addition and multiplication.

4. $A = \mathbb{Z}_n = \{0, 1, 2, \ldots, n-1\}$, the set of integers modulo n, where $+$ and $\cdot$ denote addition and multiplication taken modulo n. (For instance, if $A = \mathbb{Z}_5$, then in $\mathbb{Z}_5$, $3 + 4 = 7 = 2$, and $3 \cdot 4 = 12 = 2$).

5. $A = \mathbb{Z}[X]$, the set of polynomials in the indeterminate X with integer coefficients with addition $+$ and multiplication $\cdot$ defined in the usual way.

6. $A = \mathbb{Z} + i\sqrt{3}\mathbb{Z} = \{a + ib\sqrt{3} : a, b \in \mathbb{Z}\} \subset \mathbb{C}$. Then, A is a ring with the usual $+$ and $\cdot$ in $\mathbb{C}$.

7. (Ring of Gaussian integers). Let $A = \mathbb{Z} + i\mathbb{Z} = \{a + ib : a, b \in \mathbb{Z}\} \subset \mathbb{C}$. Then, with the usual addition $+$ and multiplication $\cdot$ in $\mathbb{C}, A$ is a ring.

8. (A ring of functions). Let $A = C[0,1]$, the set of all complex-valued continuous functions on $[0,1]$. For $t \in [0,1]$, and $f, g \in A$, set $(f+g)(t) = f(t) + g(t)$, and $(f \cdot g)(t) = f(t)g(t)$. Then, it is clear that both $f + g$ and $f \cdot g$ are in A. It is easy to check that A is a ring.

Definition 3.23. *A ring A is called commutative if for all $a, b \in A$, $ab = ba$.*

Hence if A is a non-commutative ring, there exists a pair of elements $x, y \in A$ with $xy \neq yx$.

All the rings given above in Examples 1 to 8 are commutative rings. We now present an example of a non-commutative ring.

Example 3.6

Let $A = M_2(\mathbb{Z})$, the set of all 2 by 2 matrices with integers as entries. A is a ring with the usual matrix addition $+$ and usual matrix multiplication $\cdot$. It is a non-commutative ring since $M = \left(\begin{smallmatrix} 1 & 1 \\ 0 & 0 \end{smallmatrix}\right), N = \left(\begin{smallmatrix} 0 & 0 \\ 1 & 1 \end{smallmatrix}\right)$ are in A, but $MN \neq NM$.

3.17.1.1 Unity element of a ring

Definition 3.24. *An element e of a ring A is called an identity or unity element of A if $ea = ae = a$ for all $a \in A$.*

An identity element of A, if it exists, must be unique. For, if e and f are identity elements of A, then,

$$ef = e \quad \text{as } f \text{ is an identity element of } A,$$
$$ef = f \quad \text{as } e \text{ is an identity element of } A.$$

Therefore $e = f$. Hence if a ring A has an identity element e, we can refer to it as the identity element e of A.

For the rings in Examples 1, 3 and 7 above, the number 1 is the identity element. For the ring $C[0, 1]$ of Example 8 above, the function $1 \in C[0, 1]$ defined by $1(t) = 1$ for all $t \in [0, 1]$ acts as the identity element. For the ring $M_2(\mathbb{Z})$ of Example 9, the matrix $\begin{pmatrix} 1 & 0 \\ 0 & 1 \end{pmatrix}$ acts as the identity element. A ring may not have an identity element. For instance, the ring $2\mathbb{Z}$ in Section 3.17.1 above has no identity element.

3.17.2 Units of a ring

An element a of a ring A with identity element e is called a unit in A if there exist elements b and c in A such that $ab = e = ca$.

Proposition 3.5. *If a is a unit in a ring A with identity element e, and if $ab = ca = e$, then $b = c$.*

Proof. We have, $b = eb = (ca)b = c(ab) = ce = c$. □

We denote the element $b(= c)$ described in Proposition 3.5 as the inverse of a and denote it by a^{-1}. Thus, if a is a unit in A, then there exists an element $a^{-1} \in A$ such that $aa^{-1} = a^{-1}a = e$. Clearly, a^{-1} is unique.

Proposition 3.6. *The units of a ring A (with identity element) form a group under multiplication.*

Proof. Exercise. □

3.17.2.1 Units of the ring $\mathbb{Z}_n$

Let a be a unit in the ring $\mathbb{Z}_n$. (See Example in Section 3.17.1 above). Then, there exists an $x \in \mathbb{Z}_n$ such that $ax = 1$ in $\mathbb{Z}_n$, or equivalently, $ax \equiv 1 (\bmod n)$ But this implies that $ax - 1 = bn$ for some integer b. Hence $gcd(a, n) = 1$. (Because if an integer $c > 1$ divides both a and n, then it should divide 1.) Conversely, if $(a, n) = 1$, by Euclidean algorithm, there exist integers x and y with $ax + ny = 1$, and therefore $ax \equiv 1 (\bmod n)$. This, however, means that a is a unit in $\mathbb{Z}_n$. Thus, the set U of units of $\mathbb{Z}_n$ is precisely the set of integers in $\mathbb{Z}_n$ that are relatively prime to n and we know that the order of this set is $\phi(n)$, where ϕ is the Euler function.

3.17.2.2 Zero divisors

In the ring $\mathbb{Z}$ of integers, a is a divisor of c if there exists an integer b such that $ab = c$. As $\mathbb{Z}$ is a commutative ring, we simply say that a is a divisor of c and not a left divisor or right divisor of c. Taking $c = 0$, we have the following more general definition.

Definition 3.25. *A left zero divisor in a ring A is a non-zero element a of A such that there exists a non-zero element b of A with $ab = 0$ in A. $a \in A$ is a right zero divisor in A if $ca = 0$ for some $c \in A$, $c \neq 0$.*

If A is a commutative ring, a left zero divisor a in A is automatically a right zero divisor in A and vice versa. In this case, we simply call a a zero divisor in A.

Examples relating to zero divisors

1. If $a = 2$ in $\mathbb{Z}_4$, then a is a zero divisor in $\mathbb{Z}_4$, as $2 \cdot 2 = 4 = 0$ in $\mathbb{Z}_4$.
2. In the ring $M_2(\mathbb{Z})$, the matrix $\left[\begin{smallmatrix} 1 & 1 \\ 0 & 0 \end{smallmatrix}\right]$ is a right zero divisor as
$$\begin{bmatrix} 0 & 0 \\ 0 & 1 \end{bmatrix} \begin{bmatrix} 1 & 1 \\ 0 & 0 \end{bmatrix} = \begin{bmatrix} 0 & 0 \\ 0 & 0 \end{bmatrix}$$
and $\left[\begin{smallmatrix} 0 & 0 \\ 0 & 1 \end{smallmatrix}\right]$ is not the zero matrix of $M_2(\mathbb{Z})$.
3. If p is a prime, then every non-zero element of $\mathbb{Z}_p$ is a unit. This follows from the fact that if $1 \leq a < p$, then $(a, p) = 1$. Hence no $a \in \mathbb{Z}_p$, $a \neq 0$ is a zero divisor in $\mathbb{Z}_p$.

Theorem 3.12. *The following statements are true for any ring A.*

1. $a0 = 0a$ *for any* $a \in A$.
2. $a(-b) = (-a)b = -(ab)$ *for all* $a, b \in A$.
3. $(-a)(-b) = ab$ *for all* $a, b \in A$.

Proof. Exercise. □

3.18 Integral Domains

An integral domain is an abstraction of the algebraic structure of the ring of integers.

Definition 3.26. *An integral domain A is a commutative ring with identity element having no divisors of zero.*

Examples relating to integral domains

1. The ring of integers $\mathbb{Z}$ and the ring of Gaussian integers $\mathbb{Z}+i\mathbb{Z}$ are both integral domains.

2. The ring $2\mathbb{Z}$ of even integers is not an integral domain even though it has no zero divisors (Why?).

3. Let $\mathbb{Z}\times\mathbb{Z} = \{(a,b) : a \in \mathbb{Z}, b \in \mathbb{Z}\}$. For (a,b), (c,d) in $\mathbb{Z}\times\mathbb{Z}$, define

$$\begin{aligned}(a,b) \pm (c,d) &= (a \pm c, b \pm d),\\ \text{and}(a,b)\cdot(c,d) &= (ac, bd).\end{aligned}$$

   Clearly $\mathbb{Z}\times\mathbb{Z}$ is a commutative ring with zero element $(0,0)$ and identity element $(1,1)$ but not an integral domain as it has zero divisors. For instance, $(1,0)\cdot(0,1) = (0,0)$.

3.19 Exercises

1. Prove that $\mathbb{Z}_n$, $n \geq 2$, is an integral domain iff n is a prime.

2. Give the proof of Proposition 3.6.

3. Determine the group of units of the rings:

   (i) $\mathbb{Z}$, (ii) $M_2(\mathbb{Z})$, (iii) $\mathbb{Z}+i\mathbb{Z}$, (iv) $\mathbb{Z}+i\sqrt{3}\mathbb{Z}$.

4. Let A be a ring, and $a, b_1, b_2, \ldots, b_n \in A$. Then, show that $a(b_1 + b_2 + \cdots + b_n) = ab_1 + ab_2 + \cdots ab_n$. (Hint: Apply induction on n).

5. Let A be a ring, and $a, b \in A$. Then, show that for any positive integer n,

$$n(ab) = (na)b = a(nb).$$

   (na stands for the element $a + a + \cdots$ (n times) of A).

6. Show that no unit of a ring A can be a zero divisor in A.

7. (Definition: A subset B of a ring A is a subring of A if B is a ring with respect to the binary operations $+$ and $\cdot$ of A). Prove:

   i. $\mathbb{Z}$ is a subring of $\mathbb{Q}$.

   ii. $\mathbb{Q}$ is a subring of $\mathbb{R}$.

   iii. $\mathbb{R}$ is a subring of $\mathbb{C}$.

8. Prove that any ring A with identity element and cardinality p, where p is a prime, is commutative. (Hint: Verify that the elements $1, 1+1, \ldots, 1+1+\cdots+1$ (p times) are all distinct elements of A.)

3.20 Ideals

One of the special classes of subrings of a ring is the class of ideals. Consider, for example, the set S of all multiples of 3 (positive multiples, negative multiples, and zero multiple) in the ring $\mathbb{Z}$ of integers. Then, it is easy to see that S is a subring of $\mathbb{Z}$. More than this, if $n \in \mathbb{Z}$ and $a \in S$, then $na \in S$ as na is also a multiple of 3. We then call S an ideal in $\mathbb{Z}$. We now present the formal definition of an ideal in a general ring (that is, not necessarily commutative) A.

Definition 3.27. *Let A be any ring, and $S \subset A$. S is called a left ideal in the ring A if*

i. S is a subring of A, and

ii. For $a \in A$, and $s \in S$, $as \in S$.

We call S a left ideal of A because we are multiplying s by a on the left of s. If we take sa instead of as, and if $sa \in S$ for every $a \in A$ and $s \in S$, we call S a right ideal of A. If S is both a left ideal and right ideal of A, it is a two-sided ideal of A. If A is a commutative ring, it is obvious that we need make no distinction between left ideals and right ideals.

Example 3.7

Let A be the ring of all 2×2 matrices over $\mathbb{Z}$. (Note: A is a non-commutative ring). Let

$$S = \left\{ \begin{pmatrix} a & 0 \\ b & 0 \end{pmatrix}, a, b \in \mathbb{Z} \right\}.$$

Then, as

$$\begin{pmatrix} a & 0 \\ b & 0 \end{pmatrix} \begin{pmatrix} c & 0 \\ d & 0 \end{pmatrix} = \begin{pmatrix} ac & 0 \\ bc & 0 \end{pmatrix},$$

it is easy to see that S is a subring of A. Moreover, if

$$\begin{pmatrix} a & 0 \\ b & 0 \end{pmatrix} \in S, \quad \text{and} \quad \begin{pmatrix} x & y \\ z & t \end{pmatrix} \in A,$$

we have

$$\begin{pmatrix} x & y \\ z & t \end{pmatrix} \begin{pmatrix} a & 0 \\ b & 0 \end{pmatrix} = \begin{pmatrix} xa + yb & 0 \\ za + tb & 0 \end{pmatrix} \in S.$$

Hence S is a left ideal of A. However, S is not a right ideal of A. For instance,

$$\begin{pmatrix} 5 & 0 \\ 6 & 0 \end{pmatrix} \in S, \quad \text{but} \quad \begin{pmatrix} 5 & 0 \\ 6 & 0 \end{pmatrix} \begin{pmatrix} 1 & 2 \\ 3 & 4 \end{pmatrix} = \begin{pmatrix} 5 & 10 \\ 6 & 12 \end{pmatrix} \notin S.$$

Example 3.8

Let $A = \mathbb{Z}[x]$, the ring of all polynomial in x with integer coefficients. Let $S = \langle 2, x \rangle$, the ideal generated by 2 and x in A = smallest ideal containing 2 and x in A = Set of all integer polynomials in x with even constant terms. (Note that the integer 0 is also an integer polynomial. Also remember that A is a commutative ring.)

Definition 3.28. *Let A and B be two rings. A ring homomorphism from A to B is a map $f : A \rightarrow B$ such that for all $a, b \in A$,*

i. $f(a + b) = f(a) + f(b)$, *and*

ii. $f(ab) = f(a)f(b)$.

Example 3.9

Let $A = \mathbb{Z}$ and $B = \mathbb{Z}_5$ consisting of integers modulo 5. Hence $\mathbb{Z}_5 = \{0, 1, 2, 3, 4\}$ where addition and multiplication are taken modulo 5. (For example, $2 + 3 = 0$, and $2.3 = 1$ in $\mathbb{Z}_5$.) Clearly $\mathbb{Z}_5$ is a ring. Now consider the map $f : \mathbb{Z} \rightarrow \mathbb{Z}_5$ defined by $f(a) = a_0$ where $a_0 \in \mathbb{Z}_5$, and $a \equiv a_0(mod\ 5)$. Then, f is a ring homomorphism.

In the above example, what are all the elements that are mapped to 3? They are all the numbers $n \equiv 3(mod\ 5)$ in $\mathbb{Z}$. We denote this set by $[3]$, where $[3] = \{\ldots, -12, -7, -2, 3, 8, 13, \ldots\}$. We call this set the residue class modulo 5 defined by 3 in $\mathbb{Z}$. Note that $[3] = [8]$, etc. Hence $\mathbb{Z}_5 = \{[0], [1], [2], [3], [4]\}$, where $[m] = [n]$ iff $m \equiv n(\text{mod}\, 5)$ for any integers m and n iff $m - n \in (5)$, the ideal generated by 5 in $\mathbb{Z}$. The ring $\mathbb{Z}_5$ is often referred to as the residue class ring or quotient ring modulo the ideal (5).

More generally, let S be an ideal in a commutative ring A, and let A_S denote the set of residue classes modulo the ideal S, that is, for $a \in A$, the residue class defined by $a \in A$, namely $[a] = \{a + s, s \in S\}$ then A_S is the residue class ring or quotient ring defined by the ideal S is the ring A.

3.21 Principal Ideals

Definition 3.29. *Let S be a subset of a commutative ring A. S is called a principal ideal of A if*

1. *S is an ideal of A, and*
2. *S is generated by a single element, that is $\exists\ s \in S$ such that $S = \{as : a \in A\}$.*

Definition 3.30. *A principal ideal ring (or more specifically, a principal ideal domain (P.I.D.)) is a commutative ring A without zero divisors and with unit element 1 (that is, an integral domain) and in which every ideal is principal.*

Example 3.10

The ring $\mathbb{Z}$ of integers is a PID. This is because first $\mathbb{Z}$ is an integral domain. Suppose S is an ideal of A. If $S = (0)$, the zero ideal, S is principal. So assume that $S \neq 0$. If $a \in S$, and $a \neq 0$, $(-1)a = -a \in S$. Of the two numbers a and $-a$, one is a positive integer. Let s be the least positive integer belonging to S.
Claim: $S = \langle s \rangle$, the ideal generated by s. Let b be any element of S. We have to show that $b \in \langle s \rangle$, that is, b is an integral multiple of s. By division algorithm in $\mathbb{Z}$,

$$b = qs + r, \quad 0 \leqslant r < s.$$

As $s \in S$, $qs \in S$ and hence $b - qs = r \in S$. But if $r \neq 0$, $r < s$, and this contradicts the choice of s. In other words, $b \in \langle s \rangle$, and $S = \langle s \rangle$.

Example 3.11

Let F be a field. Then, the ring $F[x]$ of polynomials in x with coefficients in F is a P.I.D.

Proof. Trivially, $F[x]$ is a P.I.D. Let S be an ideal in $F[x]$. If $S = (0)$, trivially S is principal. So assume that $S \neq (0)$: Let $s(x)$ be the monic polynomial of least degree r belonging to S. Such a polynomial $s(x)$ must exist since if $0 \neq a(x) \in S$, and if the leading coefficient of $a(x)$ is $k \in F$, $k^{-1}a(x)$ also $\in S$ (as $k^{-1} \in F \subset F[x]$) $\Rightarrow$ $k^{-1}a(x)$ is a monic polynomial in S. Moreover, such a monic polynomial in S is unique. For, if there are two such monic polynomials of the same least degree r in S, their difference is a polynomial in S of degree less than r. This contradicts the choice of r.

We now imitate the proof of Example 3.10. Let $a(x)$ be any polynomial of $F[x]$ in S. Divide $a(x)$ by $s(x)$ by Euclidean algorithm. This gives

$$a(x) = q(x)s(x) + r(x), \qquad \square$$

where either $r(x) = 0$ or deg $r(x) <$ deg $s(x)$. As $a(x)$, $s(x)$ are in S, $q(x)s(x) \in (a)$, and $a(x) - q(x)s(x) = r(x) \in S$. By the choice of $s(x)$, we should have $r(x) = 0 \Rightarrow a(x) = q(x)s(x) \Rightarrow$ every polynomial in S is a multiple of $s(x)$ in $F[x] \Rightarrow S$ is a principal ideal in $F[x] \Rightarrow F[x]$ is a principal ideal domain.

Example 3.12

The ring $\mathbb{Z}[x]$, though an integral domain, is not a P.I.D. The ideal $S = \langle 2, x\rangle$ in $\mathbb{Z}[x]$ which consists of all polynomials in x with even constant terms is not a principal ideal. For, suppose S is a principal ideal in $\mathbb{Z}[x]$. Let $S = \langle a(x)\rangle$. As $2 \in S$, S must be equal to $\langle 2\rangle$, since otherwise 2 cannot be a multiple of a non-constant polynomial. But then every polynomial in S must have even integers as coefficients, a contradiction. (For instance $2 + 3x \in S$). Hence $\mathbb{Z}[x]$ is not a P.I.D.

Definition 3.31. *Let A and B be commutative rings, and let $f : A \to B$ be a ring homomorphism, and let $K \subset A$ be the set of those elements of A that are mapped to the zero element of B. Then, K is called the kernel of the homomorphism f, and denoted by $Ker\ f$.*

Theorem 3.13. *Let A and B be two commutative rings, and let $f \to B$ be a ring homomorphism f. Then,*

i. $Kerf$ is an ideal of A, and

ii. If f is onto, then $A/Kerf \simeq B$.

Proof.

i. Let $x, y \in K$ and $a \in A$ so that $f(x) = 0 = f(y)$. Then, trivially, $x \pm y$, ax all belong to K. This is because, $f(x \pm y) = f(x) \pm f(y)$ [Note as $y + (-y) = 0$, $f(-y) = -f(y)$], and $f(ax) = f(a)f(x) = f(a) \cdot 0 = 0$. Hence K is an ideal of A.

ii. We are assuming that $f : A \to B$ is onto. Define the map $\bar{f} : A/K \to B$ by $\bar{f}(a + K) = f(a)$, $a \in A$. Note that this map is well-defined. Indeed, if $a + K = b + K$, then $a - b \in K$ and hence $f(a - b) = 0$. But $f(a - b) = f(a) - f(b) \Longrightarrow f(a) = f(b)$. Hence $\bar{f}(a + K) = f(a) = f(b) = \bar{f}(b + K)$. We now verify that f is a ring isomorphism.

A. $\bar{f}$ is $1 - 1$: $\bar{f}(a_1 + K) = \bar{f}(a_2 + K)$ for $a_1, a_2 \in A \Rightarrow f(a_1) = f(a_2) \Rightarrow f(a_1 - a_2) = 0 \in B \Rightarrow a_1 - a_2 \in K \Rightarrow a_1 + K = a_2 + K$.

B. $\bar{f}$ is onto: Let $b \in B$. Then, as f is onto, $\exists\ a \in A$ with $f(a) = b$. Then, $\bar{f}(a + K) = f(a) = b \Rightarrow \bar{f}$ is onto.

C. $\bar{f}$ is a ring homomorphism: For $a_1, a_2 \in K$, $\bar{f}((a_1+K)+(a_2+K)) = \bar{f}((a_1 + a_2) + K) = f(a_1 + a_2) = f(a_1) + f(a_2) = \bar{f}(a_1 + K) + \bar{f}(a_2 + K)$. In a similar manner, $\bar{f}((a_1 + K)(a_2 + K)) = \bar{f}((a_1a_2) + K) = f(a_1a_2) = f(a_1)f(a_2) = \bar{f}(a_1 + K)\bar{f}(a_2 + K)$. This proves that $\bar{f} : A/K \rightarrow B$ is a ring homomorphism.
Thus, $\bar{f}$ is a $1-1$, onto ring homomorphism and hence $\bar{f}$ is an isomorphism. □

Corollary 3.3. *Let F be a field and let $\phi : F[x] \rightarrow F[x]/(x^n - 1)$ be the ring homomorphism defined by:*

$$\phi(f(x)) = \overline{f(x)} = \textit{ Residue class of } f(x) \in F[x]/(x^n - 1).$$

Then, K is the principal ideal $(x^n - 1) \subset F[x]$.

Consider the special case when $F = \mathbb{R}$, and π is the ring homomorphism

$$\phi : \mathbb{R}[x] \rightarrow \mathbb{R}[x]/(x^3 - 1).$$

Then, $\phi(x^5 + x^4 + x^3 + 1) =$ the residue class $[x^2 + x + 1 + 1] = [x^2 + x + 2]$, the class defined by $x^2 + x + 2$ modulo the principal ideal $(x^3 - 1)$ in $\mathbb{R}[x]$. This is because $[x^5 + x^4 + x^3 + 1] = [(x^3 - 1)(x^2 + x + 1) + (x^2 + x + 2)] = [x^3 - 1][x^2 + x + 1] + [x^2 + x + 2] = [0][x^2 + x + 1] + [x^2 + x + 2] = [x^2 + x + 2]$. (Note: square brackets stand for residue classes.) More generally, we have:

Remark 3.1. *Every element of $\mathbb{R}[x]/(x^n - 1)$ can be identified with a real polynomial of degree at most $(n - 1)$.*

3.22 Fields

We now discuss the fundamental properties of fields and then go on to develop in the next chapter the properties of finite fields that are basic to coding theory and cryptography. If rings are algebraic abstractions of the set of integers, fields are algebraic abstractions of the sets $\mathbb{Q}$, $\mathbb{R}$ and $\mathbb{C}$ (as mentioned already).

Definition 3.32. *A field is a commutative ring with identity element in which every non-zero element is a unit.*

Hence if F is a field, and $F^* = F \setminus \{0\}$, the set of non-zero elements of F, then every element of F^* is a unit in F. Hence F^* is a group under the multiplication operation of F. Conversely, if F is a commutative ring with unit element and if F^* is a group under the multiplication operation of F, then every element of F^* is a unit under the multiplication operation of F, and hence F is a field. This observation enables one to give an equivalent definition of a field.

Definition 3.33 (Equivalent definition). *A field is a commutative ring F with unit element in which the set F^* of non-zero elements is a group under the multiplication operation of F.*

3.22.1 Examples of fields

1. $\mathbb{Q}, \mathbb{R}$ and $\mathbb{C}$ are all examples of infinite fields.

2. If F is a field, then $F(x)$, the set of all rational functions in x, is an infinite field (even if F is a finite field). Indeed, if $a(x)/b(x)$ is a non-zero rational function, so that $a(x)$ is not the zero polynomial, then its inverse is the rational function $b(x)/a(x)$.

3. If p is a prime, then the ring $\mathbb{Z}_p$ of integers modulo p is a field with p elements.

Every field is an integral domain. To see this, all we have to verify is that F has no zero divisors. Indeed, if $ab = 0$, $a \neq 0$, then as a^{-1} exists in F, we have $0 = a^{-1}(ab) = (a^{-1}a)b = b$ in F. However, not every integral domain is a field. For instance, the ring $\mathbb{Z}$ of integers is an integral domain but not a field. (Recall that the only non-zero integers which are units in $\mathbb{Z}$ are 1 and -1.)

3.23 Characteristic of a Field

Definition 3.34. *A field F is called finite if $|F|$, the cardinality of F, is finite; otherwise, F is an infinite field.*

Let F be a field whose zero and identity elements are denoted by 0_F and 1_F, respectively. A subfield of F is a subset F' of F such that F' is also a field with the same addition and multiplication operations of F. This of course means that the zero and unity elements of F' are the same as those of F. It is clear that the intersection of any family of subfields of F is again a subfield of F. Let P denote the intersection of the family of all subfields of F. Naturally, this subfield P is the smallest subfield of F. Because if P' is a subfield of F that is properly contained in P, then $P \subset P' \underset{\neq}{\subseteq} P$, a contradiction. This smallest subfield P of F is called the *prime field* of F. Necessarily, $0_F \in P$ and $1_F \in P$.

As $1_F \in P$, the elements $1_F, 1_F + 1_F = 2 \cdot 1_F$, $1_F + 1_F + 1_F = 3 \cdot 1_F$ and, in general, $n \cdot 1_F$, $n \in \mathbb{N}$, all belong to P. There are then two cases to consider:

Case 1: The elements $n \cdot 1_F$, $n \in \mathbb{N}$, are all distinct. In this case, the subfield P itself is an infinite field and therefore F is an infinite field.

Case 2: The elements $n \cdot 1_F$, $n \in \mathbb{N}$, are not all distinct. In this case, there exist $r, s \in \mathbb{N}$ with $r > s$ such that $r \cdot 1_F = s \cdot 1_F$, and therefore, $(r-s) \cdot 1_F = 0$, where $r-s$ is a positive integer. Hence, there exists a least positive integer p such that $p \cdot 1_F = 0$. We claim that p is a prime number. If not, $p = p_1 p_2$, where p_1 and p_2 are positive integers less than p. Then, $0 = p \cdot 1_F = (p_1 p_2) \cdot 1_F = (p_1 \cdot 1_F)(p_2 \cdot 1_F)$ gives, as F is a field, either $p_1 \cdot 1_F = 0$ or $p_2 \cdot 1_F = 0$. But this contradicts the choice of p. Thus, p is prime.

Definition 3.35. *The characteristic of a field F is the least positive integer p such that $p \cdot 1_F = 0$ if such a p exists; otherwise, F is said to be of characteristic zero.*

A field of characteristic zero is necessarily infinite (as its prime field already is). A finite field is necessarily of prime characteristic. However, there are infinite fields with prime characteristic. Note that if a field F has characteristic p, then $px = 0$ for each $x \in F$.

Examples of integral domains and fields

i. The fields $\mathbb{Q}$, $\mathbb{R}$ and $\mathbb{C}$ are all of characteristic zero.

ii. The field $\mathbb{Z}_p$ of integers modulo a prime p is of characteristic p.

iii. For a field F, denote by $F[X]$ the set of all polynomials in X over F, that is, polynomials whose coefficients are in F. $F[X]$ is an integral domain and the group of units of $F[X] = F^*$, the set of all non-zero elements of F.

iv. The field $\mathbb{Z}_p(X)$ of rational functions of the form $a(X)/b(X)$, where $a(X)$ and $b(X)$ are polynomials in X over $\mathbb{Z}_p$, p being a prime, and $b(X) \neq 0$, is an infinite field of (finite) characteristic p.

Theorem 3.14. *Let F be a field of (prime) characteristic p. Then, for all $x, y \in F$, $(x \pm y)^{p^n} = x^{p^n} \pm y^{p^n}$, and $(xy)^{p^n} = x^{p^n} y^{p^n}$.*

Proof. We apply induction on n. If $n = 1$, (by binomial theorem which is valid for any commutative ring with unit element).

$$\begin{aligned}(x+y)^p &= x^p + \binom{p}{1} x^{p-1} y + \cdots + \binom{p}{p-1} x y^{p-1} + y^p \\ &= x^p + y^p, \quad \text{since} \quad p \Big| \binom{p}{i}, \ 1 \leq i \leq p-1. \end{aligned} \tag{3.3}$$

So assume that

$$
\begin{aligned}
(x+y)^{p^n} &= x^{p^n} + y^{p^n}, n \in \mathbb{N}. \\
\text{Then,} \quad (x+y)^{p^{n+1}} &= \left((x+y)^{p^n}\right)^p \\
&= \left(x^{p^n} + y^{p^n}\right)^p \qquad \text{(by induction assumption)} \\
&= (x^{p^n})^p + (y^{p^n})^p \qquad \text{(by Equation 3.3)} \\
&= x^{p^{n+1}} + y^{p^{n+1}}. \qquad (3.4)
\end{aligned}
$$

Next we consider $(x-y)^{p^n}$. If $p = 2$, then $-y = y$ and so the result is valid. If p is an odd prime, change y to $-y$ in Equation 3.4. This gives

$$
\begin{aligned}
(x-y)^{p^n} &= x^{p^n} + (-y)^{p^n} \\
&= x^{p^n} + (-1)^{p^n} y^{p^n} \\
&= x^{p^n} - y^{p^n},
\end{aligned}
$$

since $(-1)^{p^n} = -1$ when p is odd. □

Chapter 4

Algebraic Structures II (Vector Spaces and Finite Fields)

4.1 Vector Spaces

In Section 3.22, we discussed some of the basic properties of fields. In the present section, we look at the fundamental properties of vector spaces. We follow up this discussion with a section on finite fields.

While the three algebraic structures—groups, rings, and fields—are natural generalizations of integers and real numbers, the algebraic structure *vector space* is a natural generalization of the 3-dimensional Euclidean space.

We start with the formal definition of a vector space. To define a vector space, we need two objects: (i) a set V of *vectors*, and (ii) a field F of *scalars*. In the case of the Euclidean 3-space, V is the set of vectors, each vector being an ordered triple (x_1, x_2, x_3) of real numbers and $F = \mathbb{R}$, the field of real numbers. The axioms for a vector space that are given in Definition 4.1 below are easily seen to be generalizations of the properties of $\mathbb{R}^3$.

Definition 4.1. *A vector space (or linear space) V over a field F is a nonvoid set V whose elements satisfy the following axioms:*

A. *V has the structure of an additive Abelian group.*

B. *For every pair of elements α and v, where $\alpha \in F$ and $v \in V$, there exists an element $\alpha v \in V$ called the product of v by α such that*

   i. *$\alpha(\beta v) = (\alpha\beta)v$ for all $\alpha, \beta \in F$ and $v \in V$, and*

   ii. *$1v = v$ for each $v \in V$ (here 1 is the identity or unity element of the field F).*

C. i. *For $\alpha \in F$, and u, v in V, $\alpha(u+v) = \alpha u + \alpha v$, that is, multiplication by elements of F is distributive over addition in V.*

   ii. *For $\alpha, \beta \in F$ and $v \in V$, $(\alpha + \beta)v = \alpha v + \beta v$, that is multiplication of elements of V by elements of F is distributive over addition in F.*

   If $F = \mathbb{R}$, V is called a real vector space*; if $F = \mathbb{C}$, then V is called* a complex vector space*. When an explicit reference to the field F is not required, we simply say that V is a vector space (omitting the words "over the field F"). The product*

$\alpha v, \alpha \in F, v \in V$, is often referred to as scalar multiplication, α being a scalar.

4.1.1 Examples of vector spaces

1. Let $V = \mathbb{R}^3$, the set of ordered triples (x_1, x_2, x_3) of real numbers. Then, $\mathbb{R}^3$ is a vector space over $\mathbb{R}$ (as mentioned earlier) and hence $\mathbb{R}^3$ is a real vector space. More generally, if $V = \mathbb{R}^n$, $n \geq 1$, the set of ordered n-tuples $(x_1, \ldots, x_n)$ of real numbers, then $\mathbb{R}^n$ is a real vector space. If $x = (x_1, \ldots, x_n)$ and $y = (y_1, \ldots, y_n) \in \mathbb{R}^n$, then $x + y = (x_1 + y_1, \ldots, x_n + y_n)$, and $\alpha x = (\alpha x_1, \ldots, \alpha x_n)$. The zero vector of $\mathbb{R}^n$ is $(0, \ldots, 0)$. $\mathbb{R}^n$ is known as the *real affine space of dimension n*.

2. Let $V = \mathbb{C}^n$, the set of ordered n-tuples $(z_1, \ldots, z_n)$ of complex numbers. Then, V is a vector space over $\mathbb{R}$ as well as over $\mathbb{C}$. We note that the real vector space $\mathbb{C}^n$ and the complex vector space $\mathbb{C}^n$ are essentially different spaces despite the fact that the underlying set of vectors is the same in both cases.

3. $\mathbb{R}$ is a vector space over $\mathbb{Q}$.

4. $F[X]$, the ring of polynomials in X over the field F, is a vector space over F.

5. The set of solutions of a homogeneous linear ordinary differential equation with real coefficients forms a real vector space. The reason is that any such differential equation has the form

$$\frac{d^n y}{dx^n} + C_1 \frac{d^{n-1} y}{dx^{n-1}} + \cdots + C_{n-1} y = 0. \tag{4.1}$$

   Clearly, if $y_1(x)$ and $y_2(x)$ are two solutions of the differential Equation 4.1, then so is $y(x) = \alpha_1 y_1(x) + \alpha_2 y_2(x), \alpha_1, \alpha_2 \in \mathbb{R}$. It is now easy to verify that the axioms of a vector space are all satisfied.

4.2 Subspaces

The notion of a subspace of a vector space is something very similar to the notions of a subgroup, subring, and subfield.

Definition 4.2. *A subspace W of a vector space V over F is a subset W of V such that W is also a vector space over F with addition and scalar multiplication as defined for V.*

Proposition 4.1. *A non-void subset W of a vector space V is a subspace of V iff for all $u, v \in W$ and $\alpha, \beta \in F$,*

$$\alpha u + \beta v \in W$$

Proof. If W is a subspace of V, then, as W is a vector space over F (with the same addition and scalar multiplication as in V), $\alpha u \in W$ and $\beta v \in W$, and therefore $\alpha u + \beta v \in W$.

Conversely, if the condition holds, then it means that W is an additive subgroup of V. Moreover, taking $\beta = 0$, we see that for each $\alpha \in F$, $u \in W$, $\alpha u \in W$. As $W \subset V$, all the axioms of a vector space are satisfied by W and hence W is a subspace of V. □

4.2.1 An example of a subspace

Let $W = \{(a, b, 0) : a, b \in \mathbb{R}\}$. Then, W is a subspace of $\mathbb{R}^3$. (To see this, apply Definition 4.2.) Geometrically, this means that the xy-plane of $\mathbb{R}^3$ (that is, the set of points of $\mathbb{R}^3$ with the z-coordinate zero) is a subspace of $\mathbb{R}^3$.

Proposition 4.2. *If W_1 and W_2 are subspaces of a vector space V, then $W_1 \cap W_2$ is also a subspace of V. More generally, the intersection of any family of subspaces of a vector space V is also a subspace of V.*

Proof. Let $u, v \in W = W_1 \cap W_2$, and $\alpha, \beta \in F$. Then, $\alpha u + \beta v$ belongs to W_1 as well as to W_2 by Proposition 4.1, and therefore to W. Hence W is a subspace of V again by Proposition 4.1. The general case is similar. □

4.3 Spanning Sets

Definition 4.3. *Let S be a subset of a vector space V over F. By the subspace spanned by S, denoted by $\langle S \rangle$, we mean the smallest subspace of V that contains S. If $\langle S \rangle = V$, we call S a spanning set of V.*

Clearly, there is at least one subspace of V containing S, namely, V. Let $\mathbf{S}$ denote the collection of all subspaces of V containing S. Then, $\cap_{W \in \mathbf{S}} W$ is also a subspace of V containing S. Clearly, it is the smallest subspace of V containing S, and hence $\langle S \rangle = \cap_{W \in \mathbf{S}} W$.

Example 4.1

We shall determine the smallest subspace W of $\mathbb{R}^3$ containing the vectors $(1, 2, 1)$ and $(2, 3, 4)$.

Clearly, W must contain the subspace spanned by $(1, 2, 1)$, that is, the line joining the origin $(0, 0, 0)$ and $(1, 2, 1)$. Similarly, W must also contain the line joining $(0,0,0)$ and $(2,3,4)$. These two distinct lines meet at the origin and hence define a unique plane through the origin, and this is the subspace spanned by the two vectors $(1,2,1)$ and $(2,3,4)$. (See Proposition 4.3 below.)

Proposition 4.3. *Let S be a subset of a vector space V over F. Then, $\langle S\rangle = L(S)$, where $L(S) = \{\alpha_1 s_1 + \alpha_2 s_2 + \cdots + \alpha_r s_r : s_i \in S,\ 1 \le i \le r$ and $\alpha_i \in F, 1 \le i \le r,\ r \in \mathbb{N}\}$ = set of all finite linear combinations of vectors of S over F.*

Proof. First, it is easy to check that $L(S)$ is a subspace of V. In fact, let $u, v \in L(S)$ so that

$$\begin{aligned} u &= \alpha_1 s_1 + \cdots + \alpha_r s_r, \quad \text{and} \\ v &= \beta_1 s_1' + \cdots + \beta_t s_t' \end{aligned}$$

where $s_i \in S$, $\alpha_i \in F$ for each i, and $s_j' \in S, \beta_j \in F$ for each j. Hence if $\alpha, \beta \in F$, then

$$\alpha u + \beta v = (\alpha\alpha_1)s_1 + \cdots + (\alpha\alpha_r)s_r + (\beta\beta_1)s_1' + \cdots + (\beta\beta_t)s_t' \in L(S).$$

Hence by Proposition 4.1, $L(S)$ is a subspace of V. Further $1 \cdot s = s \in L(S)$ for each $s \in S$, and hence $L(S)$ contains S. But by definition, $\langle S\rangle$ is the smallest subspace of V containing S. Hence $\langle S\rangle \subseteq L(S)$.

Now, let W be any subspace of V containing S. Then, any linear combination of vectors of S is a vector of W, and hence $L(S) \subseteq W$. In other words, any subspace of V that contains the set S must contain the subspace $L(S)$. Once again, as $\langle S\rangle$ is the smallest subspace of V containing S, $L(S) \subseteq \langle S\rangle$. Thus, $\langle S\rangle = L(S)$. □

Note : If $S = \{u_1, \ldots, u_n\}$ is a finite set, then $\langle S\rangle = \langle u_1, \ldots, u_n\rangle$ = subspace of linear combinations of $u_1, \ldots, u_n$ over F. In this case, we say that S generates the subspace $\langle S\rangle$ or S is a set of generators for $\langle S\rangle$. Also $L(S)$ is called the linear span of S in V.

Proposition 4.4. *Let $u_1, \ldots, u_n$ and v be vectors of a vector space V. Suppose that $v \in \langle u_1, u_2, \ldots, u_n\rangle$. Then, $\langle u_1, \ldots, u_n\rangle = \langle u_1, \ldots, u_n; v\rangle$.*

Proof. Any element $\alpha_1 u_1 + \cdots + \alpha_n u_n, \alpha_i \in F$ for each i, of $\langle u_1, \ldots, u_n\rangle$ can be rewritten as

$$\alpha_1 u_1 + \cdots + \alpha_n u_n + 0 \cdot v$$

and hence belongs to $\langle u_1, \ldots, u_n; v\rangle$. Thus,

$$\langle u_1, \ldots, u_n\rangle \subseteq \langle u_1, \ldots, u_n; v\rangle$$

Conversely, if

$$w = \alpha_1 u_1 + \cdots + \alpha_n u_n + \beta v \in \langle u_1, \ldots, u_n; v \rangle,$$

then as $v \in \langle u_1, \ldots, u_n \rangle$, $v = \gamma_1 u_1 + \cdots + \gamma_n u_n$, $\gamma_i \in F$, and therefore,

$$\begin{aligned} w &= (\alpha_1 u_1 + \cdots + \alpha_n u_n) + \beta (\gamma_1 u_1 + \cdots + \gamma_n u_n) \\ &= \sum_{i=1}^{n} (\alpha_i + \beta\gamma_i) u_i \in \langle u_1, \ldots, u_n \rangle. \end{aligned}$$

Thus, $\langle u_1, \ldots, u_n; v \rangle \subseteq \langle u_1, \ldots, u_n \rangle$ and therefore

$$\langle u_1, \ldots, u_n \rangle = \langle u_1, \ldots, u_n; v \rangle.$$

□

Corollary 4.1. *If S is any non-empty subset of a vector space V, and $v \in \langle S \rangle$, then $\langle S \cup \{v\} \rangle = \langle S \rangle$.*

Proof. $v \in \langle S \rangle$ implies, by virtue of Proposition 4.3, v is a linear combination of a finite set of vectors in S. Now the rest of the proof is as in the proof of Proposition 4.4. □

4.4 Linear Independence of Vectors

Let $V = \mathbb{R}^3$, and $e_1 = (1, 0, 0)$, $e_2 = (0, 1, 0)$ and $e_3 = (0, 0, 1)$ in $\mathbb{R}^3$. If $v = (x, y, z)$ is any vector of $\mathbb{R}^3$, then $v = xe_1 + ye_2 + ze_3$. Trivially, this is the only way to express v as a linear combination of e_1, e_2, e_3. For this reason, we call $\{e_1, e_2, e_3\}$ a base for $\mathbb{R}^3$. We now formalize these notions.

Definition 4.4. *Linear independence and linear dependence of vectors*

i. *A finite subset $S = \{v_1, \ldots, v_n\}$ of vectors of a vector space V over a field F is said to be* linearly independent *if the equation*

$$\alpha_1 v_1 + \alpha_2 v_2 + \cdots + \alpha_n v_n = 0, \quad \alpha_i \in F$$

implies that $\alpha_i = 0$ for each i.

In other words, a linearly independent set of vectors admits only the trivial linear combination between them, namely,

$$0 \cdot v_1 + 0 \cdot v_2 + \cdots + 0 \cdot v_n = 0.$$

In this case we also say that the vectors $v_1, \ldots, v_n$ are linearly independent over F. In the above equation, the zero on the right refers to the zero vector of V, while the zeros on the left refer to the scalar zero, that is, the zero element of F.

ii. *An infinite subset S of V is linearly independent in V if every finite subset of vectors of S is linearly independent.*

iii. *A subset S of V is linearly dependent over F if it is not linearly independent over F. This means that there exists a finite subset $\{v_1, \ldots, v_n\}$ of S and a set of scalars $\alpha_1, \ldots, \alpha_n$, not all zero, in F such that*

$$\alpha_1 v_1 + \cdots + \alpha_n v_n = 0.$$

If $\{v_1, \ldots, v_n\}$ is linearly independent over F, we also note that the vectors $v_1, \ldots, v_n$ are linearly independent over F.

Remark 4.1.

i. *The zero vector of V forms a linearly dependent set since it satisfies the non-trivial equation $1 \cdot 0 = 0$, where $1 \in F$ and $0 \in V$.*

ii. *Two vectors of V are linearly dependent over F iff one of them is a scalar multiple of the other.*

iii. *If $v \in V$ and $v \neq 0$, then $\{v\}$ is linearly independent (since for $\alpha \in F$, $\alpha v = 0$ implies that $\alpha = 0$).*

iv. *The empty set is always taken to be linearly independent.*

Proposition 4.5. *Any subset T of a linearly independent set S of a vector space V is linearly independent.*

Proof. First assume that S is a finite subset of V. We can take $T = \{v_1, \ldots, v_r\}$ and $S = \{v_1, \ldots, v_r; v_{r+1}, \ldots, v_n\}, n \geq r$. The relation

$$\alpha_1 v_1 + \cdots \alpha_r v_r = 0, \quad \alpha_i \in F$$

is equivalent to the condition that

$$(\alpha_1 v_1 + \cdots \alpha_r v_r) + (0 \cdot v_{r+1} + \cdots 0 \cdot v_n) = 0.$$

But this implies, as S is linearly independent over F, $\alpha_i = 0$, $1 \leq i \leq n$. Hence T is linearly independent.

If S is an infinite set and $T \subseteq S$, then any finite subset of T is a finite subset of S and hence linearly independent. Hence T is linearly independent over F. □

A restatement of Proposition 4.5 is that any superset in V of a linearly dependent subset of V is also linearly dependent.

Corollary 4.2. *If $v \in L(S)$, then $S \cup \{v\}$ is linearly dependent.*

Proof. By hypothesis, there exist $v_1, \ldots, v_n$ in S, and $\alpha_1, \ldots, \alpha_n \in F$ such that $v = \alpha_1 v_1 + \cdots \alpha_n v_n$ and therefore $\alpha_1 v_1 + \cdots \alpha_n v_n + (-1)v = 0$. Hence $\{v_1, \ldots, v_n; v\}$ is linearly dependent and so by Proposition 4.5, $S \cup \{v\}$ is linearly dependent. □

Examples of linearly independent set of vectors

1. $\mathbb{C}$ is a vector space over $\mathbb{R}$. The vectors 1 and i of $\mathbb{C}$ are linearly independent over $\mathbb{R}$. In fact, if $\alpha, \beta \in \mathbb{R}$, then

$$\alpha \cdot 1 + \beta \cdot i = 0$$

gives that $\alpha + \beta i = 0$, and therefore $\alpha = 0 = \beta$. One can check that $\{1+i, 1-i\}$ is also linearly independent over $\mathbb{R}$, while $\{2+i, 1+i, 1-i\}$ is linearly dependent over $\mathbb{R}$. The last assertion follows from the fact that if $u = 2+i$, $v = 1+i$ and $w = 1-i$, then $u+w = 3$ and $v+w = 2$ so that

$$2(u+w) = 3(v+w)$$

and this gives

$$2u - 3v - w = 0. \qquad \square$$

2. The infinite set of polynomials $S = \{1, X, X^2, \ldots\}$ in the vector space $\mathbb{R}[X]$ of polynomials in X with real coefficients is linearly independent.

Recall that an infinite set S is linearly independent iff every finite subset of S is linearly independent. So consider a finite subset $\{X^{i_1}, X^{i_2}, \ldots, X^{i_n}\}$ of S. The equation

$$\lambda_1 X^{i_1} + \lambda_2 X^{i_2} + \cdots + \lambda_n X^{i_n} = 0, \text{ the zero polynomial of } \mathbb{R}[X], \quad \text{(A)}$$

where the scalars λ_i, $1 \leq i \leq n$, all belong to $\mathbb{R}$, implies that the polynomial on the left side of Equation (A) is the zero polynomial and hence must be zero for every real value of X. In other words, every real number is a zero (root) of this polynomial. This is possible only if each λ_i is zero. Hence the set S is linearly independent over $\mathbb{R}$. $\square$

4.5 Bases of a Vector Space

Definition 4.5. *A basis (or base) of a vector space V over a field F is a subset B of V such that*

i. *B is linearly independent over F, and*

ii. *B spans V; in symbols, $\langle B \rangle = V$.*

Condition (ii) implies that every vector of V is a linear combination of (a finite number of) vectors of B while condition (i) ensures that the expression is unique. Indeed, if $u = \alpha_1 u_1 + \cdots + \alpha_n u_n = \beta_1 u_1 + \cdots + \beta_n u_n$, where the u_i's are all in B, then $(\alpha_1 - \beta_1)u_1 + \cdots + (\alpha_n - \beta_n)u_n = 0$. The linear independence of the vectors $u_1, \ldots, u_n$ means that $\alpha_i - \beta_i = 0$, that is, $\alpha_i = \beta_i$ for each i.

(*Notice that we have taken the same* $u_1, \ldots, u_n$ *in both expressions, as we can always add terms with zero coefficients. For example, if*

$$\begin{aligned} u &= \alpha_1 u_1 + \alpha_2 u_2 = \beta_1 u_1' + \beta_2 u_2', \quad \textit{then} \\ u &= \alpha_1 u_1 + \alpha_2 u_2 + 0 \cdot u_1' + 0 \cdot u_2' \\ &= 0 \cdot u_1 + 0 \cdot u_2 + \beta_1 u_1' + \beta_2 u_2'.) \end{aligned}$$

Example 4.2

The vectors $e_1 = (1,0,0)$, $e_2 = (0,1,0)$ and $e_3 = (0,0,1)$ form a basis for $\mathbb{R}^3$. This follows from the following two facts.

1. $\{e_1, e_2, e_3\}$ is linearly independent in $\mathbb{R}^3$. In fact, $\alpha_1 e_1 + \alpha_2 e_2 + \alpha_3 e_3 = 0$, $\alpha_i \in \mathbb{R}$, implies that

$$\alpha_1(1,0,0) + \alpha_2(0,1,0) + \alpha_3(0,0,1) = (\alpha_1, \alpha_2, \alpha_3) = (0,0,0),$$

   and hence $\alpha_i = 0$, $1 \leq i \leq 3$.

2. $\langle e_1, e_2, e_3 \rangle = \mathbb{R}^3$. To see this, any vector of $\langle e_1, e_2, e_3 \rangle$ is of the form $\alpha_1 e_1 + \alpha_2 e_2 + \alpha_3 e_3 = (\alpha_1, \alpha_2, \alpha_3) \in \mathbb{R}^3$ and, conversely, any $(\alpha_1, \alpha_2, \alpha_3)$ in $\mathbb{R}^3$ is $\alpha_1 e_1 + \alpha_2 e_2 + \alpha_3 e_3$ and hence belongs to $\langle e_1, e_2, e_3 \rangle$.

4.6 Dimension of a Vector Space

Definition 4.6. *By a* finite-dimensional vector space, *we mean a vector space that can be* generated *(or* spanned*) by a finite number of vectors in it.*

Our immediate goal is to establish that any finite-dimensional vector space has a finite basis and that any two bases of a finite-dimensional vector space have the same number of elements.

Lemma 4.1. *No finite-dimensional vector space can have an infinite basis.*

Proof. Let V be a finite-dimensional vector space with a finite spanning set $S = \{v_1, \ldots, v_n\}$. Suppose to the contrary that V has an infinite basis B. Then, as B is a basis, v_i is a linear combination of a finite subset B_i, $1 \leq i \leq n$ of B. Let $B' = \cup_{i=1}^n B_i$. Then, B' is also a finite subset of B. As $B' \subset B$, B' is linearly independent and further, as each $v \in V$ is a linear combination of $v_1, \ldots, v_n$, v is also a linear combination of the vectors of B'. Hence B' is also a basis for V. If $x \in B \setminus B'$, then $x \in L(B')$ and so $B' \cup \{x\}$ is a linearly dependent subset of the linearly independent set B, a contradiction. $\square$

Lemma 4.2. *A finite sequence $\{v_1, \ldots, v_n\}$ of non-zero vectors of a vector space V is linearly dependent iff for some k, $2 \leq k \leq n$, v_k is a linear combination of its preceding vectors.*

Proof. In one direction, the proof is trivial; if $v_k \in \langle v_1, \ldots, v_{k-1}\rangle$, then by Proposition 4.5, $\{v_1, \ldots v_{k-1};\, v_k\}$ is linearly dependent and so is its superset $\{v_1, \ldots, v_n\}$.

Conversely, assume that $\{v_1, v_2, \ldots, v_n\}$ is linearly dependent. As v_1 is a non-zero vector, $\{v_1\}$ is linearly independent (see (iii) of Remark 4.1). Hence, there must exist a k, $2 \leq k \leq n$ such that $\{v_1, \ldots, v_{k-1}\}$ is linearly independent while $\{v_1, \ldots, v_k\}$ is linearly dependent since at worst k can be n. Hence there exists a set of scalars $\alpha_1, \ldots, \alpha_k$, not all zero, such that

$$\alpha_1 v_1 + \cdots + \alpha_k v_k = 0.$$

Now $\alpha_k \neq 0$; for if $\alpha_k = 0$, there exists a non-trivial linear relation connecting $v_1, \ldots, v_{k-1}$ contradicting the fact that $\{v_1, \ldots, v_{k-1}\}$ is linearly independent. Thus,

$$v_k = -\alpha_k^{-1}\alpha_1 v_1 - \cdots - \alpha_k^{-1}\alpha_{k-1}v_{k-1},$$

and hence v_k is a linear combination of its preceding vectors. □

Lemma 4.2 implies, by Proposition 4.4, that under the stated conditions on v_k,

$$\langle v_1, \ldots, v_k, \ldots, v_n\rangle = \langle v_1, \ldots, v_{k-1}, v_{k+1}, \ldots, v_n\rangle = \langle v_1, \ldots, \overset{\wedge}{v_k}, \ldots, v_n\rangle,$$

where the $\wedge$ symbol upon v_k indicates that the vector v_k should be deleted.

We next prove a very important property of finite-dimensional vector spaces.

Theorem 4.1. *Any finite-dimensional vector space has a basis. Moreover, any two bases of a finite-dimensional vector space have the same number of elements.*

Proof. Let V be a finite-dimensional vector space. By Lemma 4.1, every basis of V is finite. Let $S = \{u_1, \ldots, u_m\}$, and $T = \{v_1, \ldots, v_n\}$ be any two bases of V. We want to prove that $m = n$.

Now $v_1 \in V = \langle u_1, \ldots, u_m\rangle$. Hence the set $S_1 = \{v_1;\, u_1, \ldots, u_m\}$ is linearly dependent. By Lemma 4.2, there exists a vector $u_{i_1} \in \{u_1, \ldots, u_m\}$ such that

$$\langle v_1;\, u_1, \ldots, u_m\rangle = \langle v_1;\, u_1, \ldots \overset{\wedge}{u_{i_1}}, \ldots, u_m\rangle.$$

Now consider the set of vectors

$$S_2 = \{v_2, v_1;\, u_1, \ldots, , \overset{\wedge}{u_{i_1}}, \ldots, u_m\} = \{v_2, v_1;\, u_1, \ldots u_m\} \setminus \{u_{i_1}\}.$$

As $v_2 \in V = \langle v_1;\, u_1, \ldots, \overset{\wedge}{u_{i_1}}, \ldots, u_m\rangle$, there exists a vector $u_{i_2} \in \{u_1, \ldots, \overset{\wedge}{u_{i_1}}, \ldots, u_m\}$ such that u_{i_2} is a linear combination of the vectors preceding it in the sequence S_2. (Such a vector cannot be a vector of T as every subset of T is linearly independent.) Hence if

$$\begin{aligned} S_3 &= \{v_1, v_2;\ u_1, \ldots, \overset{\wedge}{u_{i_1}}, \ldots \overset{\wedge}{u_{i_2}}, \ldots, u_m\} \\ &= \{v_1, v_2;\ u_1, \ldots, u_{i_1}, \ldots u_{i_2}, \ldots, u_m\} \setminus \{u_{i_1}, u_{i_2}\}, \\ \langle S_3 \rangle &= V. \end{aligned}$$

Thus, every time we introduce a vector from T, we are in a position to delete a vector from S. Hence $|T| \leq |S|$, that is, $n \leq m$. Interchanging the roles of the bases S and T, we see, by a similar argument, that $m \leq n$. Thus, $m = n$. $\square$

Note that we have actually shown that any finite spanning subset of a finite-dimensional vector space V does indeed contain a finite basis of V. Theorem 4.1 makes the following definition unambiguous.

Definition 4.7. *The dimension of a finite-dimensional vector space is the number of elements in any one of its bases.*

If V is of dimension n over F, we write $\dim_F V = n$ or, simply, $\dim V = n$, when F is known.

Examples for dimension of a vector space

1. $\mathbb{R}^n$ is of dimension n over $\mathbb{R}$. In fact, it is easy to check that the set of vectors

   $$S = \{e_1 = (1, 0, \ldots, 0),\ e_2 = (0, 1, 0, \ldots, 0), \ldots,\ e_n = (0, \ldots, 0, 1)\}$$

   is a basis for $\mathbb{R}^n$ over $\mathbb{R}$.

2. $\mathbb{C}^n$ is of dimension n over $\mathbb{C}$.

3. $\mathbb{C}^n$ is of dimension $2n$ over $\mathbb{R}$. In fact, if $e_k \in \mathbb{C}^n$ has 1 in the k-th position and 0 in the remaining positions, and $f_k \in \mathbb{C}^n$ has $i = \sqrt{-1}$ in the k-th position and 0 in the remaining positions, then

   $$S = \{e_1, \ldots, e_n;\ f_1, \ldots, f_n\}$$

   forms a basis of $\mathbb{C}^n$ over $\mathbb{R}$ (Verify!)

4. Let $\mathbb{P}_n(X)$ denote the set of real polynomials in X with real coefficients of degrees not exceeding n. Then, $B = \{1, X, X^2, \ldots, X^n\}$ is a basis for $\mathbb{P}_n(X)$. Hence $\dim_{\mathbb{R}} \mathbb{P}_n(X) = n + 1$.

5. The vector space $\mathbb{R}$ over $\mathbb{Q}$ is infinite-dimensional. This can be seen as follows: Suppose $\dim_{\mathbb{Q}} \mathbb{R} = n$ (finite). Then, $\mathbb{R}$ has a basis $\{v_1, \ldots, v_n\}$

over $\mathbb{Q}$. Hence

$$\mathbb{R} = \{\alpha_1 v_1 + \cdots + \alpha_n v_n : \alpha_i \in \mathbb{Q} \quad \text{for each } i\}.$$

But as $\mathbb{Q}$ is countable, the number of such linear combinations is countable. This is a contradiction as $\mathbb{R}$ is uncountable. Thus, $\mathbb{R}$ is infinite-dimensional over $\mathbb{Q}$.

Proposition 4.6. *Any* maximal *linearly independent subset of a finite-dimensional vector space V is a basis for V.*

Proof. Let $\mathbb{B}$ be a maximal linearly independent subset of V, that is, $\mathbb{B}$ is not a proper subset of $\mathbb{B}'$, where $\mathbb{B}'$ is linearly independent in V. Suppose $\mathbb{B}$ is not a basis for V. This means, by the definition of a basis for a vector space, that there exists a vector x in V such that $x \notin \langle \mathbb{B} \rangle$. Then, $\mathbb{B} \cup \{x\}$ must be a linearly independent subset of V; for, suppose $\mathbb{B} \cup \{x\}$ is linearly dependent. Then, there exist $v_1, \ldots, v_n$ in $\mathbb{B}$ satisfying a non-trivial relation

$$\alpha_1 v_1 + \cdots + \alpha_n v_n + \beta x = 0, \quad \alpha_i \in F, \quad \text{for each } i \text{ and } \beta \in F.$$

Now, $\beta \neq 0$, as otherwise, $\{v_1, \ldots, v_n\}$ would be linearly dependent over F. Thus, $x = -\beta^{-1}(\alpha_1 v_1 + \cdots + \alpha_n v_n) \in \langle \mathbb{B} \rangle$, a contradiction. Thus, $\mathbb{B} \cup \{x\}$ is linearly independent. But then the fact that $\mathbb{B} \cup \{x\} \supset \mathbb{B}$ violates the maximality of $\mathbb{B}$. Hence $\mathbb{B}$ must be a basis for V. □

4.7 Solutions of Linear Equations and Rank of a Matrix

Let

$$A = \begin{bmatrix} a_{11} & a_{12} & \cdots & a_{1n} \\ a_{21} & a_{22} & \cdots & a_{2n} \\ \vdots & \vdots & & \vdots \\ a_{m1} & a_{m2} & \cdots & a_{mn} \end{bmatrix}$$

be an m by n matrix over a field F. To be precise, we take $F = \mathbb{R}$, the field of real numbers. Let $R_1, R_2, \ldots, R_m$ be the row vectors and $C_1, C_2, \ldots, C_n$ the column vectors of A. Then, each $R_i \in \mathbb{R}^n$ and each $C_j \in \mathbb{R}^m$. The row space of A is the subspace $\langle R_1, \ldots, R_m \rangle$ of $\mathbb{R}^n$, and its dimension is the row rank of A. Clearly, (row rank of A) $\leq m$ since any m vectors of a vector space span a subspace of dimension at most m. The column space of A and the column rank of $A(\leq n)$ are defined in an analogous manner.

We now consider three elementary row transformations (or operations) defined on the row vectors of A:

i. R_{ij}: interchange of the i-th and j-th rows of A.

ii. kR_i: multiplication of the i-th row vector R_i of A by a *non-zero* scalar (real number) k.

iii. $R_i + cR_j$: addition to the i-th row of A, c times the j-th row of A, c being a scalar.

The elementary column transformations are defined in an analogous manner. The inverse of each of these three transformations is again a transformation of the same type. For instance, the inverse of $R_i + cR_j$ is obtained by adding to the new i-th row, $-c$ times R_j.

Let A^* be a matrix obtained by applying a finite sequence of elementary row transformations to a matrix A. Then, the row space of A = row space of A^*, and hence, row rank of A = dim(row space of A) = dim(row space of A^*) = row rank of A^*. Now a matrix A^* is said to be in row-reduced echelon form if:

i. The leading non-zero entry of any non-zero row (if any) of A^* is 1.

ii. The leading 1s in the non-zero rows of A^* occur in increasing order of their columns.

iii. Each column of A^* containing a leading 1 of a row of A^* has all its other entries zero.

iv. The non-zero rows of A^* precede its zero rows, if any.

Now let D be a square matrix of order n. The three elementary row (respectively column) operations considered above do not change the singular or non-singular nature of D. In other words, if D^* is a row-reduced echelon form of D, then D is singular iff D^* is singular. In particular, D is non-singular iff $D^* = I_n$, the identity matrix of order n. Hence if a row-reduced echelon form A^* of a matrix A has r non-zero rows, the maximum order of a non-singular square submatrix of A is r. This number is called the rank of A.

Definition 4.8. *The rank of a matrix A is the maximum order of a non-singular square submatrix of A. Equivalently, it is the maximum order of a non-vanishing determinant minor of A.*

Note that from our earlier discussions, rank of A = row rank of A, = column rank of A. Consequently, a set of n vectors of $\mathbf{R^n}$ is linearly independent iff the determinant formed by them is not zero; equivalently, the $n \times n$ matrix formed by them is invertible.

Example 4.3

Find the row-reduced echelon form of

$$A = \begin{bmatrix} 1 & 2 & 3 & -1 \\ 2 & 1 & -1 & 4 \\ 3 & 3 & 2 & 3 \\ 6 & 6 & 4 & 6 \end{bmatrix}.$$

As the leading entry of R_1 is 1, we perform the operations $R_2 - 2R_1$; $R_3 - 3R_1$; $R_4 - 6R_1$ (where R_i stands for the i-th row of A). This gives

$$A_1 = \begin{bmatrix} 1 & 2 & 3 & -1 \\ 0 & -3 & -7 & 6 \\ 0 & -3 & -7 & 6 \\ 0 & -6 & -14 & 12 \end{bmatrix}.$$

Next perform $(-1/3)R_2$ (that is, replace R_2 by $(-1/3)R_2$). This gives

$$A_1' = \begin{bmatrix} 1 & 2 & 3 & -1 \\ 0 & 1 & 7/3 & -2 \\ 0 & -3 & -7 & 6 \\ 0 & -6 & -14 & 12 \end{bmatrix}.$$

Now perform $R_1 - 2R_2$ (that is, replace R_1 by $R_1 - 2R_2$, etc.); $R_3 + 3R_2$; $R_4 + 6R_2$. This gives the matrix

$$A_2 = \begin{bmatrix} 1 & 0 & -5/3 & 3 \\ 0 & 1 & 7/3 & -2 \\ 0 & 0 & 0 & 0 \\ 0 & 0 & 0 & 0 \end{bmatrix}.$$

A_2 is the row-reduced echelon form of A. Note that A_2 is uniquely determined by A. Since the maximum order of a non-singular submatrix of A_2 is 2, rank of $A = 2$. Moreover, row space of $A_2 = \langle R_1, R_2(\text{of } A_2)\rangle$. Clearly R_1 and R_2 are linearly independent over $\mathbb{R}$ since for $\alpha_1, \alpha_2 \in \mathbb{R}$, $\alpha_1 R_1 + \alpha_2 R_2 = (\alpha_1, \alpha_2, -5\alpha_1/3 + 7\alpha_2/3, 3\alpha_1 - 2\alpha_2) = 0 = (0,0,0,0)$ implies that $\alpha_1 = 0 = \alpha_2$. Thus, the row rank of A is 2 and therefore the column rank of A is also 2.

Remark 4.2. *Since the last three rows of A_1 are proportional (that is, one row is a multiple of the other two), any 3×3 submatrix of A_1 will be singular. Since A_1 has a non-singular submatrix of order 2, (for example, $\left[\begin{smallmatrix} 1 & 2 \\ 0 & -3 \end{smallmatrix}\right]$), A_1 is of rank 2 and we can conclude that A is also of rank 2.*

Remark 4.3. *The word "echelon" refers to the formation of army troops in parallel divisions each with its front clear of that in advance.*

4.8 Exercises

1. Show that $\mathbb{Z}$ is not a vector space over $\mathbb{Q}$.

2. If $n \in \mathbb{N}$, show that the set of all real polynomials of degree $n \geq 1$ does not form a vector space over $\mathbb{R}$ (under usual addition and scalar multiplication of polynomials).

3. Which of the following are vector spaces over $\mathbb{R}$?

   i. $V_1 = \{(x, y, z) \in \mathbb{R}^3$ such that $y + z = 0\}$.

   ii. $V_1 = \{(x, y, z) \in \mathbb{R}^3$ such that $y + z = 1\}$.

   iii. $V_1 = \{(x, y, z) \in \mathbb{R}^3$ such that $y \geq 0\}$.

   iv. $V_1 = \{(x, y, z) \in \mathbb{R}^3$ such that $z = 0\}$.

4. Show that the dimension of the vector space of all m by n real matrices over $\mathbb{R}$ is mn. [Hint: For $m = 2$, $n = 3$, the matrices

$$\begin{bmatrix} 1 & 0 & 0 \\ 0 & 0 & 0 \end{bmatrix}, \quad \begin{bmatrix} 0 & 1 & 0 \\ 0 & 0 & 0 \end{bmatrix}, \quad \begin{bmatrix} 0 & 0 & 1 \\ 0 & 0 & 0 \end{bmatrix}, \quad \begin{bmatrix} 0 & 0 & 0 \\ 1 & 0 & 0 \end{bmatrix}$$

$$\begin{bmatrix} 0 & 0 & 0 \\ 0 & 1 & 0 \end{bmatrix}, \quad \begin{bmatrix} 0 & 0 & 0 \\ 0 & 0 & 1 \end{bmatrix}$$

   form a basis for the space of all 2 by 3 real matrices. Verify this first.]

5. Prove that a subspace of a finite-dimensional vector space is finite-dimensional.

6. Show that the vector space of real polynomials in X is infinite-dimensional over $\mathbb{R}$.

7. Find a basis and the dimension of the subspace of $\mathbb{R}^4$ spanned by the vectors $u_1 = (1, 2, 2, 0)$, $u_2 = (2, 4, 0, 1)$ and $u_3 = (4, 8, 4, 1)$.

8. Find the dimension of the subspace of $\mathbb{R}^3$ spanned by $v_1 = (2, 3, 7)$, $v_2 = (1, 0, -1)$, $v_3 = (1, 1, 2)$ and $v_4 = (0, 1, 3)$.
   [Hint: $v_1 = 3v_3 - v_2$ and $v_4 = v_3 - v_2$.]

9. State with reasons whether each of the following statements is true or false:

   A. A vector space V can have two disjoint subspaces.

   B. Every vector space of dimension n has a subspace of dimension m for each $m \leq n$.

   C. A two-dimensional vector space has exactly three subspaces.

   D. In a vector space, any two generating (that is, spanning) subsets are disjoint.

   E. If n vectors of a vector space V span a subspace U of V, then $\dim U = n$.

4.9 Solutions of Linear Equations

Consider the system of linear homogeneous equations:

$$\begin{aligned} X_1 + 2X_2 + 3X_3 - X_4 &= 0 \\ 2X_1 + X_2 - X_3 + 4X_4 &= 0 \\ 3X_1 + 3X_2 + 2X_3 + 3X_4 &= 0 \\ 6X_1 + 6X_2 + 4X_3 + 6X_4 &= 0. \end{aligned} \tag{4.2}$$

These equations are called homogeneous because if (X_1, X_2, X_3, X_4) is a solution of these equations, then so is (kX_1, kX_2, kX_3, kX_4) for any scalar k. Trivially $(0, 0, 0, 0)$ is a solution. We express these equations in the matrix form

$$AX = 0, \tag{4.3}$$

where

$$A = \begin{bmatrix} 1 & 2 & 3 & -1 \\ 2 & 1 & -1 & 4 \\ 3 & 3 & 2 & 3 \\ 6 & 6 & 4 & 6 \end{bmatrix}, \quad X = \begin{bmatrix} X_1 \\ X_2 \\ X_3 \\ X_4 \end{bmatrix}, \quad \text{and} \quad 0 = \begin{bmatrix} 0 \\ 0 \\ 0 \\ 0 \end{bmatrix}.$$

If X_1 and X_2 are any two solutions of Equation 4.3, then so is $aX_1 + bX_2$ for scalars a and b since $A(aX_1 + bX_2) = a(AX_1) + b(AX_2) = a \cdot 0 + b \cdot 0 = 0$. Thus, the set of solutions of Equation 4.3 is (as $X \in \mathbb{R}^4$) a vector subspace of $\mathbb{R}^4$, where 4 = the number of indeterminates in the Equations 4.2.

It is clear that the three elementary row operations performed on a system of homogeneous linear equations do not alter the set of solutions of the equations. Hence if A^* is the row reduced echelon form of A, the solution sets of $AX = 0$ and $A^*X = 0$ are the same.

In Example 4.3, $A^* = A_2$. Hence the equations $A^*X = 0$ are

$$\begin{aligned} X_1 - \frac{5}{3}X_3 + 3X_4 &= 0, \quad \text{and} \\ X_2 + \frac{7}{3}X_3 - 2X_4 &= 0. \end{aligned}$$

These give

$$\begin{aligned} X_1 &= \frac{5}{3}X_3 - 3X_4 \\ X_2 &= -\frac{7}{3}X_3 + 2X_4 \end{aligned}$$

so that

$$X = \begin{bmatrix} X_1 \\ X_2 \\ X_3 \\ X_4 \end{bmatrix} = X_3 \begin{bmatrix} 5/3 \\ -7/3 \\ 1 \\ 0 \end{bmatrix} + X_4 \begin{bmatrix} -3 \\ 2 \\ 0 \\ 1 \end{bmatrix}.$$

Thus, the space of solutions of $AX = 0$ is spanned by the two linearly independent vectors

$$\begin{bmatrix} 5/3 \\ -7/3 \\ 1 \\ 0 \end{bmatrix} \quad \text{and} \quad \begin{bmatrix} -3 \\ 2 \\ 0 \\ 1 \end{bmatrix}$$

and hence is of dimension 2. This number 2 corresponds to the fact that X_1, X_2, X_3, and X_4 are all expressible in terms of X_3 and X_4. Here, X_1 and X_2 correspond to the identity submatrix of order 2 of A^*, that is, the rank of A. Also, dimension $2 = 4 - 2 =$ (number of indeterminates) $-$ (rank of A). The general case is clearly similar where the system of equations is given by $a_{i1}X_1 + \cdots + a_{in}X_n = 0$, $1 \le i \le m$. These are given by the matrix equation $AX = 0$, where A is the m by n matrix (a_{ij}) of coefficients and

$$X = \begin{bmatrix} X_1 \\ \vdots \\ X_n \end{bmatrix};$$

we state it as a theorem.

Theorem 4.2. *The solution space of a system of homogeneous linear equations is of dimension $n - r$, where n is the number of unknowns and r is the rank of the matrix A of coefficients.*

4.10 Solutions of Non-homogeneous Linear Equations

A system of non-homogeneous linear equations is of the form

$$\begin{aligned} a_{11}X_1 + a_{12}X_2 + \cdots + a_{1n}X_n &= b_1 \\ \vdots \qquad \vdots \quad \vdots \qquad & \quad \vdots \\ a_{m1}X_1 + a_{m2}X_2 + \cdots + a_{mn}X_n &= b_m, \end{aligned}$$

with not all b_i being zero.
These m equations are equivalent to the single matrix equation

$$AX = B,$$

where $A = (a_{ij})$ is an m by n matrix and B is a non-zero column vector of length m.

It is possible that such a system of equations has no solution at all. For example, consider the system of equations

$$\begin{aligned} X_1 - X_2 + X_3 &= 2 \\ X_1 + X_2 - X_3 &= 0 \\ 3X_1 \qquad\qquad &= 6. \end{aligned}$$

From the last equation, we get $X_1 = 2$. This, when substituted in the first two equations, yields $-X_2 + X_3 = 0$, $X_2 - X_3 = -2$ which are mutually contradictory. Such equations are called *inconsistent* equations.

When are the equations represented by $AX = B$ consistent?

Theorem 4.3. *The equations $AX = B$ are consistent if and only if B belongs to the column space of A.*

Proof. The equations are consistent iff there exists a vector

$$X_0 = \begin{bmatrix} \alpha_1 \\ \vdots \\ \alpha_n \end{bmatrix}$$

such that $AX_0 = B$. But this happens iff $\alpha_1 C_1 + \cdots + \alpha_n C_n = B$, where $C_1, \ldots, C_n$ are the column vectors of A, that is, iff B belongs to the column space of A. □

Corollary 4.3. *The equations represented by $AX = B$ are consistent iff rank of $A =$ rank of (A, B). [(A, B) denotes the matrix obtained from A by adding one more column vector B at the end. It is called the matrix augmented by B].*

Proof. By the above theorem, the equations are consistent iff $B \in \langle C_1, \ldots, C_n \rangle$. But this is the case, by Proposition 4.4, iff $\langle C_1, \ldots, C_n \rangle = \langle C_1, \ldots, C_n, B \rangle$. The latter condition is equivalent to, column rank of $A =$ column rank of (A, B), and consequently to, rank of $A =$ rank of (A, B). □

We now ask the natural question. If the system $AX = B$ is consistent, how to solve it?

Theorem 4.4. *Let X_0 be any particular solution of the equation $AX = B$. Then, the set of all solutions of $AX = B$ is given by $\{X_0 + U\}$, where U varies over the set of solutions of the auxiliary equation $AX = 0$.*

Proof. If $AU = 0$, then $A(X_0 + U) = AX_0 + AU = B + 0 = B$, so that $X_0 + U$ is a solution of $AX = B$.

Conversely, let X_1 be an arbitrary solution of $AX = B$, so that $AX_1 = B$. Then, $AX_0 = AX_1 = B$ gives that $A(X_1 - X_0) = 0$. Setting $X_1 - X_0 = U$, we get, $X_1 = X_0 + U$, and $AU = 0$. □

As in the case of homogeneous linear equations, the set of solutions of $AX = B$ remains unchanged when we perform any finite number of elementary row transformations on A, provided we take care to perform simultaneously the same operations on the matrix B on the right. As before, we row-reduce A to its echelon form A_0 and the equation $AX = B$ gets transformed into its equivalent equation $A^* X_0 = B_0$.

4.11 LUP Decomposition

Definition 4.9. *By an LUP decomposition of a square matrix A, we mean an equation of the form*

$$PA = LU \tag{4.4}$$

where P is a permutation matrix, L a unit-lower triangular matrix, and U an upper triangular matrix.

Recall that a square matrix M is upper(resp. lower) triangular if all its entries below(resp. above) its principal diagonal are zero. It is unit upper(resp. lower) triangular if it is (i) upper(resp. lower) triangular, and (ii) all the entries of its principal diagonal are equal to 1. Hence all the matrices P, A, L and U in Equation 4.4 must be of the same order.

Suppose we have determined matrices P, L and U so that Equation 4.4 holds. The equation $AX = b$ is equivalent to $(PA)X = Pb$ in that both have the same set of solutions X. This is because P^{-1} exists and hence $(PA)X = Pb$ is equivalent to $P^{-1}(PA)X = P^{-1}(Pb)$, that is, $AX = b$. Now set $Pb = b'$. Then, $PAX = Pb$ gives

$$LUX = b'. \tag{4.5}$$

Hence if we set $UX = Y$ (a column vector), then (4.5) becomes $LY = b'$. We know from Section 4.10 as to how to solve $LY = b'$. A solution Y of this equation, when substituted in $UX = Y$, gives X, again by the same method.

4.11.1 Computing an LU decomposition

We first consider the case when $A = (a_{ij})$ is a non-singular matrix of order n. We begin by obtaining an LU decomposition for A; that is, an LUP decomposition with $P = I_n$ in Equation 4.4. The process by which we obtain the LU decomposition for A is known as *Gaussian elimination*.

Assume that $a_{11} \neq 0$. We write

$$A = \left[\begin{array}{c|ccc} a_{11} & a_{12} & \dots & a_{1n} \\ \hline a_{21} & a_{22} & \dots & a_{2n} \\ \vdots & \vdots & & \vdots \\ a_{n1} & a_{n2} & \dots & a_{nn} \end{array}\right] = \left[\begin{array}{c|ccc} a_{11} & a_{12} & \dots & a_{1n} \\ \hline a_{21} & & & \\ \vdots & & A' & \\ a_{n1} & & & \end{array}\right],$$

where A' is a square matrix of order $n-1$. We can now factor A as

$$\left[\begin{array}{c|ccc} 1 & 0 & \dots & 0 \\ \hline a_{21}/a_{11} & & & \\ \vdots & & I_{n-1} & \\ a_{n1}/a_{11} & & & \end{array}\right] \left[\begin{array}{c|ccc} a_{11} & a_{12} & \dots & a_{1n} \\ \hline 0 & & & \\ \vdots & & A' - \dfrac{vw^t}{a_{11}} & \\ 0 & & & \end{array}\right]$$

where

$$v = \begin{bmatrix} a_{21} \\ \vdots \\ a_{n1} \end{bmatrix} \quad \text{and} \quad w^t = (a_{12} \dots a_{1n}).$$

Note that vw^t is also a matrix of order $n-1$. The matrix $A_1 = A' - (vw^t/a_{11})$ is called the Schur complement of A with respect to a_{11}.

We now recursively find an LU decomposition of A. If we assume that $A_1 = L'U'$, where L' is unit lower-triangular and U' is upper-triangular, then

$$\begin{aligned} A &= \left[\begin{array}{c|c} 1 & 0 \\ \hline v/a_{11} & I_{n-1} \end{array}\right] \left[\begin{array}{c|c} a_{11} & w^t \\ \hline 0 & A_1 \end{array}\right] \\ &= \left[\begin{array}{c|c} 1 & 0 \\ \hline v/a_{11} & I_{n-1} \end{array}\right] \left[\begin{array}{c|c} a_{11} & w^t \\ \hline 0 & L'U' \end{array}\right] \\ &= \left[\begin{array}{c|c} 1 & 0 \\ \hline v/a_{11} & L' \end{array}\right] \left[\begin{array}{c|c} a_{11} & w^t \\ \hline 0 & U' \end{array}\right] \\ &= LU, \end{aligned} \tag{4.6}$$

where

$$L = \left[\begin{array}{c|c} 1 & 0 \\ \hline v/a_{11} & L' \end{array}\right], \quad \text{and} \quad U = \left[\begin{array}{c|c} a_{11} & w^t \\ \hline 0 & U' \end{array}\right].$$

The validity of the two middle equations on the right of Equation 4.6 can be verified by routine block multiplication of matrices (cf: Section 3.3.1). This method is based on the supposition that a_{11} and all the leading entries of the successive Schur complements are all non-zero. If a_{11} is zero, we interchange the first row of A with a subsequent row having a non-zero first entry. This amounts to premultiplying both sides by the corresponding permutation matrix P yielding the matrix PA on the left. We now proceed as with the case when $a_{11} \neq 0$. If a leading entry of a subsequent Schur complement is zero, once again we make interchanges of rows—not just the rows of the relevant Schur complement, but the full rows obtained from A. This again amounts to premultiplication by a permutation matrix. Since any product of permutation matrices is a permutation matrix, this process finally ends up with a matrix $P'A$, where P' is a permutation matrix of order n.

We now present two examples, one to obtain the LU decomposition when it is possible and another to determine the LUP decomposition.

Example 4.4

Find the LU decomposition of

$$A = \begin{bmatrix} 2 & 3 & 1 & 2 \\ 4 & 7 & 4 & 7 \\ 2 & 7 & 13 & 16 \\ 6 & 10 & 13 & 15 \end{bmatrix}.$$

Here,

$$a_{11} = 2, \quad v = \begin{bmatrix} 4 \\ 2 \\ 6 \end{bmatrix}, \quad w^t = [3, 1, 2].$$

Therefore

$$v/a_{11} = \begin{bmatrix} 2 \\ 1 \\ 3 \end{bmatrix}, \quad \text{and so} \quad vw^t/a_{11} = \begin{bmatrix} 6 & 2 & 4 \\ 3 & 1 & 2 \\ 9 & 3 & 6 \end{bmatrix},$$

where w^t denotes the transpose of w.
Hence the Schur complement of A is

$$A_1 = \begin{bmatrix} 7 & 4 & 7 \\ 7 & 13 & 16 \\ 10 & 13 & 15 \end{bmatrix} - \begin{bmatrix} 6 & 2 & 4 \\ 3 & 1 & 2 \\ 9 & 3 & 6 \end{bmatrix} = \begin{bmatrix} 1 & 2 & 3 \\ 4 & 12 & 14 \\ 1 & 10 & 9 \end{bmatrix}.$$

Now the Schur complement of A_1 is

$$A_2 = \begin{bmatrix} 12 & 14 \\ 10 & 9 \end{bmatrix} - \begin{bmatrix} 4 \\ 1 \end{bmatrix} [2, \ \ 3] = \begin{bmatrix} 12 & 14 \\ 10 & 9 \end{bmatrix} - \begin{bmatrix} 8 & 12 \\ 2 & 3 \end{bmatrix} = \begin{bmatrix} 4 & 2 \\ 8 & 6 \end{bmatrix}.$$

This gives the Schur complement of A_2 as

$$\begin{aligned} A_3 &= [6] - [2][2] = [2] \\ &= [1][2] = L_3 U_3, \end{aligned}$$

where $L_3 = (1)$ is lower unit triangular and $U_3 = (2)$ is upper triangular.
Tracing back we get,

$$\begin{aligned} A_2 &= \left[\begin{array}{c|c} 1 & 0 \\ \hline v_2/a_{11} & L_3 \end{array}\right] \left[\begin{array}{c|c} 4 & w_2^t \\ \hline 0 & U_3 \end{array}\right] \\ &= \left[\begin{array}{c|c} 1 & 0 \\ \hline 2 & 1 \end{array}\right] \left[\begin{array}{c|c} 4 & 2 \\ \hline 0 & 2 \end{array}\right] = L_2 U_2. \end{aligned}$$

This gives,

$$A_1 = \left[\begin{array}{c|cc} 1 & 0 & 0 \\ \hline 4 & 1 & 0 \\ 1 & 2 & 1 \end{array}\right] \left[\begin{array}{c|cc} 1 & 2 & 3 \\ \hline 0 & 4 & 2 \\ 0 & 0 & 2 \end{array}\right].$$

Consequently,

$$A = \left[\begin{array}{c|ccc} 1 & 0 & 0 & 0 \\ \hline 2 & 1 & 0 & 0 \\ 1 & 4 & 1 & 0 \\ 3 & 1 & 2 & 1 \end{array}\right] \left[\begin{array}{c|ccc} 2 & 3 & 1 & 2 \\ \hline 0 & 1 & 2 & 3 \\ 0 & 0 & 4 & 2 \\ 0 & 0 & 0 & 2 \end{array}\right] = LU,$$

where

$$L = \left[\begin{array}{c|ccc} 1 & 0 & 0 & 0 \\ \hline 2 & 1 & 0 & 0 \\ 1 & 4 & 1 & 0 \\ 3 & 1 & 2 & 1 \end{array}\right] \quad \text{is unit lower-triangular, and}$$

$$U = \left[\begin{array}{c|ccc} 2 & 3 & 1 & 2 \\ \hline 0 & 1 & 2 & 3 \\ 0 & 0 & 4 & 2 \\ 0 & 0 & 0 & 2 \end{array}\right] \quad \text{is upper-triangular.}$$

Note that A is non-singular as L and U are.

Example 4.5

Find the LUP decomposition of

$$A = \begin{bmatrix} 2 & 3 & 1 & 2 \\ 4 & 6 & 4 & 7 \\ 2 & 7 & 13 & 16 \\ 6 & 10 & 13 & 15 \end{bmatrix}.$$

Suppose we proceed as before: The Schur complement of A is

$$A_1 = \begin{bmatrix} 6 & 4 & 7 \\ 7 & 13 & 16 \\ 10 & 13 & 15 \end{bmatrix} - \begin{bmatrix} 2 \\ 1 \\ 3 \end{bmatrix} \begin{bmatrix} 3 & 1 & 2 \end{bmatrix} = \begin{bmatrix} 6 & 4 & 7 \\ 7 & 13 & 16 \\ 10 & 13 & 15 \end{bmatrix} - \begin{bmatrix} 6 & 2 & 4 \\ 3 & 1 & 2 \\ 9 & 3 & 6 \end{bmatrix}$$

$$= \begin{bmatrix} 0 & 2 & 3 \\ 4 & 12 & 14 \\ 1 & 10 & 9 \end{bmatrix}.$$

Since the leading entry is zero, we interchange the first row of A_1 with some other row. Suppose we interchange the first and third rows of A_1. This amounts to considering the matrix PA instead of A_1, where

$$P = \begin{bmatrix} 1 & 0 & 0 & 0 \\ 0 & 0 & 0 & 1 \\ 0 & 0 & 1 & 0 \\ 0 & 1 & 0 & 0 \end{bmatrix}.$$

Note that the first row of A_1 corresponds to the second row of A and the last row of A_1 to the fourth row of A. This means that the Schur complement of PA (instead of A) is

$$A_1' = \begin{bmatrix} 1 & 10 & 9 \\ 4 & 12 & 14 \\ 0 & 2 & 3 \end{bmatrix}.$$

We now proceed with A_1' as before. The Schur complement of A_1' is

$$A_2 = \begin{bmatrix} 12 & 14 \\ 2 & 3 \end{bmatrix} - \begin{bmatrix} 4 \\ 0 \end{bmatrix} \begin{bmatrix} 10 & 9 \end{bmatrix} = \begin{bmatrix} 12 & 14 \\ 2 & 3 \end{bmatrix} - \begin{bmatrix} 40 & 36 \\ 0 & 0 \end{bmatrix} = \begin{bmatrix} -28 & -22 \\ 2 & 3 \end{bmatrix}.$$

The Schur complement of A_2 is

$$A_3 = [3] - \left[\tfrac{2}{-28}\right][-22] = \left[3 - \tfrac{11}{7}\right] = \left[\tfrac{10}{7}\right] = (1)(10/7) = L_3U_3,$$

where $L_3 = [1]$ is unit lower triangular, and $U_3 = [10/7]$ is upper triangular. Hence

$$A_2 = \begin{bmatrix} 1 & 0 \\ \frac{-2}{28} & 1 \end{bmatrix} \begin{bmatrix} -28 & -22 \\ 0 & \frac{10}{7} \end{bmatrix}.$$

This gives

$$A_1' = \begin{bmatrix} 1 & 0 & 0 \\ 4 & 1 & 0 \\ 0 & \frac{-1}{14} & 1 \end{bmatrix} \begin{bmatrix} 1 & 10 & 9 \\ 0 & -28 & -22 \\ 0 & 0 & \frac{10}{7} \end{bmatrix}.$$

Thus,

$$\begin{bmatrix} 1 & 0 & 0 & 0 \\ 3 & 1 & 0 & 0 \\ 1 & 4 & 1 & 0 \\ 2 & 0 & \frac{-1}{14} & 1 \end{bmatrix} \begin{bmatrix} 2 & 3 & 1 & 2 \\ 0 & 1 & 10 & 9 \\ 0 & 0 & -28 & -22 \\ 0 & 0 & 0 & \frac{10}{7} \end{bmatrix} = L'U,$$

where L' and U are the first and second matrices in the product. Notice that we have interchanged the second and fourth rows of A while computing L'. Interchanging the second and fourth rows of L', we get

$$A = \begin{bmatrix} 1 & 0 & 0 & 0 \\ 2 & 0 & \frac{-1}{14} & 1 \\ 1 & 4 & 1 & 0 \\ 3 & 1 & 0 & 0 \end{bmatrix} \begin{bmatrix} 2 & 3 & 1 & 2 \\ 0 & 1 & 10 & 9 \\ 0 & 0 & -28 & -22 \\ 0 & 0 & 0 & \frac{10}{7} \end{bmatrix} = LU.$$

We have assumed, to start with, that A is a non-singular matrix. Even if A is invertible, it is possible that A has no LU decomposition. For example, the non-singular matrix $\begin{bmatrix} 0 & 1 \\ 1 & 0 \end{bmatrix}$ has no LU decomposition. It is easy to check that if A is singular, it is possible that not only a column of a Schur complement but even the full column corresponding to it may be zero. In that case, we have to interchange columns. This would result in a matrix of the form AP rather than PA. It is also possible that we get the form P_1AP_2 where P_1 and P_2 are both permutation matrices.

Example 4.6

Solve the system of linear equations using the LUP decomposition method:

$$\begin{aligned} 2X_1 + 3X_2 + X_3 + 2X_4 &= 10 \\ 4X_1 + 6X_2 + 4X_3 + 7X_4 &= 25 \\ 2X_1 + 7X_2 + 13X_3 + 16X_4 &= 40 \\ 6X_1 + 10X_2 + 13X_3 + 15X_4 &= 50. \end{aligned} \tag{4.7}$$

These equations are equivalent to $AX = B$, where A is the matrix of Example 4.5 and

$$B = \begin{bmatrix} 10 \\ 25 \\ 40 \\ 50 \end{bmatrix}.$$

If we interchange any pair of rows of $(A|B)$, it amounts to interchanging the corresponding equations. However, this will in no way alter the solution. Hence the solutions of Equation 4.7 are the same as solutions of $PAX = PB$, where P is the permutation matrix obtained in Example 4.5. Thus, the solutions are the same as the solutions of $LUX = B$, where L and M are again those obtained in Example 4.5.

Now set $UX = Y$ so that the given equations become $LY = B$, where

$$Y = \begin{bmatrix} Y_1 \\ Y_2 \\ Y_3 \\ Y_4 \end{bmatrix}.$$

This gives

$$\begin{bmatrix} 1 & 0 & 0 & 0 \\ 2 & 0 & \frac{-1}{14} & 1 \\ 1 & 4 & 1 & 0 \\ 3 & 1 & 0 & 0 \end{bmatrix} \begin{bmatrix} Y_1 \\ Y_2 \\ Y_3 \\ Y_4 \end{bmatrix} = \begin{bmatrix} 10 \\ 50 \\ 40 \\ 25 \end{bmatrix}.$$

These are equivalent to

$$\begin{aligned} 3Y_1 &= 10 \\ 2Y_1 - \frac{1}{14}Y_3 + Y_4 &= 25 \\ Y_1 + 4Y_2 + Y_3 &= 40 \\ 3Y_1 + Y_2 &= 50, \end{aligned}$$

and we get $Y_1 = 10$, $Y_2 = 20$, $Y_3 = -50$ and $Y_4 = 10/7$. Substituting these values in $UX = Y$, we get

$$\begin{bmatrix} 2 & 3 & 1 & 2 \\ 0 & 1 & 10 & 9 \\ 0 & 0 & -28 & -22 \\ 0 & 0 & 0 & \frac{10}{7} \end{bmatrix} \begin{bmatrix} X_1 \\ X_2 \\ X_3 \\ X_4 \end{bmatrix} = \begin{bmatrix} 10 \\ 20 \\ -50 \\ \frac{10}{7} \end{bmatrix}.$$

These give

$$\begin{aligned} 2X_1 + 3X_2 + X_3 + 2X_4 &= 10 \\ X_2 + 10X_3 + 9X_4 &= 20 \\ -28X_3 - 22X_4 &= -50 \\ \frac{10}{7}X_4 &= \frac{10}{7} \end{aligned}$$

Solving backward, we get $X_1 = 2$, $X_2 = X_3 = X_4 = 1$.

4.12 Exerciscs

1. Examine if the following equations are consistent.

$$\begin{aligned} X_1 + X_2 + X_3 + X_4 &= 0 \\ 2X_1 - X_2 + 3X_3 + 4X_4 &= 1 \\ 3X_1 + 4X_3 + 5X_4 &= 2. \end{aligned}$$

2. Solve the system of homogeneous linear equations:

$$\begin{aligned} 4X_1 + 4X_2 + 3X_3 - 5X_4 &= 0 \\ X_1 + X_2 + 2X_3 - 3X_4 &= 0 \\ 2X_1 + 2X_2 - X_3 &= 0 \\ X_1 + X_2 + 2X_3 - 2X_4 &= 0. \end{aligned}$$

Show that the solution space is of dimension 2.

3. Solve:

$$\begin{aligned} X_1 + X_2 + X_3 + X_4 &= 0 \\ X_1 + 3X_2 + 2X_3 + 4X_4 &= 0 \\ 2X_1 + X_3 - X_4 &= 0. \end{aligned}$$

4. Solve by finding LUP decomposition:

i.

$$\begin{aligned} 2X_1 + 3X_2 - 5X_3 + 4X_4 &= -8 \\ 3X_1 + X_2 - 4X_3 + 5X_4 &= -8 \\ 7X_1 + 3X_2 - 2X_3 + X_4 &= 56 \\ 4X_1 + X_2 - X_3 + 3X_4 &= 20. \end{aligned}$$

ii.

$$\begin{aligned} 3X_1 - 2X_2 + X_3 &= 7 \\ X_1 + X_2 + X_3 &= 12 \\ -X_1 + 4X_2 - X_3 &= 3. \end{aligned}$$

iii.

$$\begin{aligned} 2X_1 + 4X_2 - 5X_3 + X_4 &= 8 \\ 4X_1 + -5X_3 - X_4 &= 16 \\ -4X_1 + 2X_2 + X_4 &= -5 \\ 6X_1 + 4X_2 - 10X_3 + 7X_4 &= 13. \end{aligned}$$

4.13 Finite Fields

In this section, we discuss the basic properties of finite fields. Finite fields are fundamental to the study of codes and cryptography.

Recall that a field F is finite if its cardinality $|F|$ is finite. $|F|$ is the order of F. The characteristic of a finite field F, as seen in Section 3.23, is a prime number p, and the prime field P of F is a field of p elements. P consists of the p elements 1_F, $2 \cdot 1_F = 1_F + 1_F, \ldots, p \cdot 1_F = 0_F$. Clearly, F is a vector space over P. If the dimension of the finite field F over P is n, then n is finite. Hence F has a basis $\{u_1, \ldots, u_n\}$ of n elements over P. This means that each element $v \in F$ is a unique linear combination of $u_1, \ldots, u_n$, say,

$$v = \alpha_1 u_1 + \alpha_2 u_2 + \cdots + \alpha_n u_n, \quad \alpha_i \in \mathbb{P}, \quad 1 \leq i \leq n.$$

For each i, α_i can take $|P| = p$ values, and so there are $p \cdot p \cdots (n \text{ times}) = p^n$ distinct elements in F. Thus, we have proved the following result.

Theorem 4.5. *The order of a finite field is a power of a prime number.* □

Finite fields are known as Galois fields after the French mathematician Évariste Galois (1811–1832) who first studied them. A finite field of order q is denoted by $GF(q)$.

We now look at the converse of Theorem 4.5. Given a prime power p^n (where p is a prime), does there exist a field of order p^n? The answer to this question is in the affirmative. We give below two different constructions that yield a field of order p^n.

Theorem 4.6. *Given p^n (where p is a prime), there exists a field of p^n elements.*

Construction 1: Consider the polynomial $X^{p^n} - X \in Z_p[X]$ of degree p^n. (Recall that $Z_p[X]$ stands for the ring of polynomials in X with coefficients from the field Z_p of p elements.) The derivative of this polynomial is

$$p^n X^{p^n - 1} - 1 = -1 \in Z_p[X],$$

and is therefore relatively prime to $X^{p^n} - X$. Hence the p^n roots of $X^{p^n} - X$ are all distinct. (Here, though no concept of the limit is involved, the notion of the derivative has been employed as though it is a real polynomial.) It is known [12] that the roots of this polynomial lie in an extension field $K \supset \mathbb{Z}_p$. K is also of characteristic p. If a and b are any two roots of $X^{p^n} - X$, then

$$a^{p^n} = a, \quad \text{and} \quad b^{p^n} = b.$$

Now by Theorem 3.14

$$(a \pm b)^{p^n} = a^{p^n} \pm b^{p^n},$$

and, by the commutativity of multiplication in K,

$$a^{p^n} b^{p^n} = (ab)^{p^n},$$

and so $a \pm b$ and ab are also roots of $X^{p^n} - X$. Moreover, if a is a non-zero root of $X^{p^n} - X$, then so is a^{-1} since $(a^{-1})^{p^n} = (a^{p^n})^{-1} = a^{-1}$. Also the associative and distributive laws are valid for the set of roots since they are all elements of the field K. Finally 0 and 1 are also roots of $X^{p^n} - X$. In other words, the p^n roots of the polynomial $X^{p^n} - X \in Z_p[X]$ form a field of order p^n. □

Construction 2: Let $f(X) = X^n + a_1X^{n-1} + \cdots + a_n \in \mathbb{Z}_p[X]$ be a polynomial of degree n irreducible over $\mathbb{Z}_p$. (The existence of such an irreducible polynomial (with leading coefficient 1) of degree n is guaranteed by a result (see [12]) in Algebra.) Let F denote the ring of polynomials in $\mathbb{Z}_p[X]$ reduced modulo $f(X)$ (that is, if $g(X) \in \mathbb{Z}_p[X]$, divide $g(X)$ by $f(X)$ and take the remainder $g_1(X)$ which is 0 or of degree less than n). Then, every non-zero polynomial in F is a polynomial of $\mathbb{Z}_p[X]$ of degree at most $n-1$. Moreover, if $a_0X^{n-1} + \cdots + a_{n-1}$ and $b_0X^{n-1} + \cdots + b_{n-1}$ are two polynomials in F of degrees at most $n-1$, and if they are equal, then,

$$(a_0 - b_0)X^{n-1} + \cdots + (a_{n-1} - b_{n-1})$$

is the zero polynomial of F, and hence is a multiple of $f(X)$ in $\mathbb{Z}_p[X]$. As degree of f is n, this is possible only if $a_i = b_i$, $0 \le i \le n-1$. Now if $a_0X^{n-1} + a_1X^{n-2} + \cdots + a_{n-1}$ is any polynomial of F, $a_i \in \mathbb{Z}_p$ and hence a_i has p choices. Hence the number of polynomials of the form

$$a_0X^{n-1} + \cdots + a_{n-1} \in \mathbb{Z}_p[X]$$

is p^n.

We now show that F is a field. Clearly, F is a commutative ring with unit element $1 (= 0 \cdot X^{n-1} + \cdots + 0 \cdot X + 1)$. Hence we need only verify that if $a(X) \in F$ is not zero, then there exists $b(X) \in F$ with $a(X)b(X) = 1$. As $a(X) \neq 0$, and $f(X)$ is irreducible over $\mathbb{Z}_p$, the gcd $(a(X), f(X)) = 1$. So by Euclidean algorithm, there exist polynomials $C(X)$ and $g(X)$ in $\mathbb{Z}_p[X]$ such that

$$a(X)C(X) + f(X)g(X) = 1 \tag{4.8}$$

in $\mathbb{Z}_p[X]$. Now there exists $C_1(X) \in F$ with $C_1(X) \equiv C(X) (\bmod f(X))$. This means that there exists a polynomial $h(X)$ in $\mathbb{Z}_p[X]$ with $C(X) - C_1(X) = h(X)f(X)$, and hence $C(X) = C_1(X) + h(X)f(X)$. Substituting this in Equation 4.8 and taking modulo $f(X)$, we get, $a(X)C_1(X) = 1$ in F. Hence $a(X)$ has $C_1(X)$ as inverse in F. Thus, every non-zero element of F has a multiplicative inverse in F, and so F is a field of p^n elements. □

We have constructed a field of p^n elements in two different ways—one, as the field of roots of the polynomial $X^{p^n} - X \in \mathbb{Z}_p[X]$, and the other, as the field of polynomials in $\mathbb{Z}_p[X]$ reduced modulo the irreducible polynomial $f(X)$ of degree n over $\mathbb{Z}_P$. Essentially, there is not much of a difference between the two constructions, as our next theorem shows (for a proof, see [18]).

Theorem 4.7. *Any two finite fields of the same order are isomorphic under a field isomorphism.* □

Example 4.7

Take $p = 2$ and $n = 3$. The polynomial $X^3 + X + 1$ of degree 3 is irreducible over $\mathbb{Z}_2$. (If it is reducible, one of the factors must be of degree 1, and it must be either X or $X + 1 = X - 1 \in \mathbb{Z}_2[X]$. But 0 and 1 are not roots of $X^3 + X + 1 \in \mathbb{Z}_2[X]$.) The $2^3 = 8$ polynomials over $\mathbb{Z}_2$ reduced modulo $X^3 + X + 1$ are

$$0, 1, X, X + 1, X^2, X^2 + 1, X^2 + X, X^2 + X + 1$$

and they form a field. (Note that $X^3 = X + 1$, $X^3 + X = 1$ and $X^3 + X + 1 = 0$.) We have, for instance, $\left(X^2 + X + 1\right) + (X + 1) = X^2$ and $(X^2 + 1)(X + 1) = X^3 + X^2 + X + 1 = X^2$. Also $(X + 1)^2 = X^2 + 1$.

We know that if F is a field, the set F^* of non-zero elements of F is a group. In the case when F is a finite field, F^* has an additional algebraic structure.

Theorem 4.8. *If F is a finite field, F^* (the set of non-zero elements of F) is a cyclic group.*

Proof. We know (as F is a field), F^* is a group. Hence, we need only show that F^* is generated by a single element.

Let α be an element of the group F^* of maximum order, say, k. Necessarily, $k \leq q-1$, where $q = |F|$. Choose $\beta \in F^*$, $\beta \neq \alpha, 1$. Let $o(\beta)$ (= the order of β) $= l$. Then, $l > 1$. We first show that $l|k$. Now $o(\beta^{(k,l)}) = l/(k,l)$, where (k,l) denotes the gcd of k and l. Further, as $(o(\alpha), o(\beta^{(k,l)})) = (k, (l/(k,l))) = 1$, we have $o(\alpha\beta^{(k,l)}) = o(\alpha) \cdot o(\beta^{(k,l)}) = k \cdot (l/(k,l)) = [k,l]$, the lcm of k and l. But, by our choice, the maximum order of any element of F^* is k. Therefore $[k,l] = k$ which implies that $l|k$. But $l = o(\beta)$. Therefore $\beta^k = 1$. Thus, for each of the $q - 1$ elements z of F^*, $z^k = 1$ and so z is a root of $x^k - 1$. This means that, as $|F^*| = q - 1$, $k = q - 1$. Thus, $o(\alpha) = q - 1$ and so F^* is the cyclic group generated by α. □

Definition 4.10. *(Primitive element, Minimal polynomial)*

i. *A primitive element of a finite field F is a generator of the cyclic group F^*.*

ii. *A monic polynomial is a polynomial with leading coefficient 1. For example, $X^2 + 2X + 1 \in \mathbb{R}[X]$ is monic while $2X^2 + 1 \in \mathbb{R}[X]$ is not.*

iii. *Let F be a finite field with prime field P. A primitive polynomial in $F[X]$ over P is the minimal polynomial in $P[X]$ of a primitive element of F. A minimal polynomial in $P[X]$ of an element $\alpha \in F$ is a monic polynomial of least degree in $P[X]$ having α as a root.*

Clearly the minimal polynomial of any element of F in $P[X]$ is irreducible over P.

Let α be a primitive element of F, and $f(X) = X^n + a_1X^{n-1} + \cdots + a_n \in \mathbb{P}[X]$ be the primitive polynomial of α. Then, any polynomial in $P[\alpha]$ of degree n or more can be reduced to a polynomial in $P[\alpha]$ of degree at most $n-1$. Moreover, no two distinct polynomials of $P[\alpha]$ of degree at most $n-1$ can be equal; otherwise, α would be a root of a polynomial of degree less than n over P. Hence all the polynomials of the form

$$a_0 + a_1\alpha + \cdots + a_{n-1}\alpha^{n-1},$$

in $P[\alpha]$ are all distinct and so $|P[\alpha]| = p^n$ where $p = |P|$. These p^n elements constitute a subfield F' of F and $\alpha \in F'$. But then $F \subseteq F'$ and hence $F = F'$. Thus, $|F| = |F'| = p^n$. As α is a primitive element of F, this means that

$$F = \left\{0;\ \alpha, \alpha^2, \ldots, \alpha^{p^n-1} = 1\right\}.$$

Example 4.8

Consider the polynomial $X^4 + X + 1 \in \mathbb{Z}_2[X]$. This is irreducible over $\mathbb{Z}_2$ (check that it can have no linear or quadratic factor in $\mathbb{Z}_2[X]$). Let α be a root (in an extension field of $\mathbb{Z}_2$) of this polynomial so that $\alpha^4 + \alpha + 1 = 0$. This means that $\alpha^4 = \alpha + 1$.

We now prove that α is a primitive element of a field of 16 elements over $\mathbb{Z}_2$ by checking that the 15 powers $\alpha, \alpha^2, \ldots, \alpha^{15}$ are all distinct and that $\alpha^{15} = 1$. Indeed, we have

$$\begin{aligned}
\alpha^1 &= \alpha \\
\alpha^2 &= \alpha^2 \\
\alpha^3 &= \alpha^3 \\
\alpha^4 &= \alpha + 1 (as\ \alpha^4 + \alpha + 1 = 0 \rightarrow \alpha^4 = \alpha + 1, as 1 = -1) \\
\alpha^5 &= \alpha\alpha^4 = \alpha(\alpha+1) = \alpha^2 + \alpha \\
\alpha^6 &= \alpha\alpha^5 = \alpha^3 + \alpha^2 \\
\alpha^7 &= \alpha\alpha^6 = \alpha^4 + \alpha^3 = \alpha^3 + \alpha + 1 \\
\alpha^8 &= \alpha\alpha^7 = \alpha^4 + (\alpha^2 + \alpha) = (\alpha + 1) + (\alpha^2 + \alpha) = \alpha^2 + 1 \\
\alpha^9 &= \alpha\alpha^8 = \alpha^3 + \alpha \\
\alpha^{10} &= \alpha\alpha^9 = \alpha^4 + \alpha^2 = \alpha^2 + \alpha + 1 \\
\alpha^{11} &= \alpha\alpha^{10} = \alpha^3 + \alpha^2 + \alpha \\
\alpha^{12} &= \alpha\alpha^{11} = \alpha^4 + (\alpha^3 + \alpha^2) = (\alpha + 1) + (\alpha^3 + \alpha^2) = \alpha^3 + \alpha^2 + \alpha + 1
\end{aligned}$$

$$\begin{aligned}\alpha^{13} &= \alpha\alpha^{12} = \alpha^4 + (\alpha^3 + \alpha^2 + \alpha) = (\alpha + 1) + (\alpha^3 + \alpha^2 + \alpha)\\ &= \alpha^3 + \alpha^2 + 1\\ \alpha^{14} &= \alpha\alpha^{13} = \alpha^4 + (\alpha^3 + \alpha) = (\alpha + 1) + (\alpha^3 + \alpha) = \alpha^3 + 1\\ \alpha^{15} &= \alpha\alpha^{14} = \alpha^4 + \alpha = (\alpha + 1) + \alpha = 1\end{aligned}$$

Thus, $F = \{0\} \cup F^* = \{0, \alpha, \alpha^2, \ldots, \alpha^{15} = 1\}$ and so α is a primitive element of $F = GF(2^4)$.

We observe that a polynomial irreducible over a field F need not be primitive over F. For instance, the polynomial $f(X) = X^4+X^3+X^2+X+1 \in \mathbb{Z}_2[X]$ is irreducible over $\mathbb{Z}_2$ but it is not primitive. To check that $f(X)$ is irreducible, verify that $F(X)$ has no linear or quadratic factor over $\mathbb{Z}_2$. Next, for any root α of $f(X)$, check that $\alpha^5 = 1$ so that $o(\alpha) < 15$, and $f(X)$ is not primitive over $\mathbb{Z}_2$. (Recall that if $f(X)$ were a primitive polynomial, some root of $f(X)$ should be a primitive element of $GF(2^4)$.)

4.14 Factorization of Polynomials Over Finite Fields

Let α be a primitive element of the finite field $F = GF(p^n)$, where p is a prime. Then, $F^* = \{\alpha, \alpha^2, \ldots, \alpha^{p^n-1} = 1\}$, and for any $x \in F$, $x^{p^n} = x$. Hence for each i, $1 \le i \le p^n - 1$,

$$(\alpha^i)^{p^n} = \alpha^i.$$

This shows that there exists a least positive integer t such that $\alpha^{i \cdot p^{t+1}} = \alpha^i$. Then, set

$$C_i = \{i, pi, p^2 i, \ldots, p^t i\}, \quad 0 \le i \le p^n - 1.$$

The sets C_i are called the *cyclotomic cosets* modulo p defined with respect to F and α. Now, corresponding to the coset C_i, $0 \le i \le p^n - 1$, consider the polynomial

$$f_i(X) = (X - \alpha^i)(X \quad \alpha^{i \cdot p})(X - \alpha^{i \cdot p^2}) \cdots (X - \alpha^{i \cdot p^t}).$$

The coefficients of f_i are elementary symmetric functions of $\alpha^i, \alpha^{ip}, \ldots, \alpha^{ip^t}$ and if β denotes any of these coefficients, then β satisfies the relation $\beta^p = \beta$. Hence $\beta \in \mathbb{Z}_p$ and $f_i(X) \in \mathbb{Z}_p[X]$ for each i, $0 \le i \le p^n - 1$. Each element of C_i determines the same cyclotomic coset, that is, $C_i = C_{ip} = C_{ip^2} = \cdots = C_{ip^t}$. Moreover, if $j \notin C_i$, $C_i \cap C_j = \phi$. This gives a factorization of $X^{p^n} - X$ into irreducible factors over $\mathbb{Z}_p$. In fact, $X^{p^n} - X = X(X^{p^n-1} - 1)$, and

$$\begin{aligned} X^{p^n-1} - 1 &= (x-\alpha)(X-\alpha^2)\cdots(X-\alpha^{p^n-1}) \\ &= \prod_i \left(\prod_{j \in C_i} (X-\alpha^j) \right), \end{aligned}$$

where the first product is taken over all the distinct cyclotomic cosets. Furthermore, each polynomial $f_i(X)$ is irreducible over $\mathbb{Z}_p$ as shown below. To see this, assume that

$$\begin{aligned} g(X) &= a_0 + a_1 X + \cdots + a_k X^k \in F[X]. \\ \text{Then,} \quad (g(X))^p &= a_0^p + a_1^p X^p + \cdots + a_k^p (X^k)^p \text{ (Refer Section 3.23)} \\ &= a_0 + a_1 X^p + \cdots + a_k X^{kp} \\ &= g(X^p). \end{aligned}$$

Consequently, if β is a root of g, $g(\beta) = 0$, and therefore $0 = (g(\beta))^p = g(\beta^p)$, that is, β^p is also a root of $g(X)$. Hence if $j \in C_i$ and α^j is a root of $f_i(X)$, then all the powers α^k, $k \in C_i$, are roots of $f_i(X)$. Hence any non-constant irreducible factor of $f_i(X)$ over $\mathbb{Z}_p$ must contain all the terms $(X - \alpha^j)$, $j \in C_i$ as factors. In other words, $g(X)$ is irreducible over $\mathbb{Z}_p$.

Thus, the determination of the cyclotomic cosets yields a simple device to factorize $X^{p^n} - X$ into irreducible factors over $\mathbb{Z}_P$. We illustrate this fact by an example.

Example 4.9

Factorize $X^{2^4} - X$ into irreducible factors over $\mathbb{Z}_2$.

Let α be a primitive element of the field $GF(2^4)$. As a primitive polynomial of degree 4 over $\mathbb{Z}_2$ having α as a root, we can take (see Example 4.8) $X^4 + X + 1$.

The cyclotomic cosets modulo 2 w.r.t. $GF(2^4)$ and α are

$$\begin{aligned} C_0 &= \{0\} \\ C_1 &= \{1, 2, 2^2 = 4, 2^3 = 8\} \quad \text{(Note: } 2^4 = 16 \equiv 1 (\bmod 15)) \\ C_3 &= \{3, 6, 12, 9\} \\ C_5 &= \{5, 10\} \\ C_7 &= \{7, 14, 13, 11\}. \end{aligned}$$

Note that $C_2 = C_1 = C_4$, and so on. Thus,

$$X^{16} - X = X(X^{15} - 1) = X \prod_{i=1}^{15} (X - \alpha^i) \tag{4.9}$$

$$= X\left(X - \alpha^0\right) \left[\prod_{i \in C_1} \left(X - \alpha^i\right)\right] \left[\prod_{i \in C_3} \left(X - \alpha^i\right)\right]$$

$$\times \left[\prod_{i \in C_5} \left(X - \alpha^i\right)\right] \left[\prod_{i \in C_7} \left(X - \alpha^i\right)\right]$$

$$= X\left(X+1\right)\left(X^4 + X + 1\right)\left(X^4 + X^3 + X^2 + X + 1\right)$$

$$\times \left(X^2 + X + 1\right)\left(X^4 + X^3 + 1\right).$$

In computing the products, we have used the relation $\alpha^4 + \alpha + 1 = 0$, that is, $\alpha^4 = \alpha + 1$. Hence, for instance,

$$\prod_{i \in C_5} \left(X - \alpha^i\right) = \left(X - \alpha^5\right)\left(X - \alpha^{10}\right)$$

$$= X^2 - \left(\alpha^5 + \alpha^{10}\right) X + \alpha^{15}$$

$$= X^2 + \left[\left(\alpha^2 + \alpha\right) + \left(\alpha^2 + \alpha + 1\right)\right] X + \alpha^{15}$$

(see Example 4.8)

$$= X^2 + X + 1.$$

The six factors on the right of Equation 4.9 are all irreducible over $\mathbb{Z}_2$. The minimal polynomials of α, α^3 and α^7 are all of degree 4 over $\mathbb{Z}_2$. However, while α and α^7 are primitive elements of $GF(2^4)$ (so that the polynomials $X^4 + X + 1$ and $X^4 + X^3 + 1$ are primitive), α^3 is not (even though its minimal polynomial is also of degree 4).

Primitive polynomials are listed in [19]

4.15 Exercises

1. Construct the following fields:
   $GF(2^4)$, $GF(2^5)$ and $GF(3^2)$.

2. Show that $GF(2^5)$ has no $GF(2^3)$ as a subfield.

3. Factorize $X^{2^3} + X$ and $X^{2^5} + X$ over $\mathbb{Z}_2$.

4. Factorize $X^{3^2} - X$ over $\mathbb{Z}_3$.

5. Using Theorem 4.8, prove Fermat's Little Theorem that for any prime p, $a^{p-1} \equiv 1 \pmod{p}$, if $a \not\equiv 0 \pmod{p}$.

4.16 Mutually Orthogonal Latin Squares

In this section, we show, as an application of finite fields, the existence of $n-1$ mutually orthogonal Latin Squares (MOLSs) of order n.

A Latin square of order n is a double array L of n rows and n columns in which the entries belong to a set S of n elements such that no two entries of the same row or column of L are equal. Usually, we take S to be the set $\{1, 2, \ldots, n\}$ but this is not always essential.

For example,

$$\begin{bmatrix} 1 & 2 \\ 2 & 1 \end{bmatrix} \quad \text{and} \quad \begin{bmatrix} 1 & 2 & 3 \\ 2 & 3 & 1 \\ 3 & 1 & 2 \end{bmatrix}$$

are Latin squares of orders 2 and 3, respectively. Let $L_1 = (a_{ij})$, and $L_2 = (b_{ij})$ be two Latin squares of order n with entries in S. We say that L_1 and L_2 are orthogonal Latin squares if the n^2 ordered pairs (a_{ij}, b_{ij}) are all distinct. For example,

$$L_1 = \begin{bmatrix} 1 & 2 & 3 \\ 2 & 3 & 1 \\ 3 & 1 & 2 \end{bmatrix}, \quad \text{and} \quad L_2 = \begin{bmatrix} 1 & 2 & 3 \\ 3 & 1 & 2 \\ 2 & 3 & 1 \end{bmatrix}$$

are orthogonal Latin squares of order 3 since the nine ordered pairs $(1,1)$, $(2,2)$, $(3,3)$; $(2,3)$, $(3,1)$, $(1,2)$; $(3,2)$, $(1,3)$, $(2,1)$ are all distinct. However, if $M_1 = \left[\begin{smallmatrix} 1 & 2 \\ 2 & 1 \end{smallmatrix}\right]$ and $M_2 = \left[\begin{smallmatrix} 2 & 1 \\ 1 & 2 \end{smallmatrix}\right]$, then the four ordered pairs $(1,2)$, $(2,1)$, $(2,1)$ and $(1,2)$ are not all distinct. Hence M_1 and M_2 are not orthogonal. The study of orthogonal Latin squares started with Euler, who had proposed the following problem of 36 officers (See [6]). The problem asks for an arrangement of 36 officers of 6 ranks and from 6 regiments in a square formation of size 6 by 6. Each row and each column of this arrangement is to contain only one officer of each rank and only one officer from each regiment. We label the ranks and the regiments from 1 through 6, and assign to each officer an ordered pair of integers in 1 through 6. The first component of the ordered pair corresponds to the rank of the officer and the second component his regiment. Euler's problem then reduces to finding a pair of orthogonal Latin squares of order 6. Euler conjectured in 1782 that there exists no pair of orthogonal Latin squares of order $n \equiv 2 \pmod 4$. Euler himself verified the conjecture for $n = 2$, while Tarry in 1900 verified it for $n = 6$ by a systematic case-by-case analysis. But the most significant result with regard to the Euler conjecture came from Bose, Shrikande and Parker who disproved the Euler conjecture by establishing that if $n \equiv 2 \pmod 4$ and $n > 6$, then there exists a pair of orthogonal Latin squares of order n (see also [17,35]).

A set $\{L_1, \ldots, L_t\}$ of t Latin squares of order n on S is called a set of MOLS (Mutually Orthogonal Latin squares) if L_i and L_j are orthogonal whenever $i \neq j$. It is easy to see [17] that the number t of MOLS of order n is bounded by $n-1$. Further, any set of $n-1$ MOLS of order n is known to be equivalent to the existence of a finite projective plane of order n [17]. A long standing conjecture is that if n is not a prime power, then there exists no complete set of MOLS of order n.

We now show that if n is a prime power, there exists a set of $n-1$ MOLS of order n. (Equivalently, this implies that there exists a projective plane of any prime power order, though we do not prove this here) (see [17] for more details on finite projective planes).

Theorem 4.9. *Let $n = p^k$, where p is a prime and k is a positive integer. Then, for $n \geq 3$, there exists a complete set of MOLS of order n.*

Proof. By Theorem 4.6, we know that there exists a finite field $GF(p^k) = GF(n) = F$, say. Denote the elements of F by $a_0 = 0$, $a_1 = 1$, $a_2, \ldots, a_{n-1}$. Define the $n-1$ matrices $A_1, \ldots, A_{n-1}$ of order n by

$$A_t = (a_{ij}^t), \quad 0 \leq i, j \leq n-1; \quad \text{and} \quad 1 \leq t \leq n-1,$$

$a_{ij}^t = a_t a_i + a_j$ (here a_{ij}^t stands for the (i,j)-th entry of the matrix A_t). The entries a_{ij}^t are all elements of the field F. We claim that each A_t is a Latin square. Suppose, for instance, two entries of some i-th row of A_t, say a_{ij}^t and a_{il}^t are equal. This implies that

$$a_t a_i + a_j = a_t a_i + a_l,$$

and hence $a_j = a_l$. Consequently $j = l$. Thus, all the entries of the i-th row of A_t are distinct. For a similar reason, no two entries of the same column of A_t are equal. Hence A_t is a Latin square for each t.

We next claim that $\{A_1, \ldots, A_{n-1}\}$ is a set of MOLS. Suppose $1 \leq r < u \leq n-1$. Then, A_r and A_u are orthogonal. For suppose that

$$\left(a_{ij}^r, a_{ij}^u\right) = \left(a_{i'j'}^r, a_{i'j'}^u\right). \tag{4.10}$$

This means that

$$\begin{aligned} a_r a_i + a_j &= a_r a_{i'} + a_{j'}, \\ \text{and} \quad a_u a_i + a_j &= a_u a_{i'} + a_{j'}. \end{aligned}$$

Subtraction gives

$$(a_r - a_u)a_i = (a_r - a_u)a_i'$$

and hence, as $a_r \neq a_u$, $a_i = a_{i'}$. Consequently, $i = i'$ and $j = j'$. Thus, A_r and A_u are orthogonal. □

Chapter 5

Introduction to Coding Theory

ASCII (American Standard Code for Information Interchange) was designed by the American Standards Association (ASA) subcommittee, in order to facilitate the general interchange of information among information processing systems, communication systems, and associated equipment. An 8-bit set was considered but the need for more than 128 codes in general applications was not yet evident.

ASA subcommittee

5.1 Introduction

Coding theory has its origin in communication engineering. With Shannon's seminal paper of 1948 [22], it has been greatly influenced by mathematics with a variety of mathematical techniques to tackle its problems. Algebraic coding theory uses a great deal of matrices, groups, rings, fields, vector spaces, algebraic number theory and, not to speak of, algebraic geometry. In algebraic coding, each message is regarded as a block of symbols taken from a finite alphabet. On most occasions, this alphabet is $\mathbb{Z}_2 = \{0, 1\}$. Each message is then a finite string of 0s and 1s. For example, 00110111 is a message. Usually, the messages get transmitted through a communication channel. It is quite possible that such channels are subjected to noises, and consequently, the messages get changed. The purpose of an error correcting code is to add redundancy symbols to the message, based of course on some rule so that the original message could be retrieved even though it is garbled. Each message is also called a codeword and the set of codewords is a code.

Any communication channel looks as in Figure 5.1. The first box of the channel indicates the message. It is then transmitted to the encoder, which adds a certain number of redundancy symbols. In Figure 5.1, these redundancy symbols are 001 which when added to the message 1101 give the coded message 1101001. Because of channel noise, the coded message gets distorted and the received message is 0101001. This message then enters the decoder. The decoder applies the decoding algorithm and retrieves the coded message using the added redundancy symbols. From this, the original message is read off in the last box (see Figure 5.1). The decoder has thus corrected a single error, that is, error in one place.

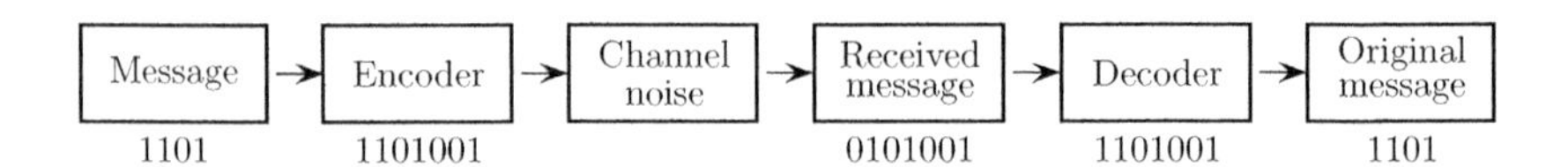

FIGURE 5.1: Communication channel.

The efficiency of a code is the number of errors it can correct. A code is perfect if it can correct all of its errors. It is *k-error-correcting* if it can correct k or fewer errors. The aim of coding theory is to devise efficient codes. Its importance lies in the fact that erroneous messages could prove to be disastrous.

It is relatively easier to detect errors than to correct them. Sometimes, even detection may prove to be helpful as in the case of a feedback channel, that is, a channel that has a provision for retransmission of the messages. Suppose the message 1111 is sent through a feedback channel. If a single error occurs and the received message is 0111, we can ask for a feedback twice and *may* get 1111 on both occasions. We then conclude that the received message should have been 1111. Obviously, this method is not perfect as the original message could have been 0011. On the other hand, if the channel is two-way, that is, it can detect errors so that the receiver knows the places where the errors have occurred and also contains the provision for feedback, then it can prove to be more effective in decoding the received message.

5.2 Binary Symmetric Channels

One of the simplest channels is the binary symmetric channel (BSC). This channel has no memory and it simply transmits two symbols 0 and 1. It has the property that each transmitted symbol has the probability $p(\leq 1/2)$ of being received in error, so that the probability that a transmitted symbol is received correctly is $q = 1 - p$. This is pictorially represented in Figure 5.2. Before considering an example of a BSC, we first give the formal definition of a code.

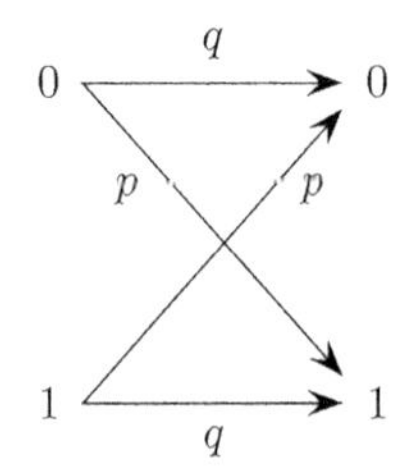

FIGURE 5.2: Binary symmetric channel.

Definition 5.1. *A code C of length n over a finite field F is a set of vectors in F^n, the space of ordered n-tuples over F. Any element of C is called a codeword of C.*

As an example, the set of vectors, $C = \{10110, 00110, 11001, 11010\}$ is a code of length 5 over the field $\mathbb{Z}_2$. This code C has 4 codewords.

Suppose a binary codeword of length 4 (that is, a 4-digit codeword) is sent through a BSC with probability $q = 0.9$. Then, the probability that the sent word is received correctly is $q^4 = (0.9)^4 = 0.6561$.

We now consider another code, namely, the Hamming (7, 4)-code. (See Section 5.5 below). This code has as its words the binary vectors 1000001, 0100011, 0010010, 0001111, of length 7 and all of their linear combinations over the field $\mathbb{Z}_2$.

The first four positions are information positions and the last three are the redundancy positions. There are in all $2^4 = 16$ codewords in the code. We shall see later that this code can correct one error. Hence the probability that a received vector yields the transmitted vector is $q^7 + 7pq^6$, where the first term corresponds to the case of no error and the term $7pq^6$ corresponds to a single error in each of the seven possible positions. If $q = 0.9$, $q^7 + 7pq^6 = 0.4783 + 0.3720 = 0.8503$, which is quite large compared to the probability 0.6561 arrived at earlier in the case of a BSC.

Hamming code is an example of a class of codes called *Linear Codes*. We now present some basic facts about linear codes.

5.3 Linear Codes

Definition 5.2. *An $[n, k]$-linear code C over a finite field F is a k-dimensional subspace of F^n, the space of ordered n-tuples over F.*

If F has q elements, that is, $F = GF(q)$, the $[n, k]$-code will have q^k codewords. The codewords of C are all of length n as they are n-tuples over F. k is called the *dimension* of C. C is a binary code if $F = \mathbb{Z}_2$.

A linear code C is best represented by any one of its generator matrices.

Definition 5.3. *A generator matrix of a linear code C over F is a matrix whose row vectors form a basis for C over F.*

If C is an $[n, k]$-linear code over F, then a generator matrix of C is a k by n matrix G over F whose row vectors form a basis for C.

For example, consider the binary code C_1 with generator matrix

$$G_1 = \begin{bmatrix} 1 & 0 & 0 & 1 & 1 \\ 0 & 1 & 0 & 1 & 0 \\ 0 & 0 & 1 & 0 & 1 \end{bmatrix}.$$

Clearly, all the three row vectors of G_1 are linearly independent over $\mathbb{Z}_2$. Hence C_1 has $2^3 = 8$ codewords. The first three columns of G_1 are linearly independent over $\mathbb{Z}_2$. Therefore, the first three positions of any codeword of C_1 may be taken as information positions, and the remaining two as redundancy positions. In fact, the positions corresponding to any three linearly independent columns of G_1 may be taken as information positions and the rest redundancies. Now any word X of C_1 is given by

$$X = x_1R_1 + x_2R_2 + x_3R_3, \tag{5.1}$$

where x_1, x_2, x_3 are all in $\mathbb{Z}_2$ and R_1, R_2, R_3 are the three row vectors of G_1 in order. Hence by Equation 5.1, $X = (x_1, x_2, x_3, x_1 + x_2, x_1 + x_3)$. If we take $X = (x_1, x_2, x_3, x_4, x_5)$, we have the relations

$$\begin{aligned} x_4 &= x_1 + x_2, \quad \text{and} \\ x_5 &= x_1 + x_3. \end{aligned} \tag{5.2}$$

In other words, the first redundancy coordinate of any codeword is the sum of the first two information coordinates of that word, while the next redundancy coordinate is the sum of the first and third information coordinates.

Equations 5.2 are the *parity-check* equations of the code C_1. They can be rewritten as

$$\begin{aligned} x_1 + x_2 - x_4 &= 0, \quad \text{and} \\ x_1 + x_3 - x_5 &= 0. \end{aligned} \tag{5.3}$$

In the binary case, Equations 5.3 become

$$\begin{aligned} x_1 + x_2 + x_4 &= 0, \quad \text{and} \\ x_1 + x_3 + x_5 &= 0. \end{aligned} \tag{5.4}$$

In other words, the vector $X = (x_1, x_2, x_3, x_4, x_5) \in C_1$ iff its coordinates satisfy Equations 5.4. Equivalently, $X \in C_1$ iff it is orthogonal to the two vectors 11010 and 10101. If we take these two vectors as the row vectors of a matrix H_1, then H_1 is the 2 by 5 matrix:

$$H_1 = \begin{bmatrix} 1 & 1 & 0 & 1 & 0 \\ 1 & 0 & 1 & 0 & 1 \end{bmatrix}.$$

H_1 is called a parity-check matrix of the code C_1. The row vectors of H_1 are orthogonal to the row vectors of G_1. (Recall that two vectors $X = (x_1, \ldots, x_n)$ and $Y = (y_1, \ldots, y_n)$ of the same length n are orthogonal if their inner product (= scalar product $\langle X, Y \rangle = x_1y_1 + \cdots + x_ny_n$) is zero). Now if a vector v is orthogonal to $u_1, \ldots, u_k$, then it is orthogonal to any linear combination of $u_1, \ldots, u_k$. Hence the row vectors of H_1, which are orthogonal to the row vectors of G_1, are orthogonal to all the vectors of the row space of G_1, that is, to all the vectors of C_1. Thus,

$$C_1 = \left\{X \in \mathbb{Z}_2^5 : H_1X^t = 0\right\} = \text{Null space of the matrix } H_1,$$

where X^t is the transpose of X. The orthogonality relations $H_1X^t = 0$ give the parity-check conditions for the code C_1. These conditions fix the redundancy positions, given the message positions of any codeword. A similar result holds good for any linear code. Thus, any linear code over a field F is either the row space of one of its generator matrices or the null space of its corresponding parity-check matrix.

Note that if we form any k linear combinations of the generator matrix of a linear code which are also linearly independent over the base field, the resulting k words also form a basis for $\mathcal{C}$. For instance, if

$$G = \begin{pmatrix} 1 & 0 & 1 & 1 \\ 0 & 1 & 1 & 1 \end{pmatrix}$$

is a generator matrix of a binary linear code $\mathcal{C}$ of length 4, the matrix

$$G' = \begin{pmatrix} 1 & 0 & 1 & 1 \\ 1 & 1 & 0 & 0 \end{pmatrix}$$

is also a generator matrix of $\mathcal{C}$. The reason is that every row vector of $G' \in \mathcal{C}$ and $\text{rank}(G) = \text{rank}(G') = 2$.

So far, we have been considering binary linear codes. We now consider linear codes over an arbitrary finite field F. As mentioned in Definition 5.2, an $[n, k]$ linear code $\mathcal{C}$ over F is a k-dimensional subspace of F^n, the space of all ordered n-tuples over F. If $\{u_1, \ldots, u_k\}$ is a basis of $\mathcal{C}$ over F, every word of $\mathcal{C}$ is a unique linear combination

$$\alpha_1 u_1 + \cdots + \alpha_k u_k,\ \alpha_i \in F \quad \text{for each}\, i.$$

Since α_i can take q values for each i, $1 \le i \le k$, $\mathcal{C}$ has $q \cdot q \cdots q(k\,\text{times}) = q^k$ codewords.

Let G be the k by n matrix over F having $u_1, \ldots, u_k$ of F^n as its row vectors. Then, as G has k ($=$ dimension of $\mathcal{C}$) rows and all the k rows form a linearly independent set over F, G is a generator matrix of $\mathcal{C}$. Consequently, $\mathcal{C}$ is the row space of G over F. The null space of $\mathcal{C}$ is the space of vectors $X \in F^n$ which are orthogonal to all the words of $\mathcal{C}$. In other words, it is the dual space $\mathcal{C}^\perp$ of $\mathcal{C}$. As $\mathcal{C}$ is of dimension k over F, $\mathcal{C}^\perp$ is of dimension $n - k$ over F. Let $\{X_1, \ldots, X_{n-k}\}$ be a basis of $\mathcal{C}^\perp$ over F. If H is the matrix whose row vectors are $X_1, \ldots, X_{n-k}$, then H is a parity-check matrix of $\mathcal{C}$. It is an $(n-k)$ by n matrix. Thus,

$$\begin{aligned} \mathcal{C} &= \text{row space of } G \\ &= \text{null space of } H \\ &= \left\{X \in F^n : HX^t = 0\right\}. \end{aligned}$$

Theorem 5.1. *Let $G = (I_k|A)$ be a generator matrix of a linear code $\mathcal{C}$ over F, where I_k is the identity matrix of order k over F, and A is a k by $(n-k)$ matrix over F. Then, a generator matrix of $\mathcal{C}^\perp$ is given by*

$$H = \left(-A^t | I_{n-k}\right)$$

over F.

Proof. Each row of H is orthogonal to all the rows of G since (by block multiplication, see Chapter 3, Section 3.3.1),

$$GH^t = [I_k|A]\begin{bmatrix}-A\\ I_{n-k}\end{bmatrix} = -A + A = 0.$$ □

Recall that in the example following Definition 5.3, $k = 3$, $n = 5$, and

$$G_1 = (I_3|A), \quad \text{where } A = \begin{bmatrix}1 & 1\\ 1 & 0\\ 0 & 1\end{bmatrix} \quad \text{while } H_1 = \left(-A^t|I_2\right) = \left(A^t|I_2\right) \text{ over } \mathbb{Z}_2.$$

Corollary 5.1. *$G = [I_k|A]$ is a generator matrix of a linear code $\mathcal{C}$ of length n iff $H = [-A^t|I_{n-k}]$ is a parity-check matrix of $\mathcal{C}$.*

5.4 Minimum Distance of a Code

Definition 5.4. *The weight $wt(v)$ of a codeword v of a code $\mathcal{C}$ is the number of non-zero coordinates in v. The minimum weight of $\mathcal{C}$ is the least of the weights of its* non-zero *codewords. The weight of the zero vector of $\mathcal{C}$ is naturally zero.*

Example 5.1

As an example, consider the binary code $\mathcal{C}_2$ with generator matrix

$$G_2 = \left[\begin{array}{cc|ccc}1 & 0 & 1 & 1 & 0\\ 0 & 1 & 1 & 0 & 1\end{array}\right] = [I_2|A].$$

$\mathcal{C}_2$ has four codewords. Its three non-zero words are $u_1 = 10110$, $u_2 = 01101$, and $u_3 = u_1 + u_2 = 11011$. Their weights are 3, 3 and 4, respectively. Hence the minimum weight of $\mathcal{C}_2$ is 3. A parity-check matrix of $\mathcal{C}_2$, by Theorem 5.1, is (as $F = \mathbb{Z}_2$)

$$H_2 = \left[-A^t|I_{5-2}\right] = \left[A^t|I_3\right] = \left[\begin{array}{cc|ccc}1 & 1 & 1 & 0 & 0\\ 1 & 0 & 0 & 1 & 0\\ 0 & 1 & 0 & 0 & 1\end{array}\right].$$

Definition 5.5. *Let $X, Y \in F^n$. The* distance $d(X, Y)$, *also called the* Hamming distance *between X and Y, is defined to be the number of places in which X and Y differ. Accordingly, $d(X, Y) = wt(X - Y)$.*

If X and Y are codewords of a linear code $\mathcal{C}$, then $X-Y$ is also in $\mathcal{C}$ and has non-zero coordinates only at the places where X and Y differ. Accordingly, if X and Y are words of a linear code $\mathcal{C}$, then

$$d(X,Y) = \text{wt}\,(X-Y). \tag{5.5}$$

As a consequence, we have our next theorem.

Theorem 5.2. *The minimum distance of a* linear *code $\mathcal{C}$ is the minimum weight of a non-zero codeword of $\mathcal{C}$ and hence it is equal to the minimum weight of $\mathcal{C}$.*

Thus, for the linear code $\mathcal{C}_2$ of Example 5.1, the minimum distance is 3.

The function $d(X,Y)$ defined in Definition 5.5 does indeed define a distance function (that is, a metric) on F^n. That is to say, it has the following three properties:

For all X, Y, Z in F^n,

i. $d(X,Y) \geq 0$, and $d(X,Y) = 0$ iff $X = Y$,

ii. $d(X,Y) = d(Y,X)$, and

iii. $d(X,Z) \leq d(X,Y) + d(Y,Z)$.

We now give another interpretation for the minimum weight of a linear code $\mathcal{C}$ over $\mathbb{F}_q^*$

Theorem 5.3. *Let H be a parity-check matrix of a linear code $\mathcal{C}$ over $\mathbb{F}_q$. The minimum distance of $\mathcal{C}$ is d if and only if every set of $d-1$ columns of H is linearly independent and some set of d columns of H is linearly dependent.*

Proof. Recall that $\mathcal{C} = \{X \in \mathbb{F}_q^n : HX^t = 0\} = \{X \in \mathbb{F}_q^n : XH^t = 0\} \Leftrightarrow x_1C_1 + x_2C_2 + \cdots + x_nC_n = 0$ where $C_1, C_2, \ldots, C_n$ are the column vectors of H and $X = (x_1, x_2, \ldots, x_n)$. If $wt(X) = k$, (for instance, if $n = 5$, $X = (2,0,3,0,-4)$ and $2C_1 + 3C_3 - 4C_5 = 0$, it means that $(2,0,3,0,-4)H^t = 0$ and hence $(2,0,3,0,-4)$ belongs to code $\mathcal{C}$, and it is of weight 3) then there are k of the coefficients among $x_1, x_2, \ldots, x_n$ that are not zero, and hence there is a linear dependence relation among the corresponding k column vectors, and in this case $wt(X) = k$. As X is a codeword, $d(\mathcal{C}) = d$ is the minimum k such that there exists a linear dependence relation between d column vectors of H. □

5.5 Hamming Codes

Hamming codes are binary linear codes. They can be defined either by their generator matrices or by their parity-check matrices. We prefer the latter. Let us start by defining the [7, 4]-Hamming code $\mathbb{H}_3$. The seven column vectors of its parity-check matrix H of $\mathbb{H}_3$ are the binary representations of the numbers 1 to 7 written in such a way that the last three of its column vectors form I_3, the identity matrix of order 3. Thus,

$$H = \left[\begin{array}{cccc|ccc} 1 & 1 & 1 & 0 & 1 & 0 & 0 \\ 1 & 1 & 0 & 1 & 0 & 1 & 0 \\ 1 & 0 & 1 & 1 & 0 & 0 & 1 \end{array}\right].$$

The columns of H are the binary representations of the numbers 7, 6, 5, 3; 4, 2, 1, respectively ($7 = 2^2 + 2^1 + 2^0$ etc.,). As H is of the form $[-A^t|I_3]$, the generator matrix of $\mathbb{H}_3$ is given by

$$G = [I_{7-3}|A] = [I_4|A] = \left[\begin{array}{cccc|ccc} 1 & 0 & 0 & 0 & 1 & 1 & 1 \\ 0 & 1 & 0 & 0 & 1 & 1 & 0 \\ 0 & 0 & 1 & 0 & 1 & 0 & 1 \\ 0 & 0 & 0 & 1 & 0 & 1 & 1 \end{array}\right].$$

$\mathbb{H}_3$ is of length $2^3 - 1 = 7$, and dimension $4 = 7 - 3 = 2^3 - 1 - 3$. What is the minimum distance of $\mathbb{H}_3$? One way of finding it is to list all the $2^4 - 1$ non-zero codewords (see Theorem 5.2). However, a better way of determining it is the following. The first row of G is of weight 4, while the remaining rows are of weight 3. The sum of any two or three of these row vectors as well as the sum of all the four row vectors of G are all of weight at least 3. Hence the minimum distance of $\mathbb{H}_3$ is 3.

5.6 Standard Array Decoding

We now write the coset decomposition of $\mathbb{Z}_2^7$ with respect to the subspace $\mathbb{H}_3$. (Recall that $\mathbb{H}_3$ is a subgroup of the additive group $\mathbb{Z}_2^7$). As $\mathbb{Z}_2^7$ has 2^7 vectors, and $\mathbb{H}_3$ has 2^4 codewords, the number of cosets of $\mathbb{H}_3$ in F^7 is $2^7/2^4 = 2^3$. (See Chapter 3). Each coset is of the form $X + \mathbb{H}_3 = \{X + v : v \in \mathbb{H}_3\}$ where $X \in \mathbb{F}^7$. Any two cosets are either identical or disjoint. The vector X is a representative of the coset $X + \mathbb{H}_3$. The zero vector is a representative of the coset $\mathbb{H}_3$. If X and Y are each of weight 1, the coset $X + \mathbb{H}_3 \neq Y + \mathbb{H}_3$, since $X - Y$ is of weight at most 2 and hence does not belong to $\mathbb{H}_3$ (as $\mathbb{H}_3$ is of minimum weight 3). Hence the seven vectors of weight 1 in $\mathbb{Z}_2^7$ together with the zero vector define $8 = 2^3$ pairwise disjoint cosets exhausting all the $2^3 \times 2^4 = 2^7$ vectors of $\mathbb{Z}_2^7$. These eight vectors (namely, the seven vectors of weight one and the zero vector) are called coset leaders.

We now construct a double array (Figure 5.3) of vectors of $\mathbb{Z}_2^7$ with the cosets defined by the eight coset leaders (mentioned above). The first row is the coset defined by the zero vector, namely, $\mathbb{H}_3$.

Figure 5.3 gives the *standard array* for the code $\mathbb{H}_3$. If u is the message vector (that is, codeword) and v is the received vector, then $v - u = e$ is

Coset leader	$\mathcal{C} = \mathbb{H}_3$														
0000000	1000111	0100110	0010101	0001011	1100001	1010010	1001100	0110011	0101101	0011010	0111000	1011001	1101010	1110100	1111111
1000000	0000111	1100110	1010101	1001011	0100001	0010010	0001100	1110011	1101101	1011010	1111000	0011001	0101010	0110100	0111111
0100000	1100111	0000110	0110101	0101011	1000001	1110010	1101100	0010011	0001101	0111010	0011000	1111001	1001010	1010100	1011111
0010000	1010111	0110110	0000101	0011011	1110001	1000010	1011100	0100011	0111101	0001010	0101000	1001001	1111010	1100100	1101111
0001000	1001111	0101110	0011101	0000011	1101001	1011010	1000100	0111011	0100101	0010010	0110000	1010001	1100010	1111100	1110111
0000100	1000011	0100010	0010001	0001111	1100101	1010110	1001000	0110111	0101001	0011110	0111100	1011101	1101110	1110000	1111011
0000010	1000101	0100100	0010111	0001001	1100011	1010000	1001110	0110001	0101111	0011000	0111010	1011011	1101000	1110110	1111101
0000001	1000110	0100111	0010100	0001010	1100000	1010011	1001101	0110010	0101100	0011011	0111001	1011000	1101011	1110101	1111110

FIGURE 5.3: Standard array decoding of $\mathbb{Z}_2^7$.

the error vector. If we assume that v has one or no error, then e is of weight 1 or 0. Accordingly e is a coset leader of the standard array. Hence to get u from v, we subtract e from v. In the binary case (as $-e = e$), $u = v + e$. For instance, if in Figure 5.3, $v = 1100110$, then v is present in the second coset for which the leader is $e = 1000000$. Hence the message is $u = v + e = 0100110$. This incidentally shows that $\mathbb{H}_3$ can correct single errors. However, if for instance, $u = 0100110$ and $v = 1000110$, then $e = 1100000$ is of weight 2 and is not a coset leader of the standard array of Figure 5.3. In this case, the standard array decoding of $\mathbb{H}_3$ will not work as it would wrongly decode v as $1000110 - 0000001 = 1000111 \in \mathbb{H}_3$. (Notice that v is present in the last row of Figure 5.3). The error is due to the fact that v has two errors and not just one. Standard array decoding is therefore *maximum likelihood decoding*.

The general Hamming code $\mathbb{H}_m$ is defined analogous to $\mathbb{H}_3$. Its parity-check matrix H has the binary representations of the numbers 1, 2, ..., $2^m - 1$ as its column vectors. Each such vector is a vector of length m. Hence H is an m by $2^m - 1$ binary matrix and the dimension of $\mathbb{H}_m$ is $(2^m - 1) - m =$ (number of columns in H) $-$ (number of rows in H). In other words, $\mathbb{H}_m$ is a $[2^m - 1, 2^m - 1 - m]$ linear code over $\mathbb{Z}_2$. Notice that H has rank m since H contains I_m as a submatrix.

The minimum distance of $\mathbb{H}_m$ is 3, $m \geq 2$. This can be seen as follows. Recall that $\mathbb{H}_m = \left\{X \in \mathbb{Z}_2^{2^m - 1} : HX^t = 0\right\}$. Let i, j, k denote respectively the numbers of the columns of H in which the m-tuples (that is, vectors of length m) $0 \ldots 011, 00 \ldots 0101$ and $0 \ldots 0110$ are present (Figure 5.4).

Let v be the binary vector of length $2^m - 1$ which has 1 in the i-th, j-th and k-th positions and zero at other positions. Clearly, v is orthogonal to all the row vectors of H and hence belongs to $\mathbb{H}_m$. Hence $\mathbb{H}_m$ has a word of weight 3. Further, $\mathbb{H}_m$ has no word of weight 2 or 1. Suppose $\mathbb{H}_m$ has a word u of weight 2. Let i, j be the two positions where u has 1. As $Hu^T = 0$, by the rule of matrix multiplication, $C_i + C_j = 0$, where C_i, C_j are the i-th and j-th columns of H, respectively. This means that $C_i = -C_j = C_j$ (the code being binary). But this contradicts the fact that the columns of H are all distinct. For a similar reason, H_m has no codeword of weight 1. Indeed, assume that H_m has a codeword v of weight 1. Let v have 1 in its, say, k-th position and 0 elsewhere. As $v \in H_m$, $Hv = 0$, and this gives that the k-th column of H is

$$H: \begin{bmatrix} \cdots & 0 & \cdots & 0 & \cdots & 0 & \cdots \\ & \vdots & & \vdots & & \vdots & \\ & 0 & & 1 & & 1 & \\ \cdots & 1 & \cdots & 0 & \cdots & 0 & \cdots \\ \cdots & 1 & \cdots & 1 & \cdots & 0 & \cdots \end{bmatrix}$$

FIGURE 5.4: Form of the parity-check matrix of $\mathbb{H}_m (m \geq 2)$.

the zero vector, a contradiction. Thus, H_m has no codeword of weight 1 or 2 but it has a word of weight 3. Hence the minimum weight of H_m is 3.

5.7 Sphere Packings

As before, let F^n denote the vector space of all ordered n-tuples over the field F. Recall (Section 5.4) that F^n is a metric space with the Hamming distance between vectors of F^n as a metric.

Definition 5.6. *In F^n, the sphere with center X and radius r is the set*

$$S(X, r) = \{Y \in F^n : d(X, Y) \leq r\} \subseteq F^n.$$

Definition 5.7. *An r-error-correcting linear code C is perfect if the spheres of radius r with the codewords of C as centers are pairwise disjoint and their union is F^n.*

The above definition is justified because if v is a received vector that has at most r errors, then v is at a distance at most r from a unique codeword u of C, and hence v belongs to the unique sphere $S(u, r)$; then, v will be decoded as u.
Out next theorem shows that the minimum distance d of a linear code stands as a good error-correcting measure of the code.

Theorem 5.4. *If C is a linear code of minimum distance d, then C can correct $t = \lfloor (d-1)/2 \rfloor$ or fewer errors.*

Proof. It is enough to show that the spheres of radius t centered at the codewords of C are pairwise disjoint. Indeed, if u and v are in C, and if $z \in B(u, t) \cap B(v, t)$, that is, if the two spheres intersect, (see Figure 5.5), then

$$\begin{aligned} d(u, v) &\leq d(u, z) + d(z, v) \\ &\leq t + t = 2t \leq d - 1 < d, \end{aligned}$$

a contradiction to the fact that d is the minimum distance of C. Hence the two spheres must be disjoint. □

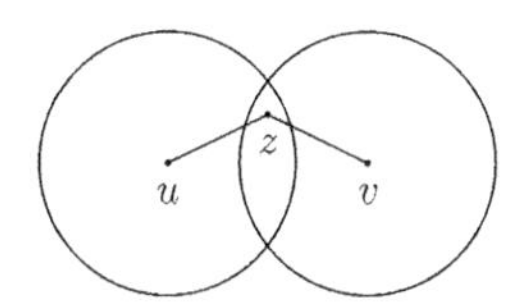

FIGURE 5.5: Case when two spheres intersect.

Corollary 5.2. *The Hamming code $\mathbb{H}_m$ is single error correcting.*

Proof. As $d = 3$ for $\mathbb{H}_m$, $\lfloor (d-1)/2 \rfloor = 1$. Now apply Theorem 5.4. □

Theorem 5.5. *The Hamming code $\mathbb{H}_m$ is a single-error-correcting perfect code.*

Proof. $\mathbb{H}_m$ is a code of dimension $2^m - 1 - m$ over $\mathbb{Z}_2$ and hence has $2^{(2^m-1-m)}$ words. Now, if v is any codeword of $\mathbb{H}_m$, $S(v,1)$ contains v (which is at distance zero from v) and the $2^m - 1$ codewords got from v (which is of length $2^m - 1$) by altering each position once at a time. Thus, $S(v,1)$ contains $1 + (2^m - 1) = 2^m$ words of $\mathbb{H}_m$ (recall that $\mathbb{H}_m$ is single error correcting). As $d = 3$ for $\mathbb{H}_m$, the spheres $s(v,1)$, $v \in \mathbb{H}_m$ are pairwise disjoint. The cardinality of the union of the spheres $S(v,1)$ as v varies over $\mathbb{H}_m$ is $2^{2^m-1-m} \cdot 2^m = 2^{2^m-1}$ = the number of vectors in F^n, where $n = 2^m - 1$. Thus the spheres $S(v,1)$, $v \in H_m$ are pairwise disjoint and cover the whole space F^n. Hence $\mathbb{H}_m$ is perfect. □

5.8 Extended Codes

Let C be a binary linear code of length n. We can extend this code by adding an overall parity check at the end. This means, we add a zero at the end of each word of even weight in C, and add 1 at the end of every word of odd weight. This gives an extended code C' of length $n+1$.

For looking at some of the properties of C', we need a lemma.

Lemma 5.1. *Let w denote the weight function of a binary code C. Then,*

$$d(X,Y) = w(X+Y) = w(X) + w(Y) - 2(X \star Y) \tag{5.6}$$

where $X \star Y$ is the number of common 1s in X and Y.

Proof. Let X and Y have common 1s in the $i_1, i_2, \ldots, i_p$-th positions so that $X \star Y = p$. Let X have 1s in the $i_1, \ldots, i_p$ and $j_1, \ldots, j_q$-th positions, and Y in the $i_1, \ldots, i_p$ and $l_1, \ldots, l_r$-th positions (where no j_k is an l_t). Then, $w(X) = p + q$, $w(Y) = p + r$ and $w(X+Y) = q + r$. The proof is now clear. □

Coming back to the extended code C', by definition, every codeword of C' is of even weight. Hence C' is an even weight code.

Theorem 5.6. *Suppose d is odd. Then, a binary $[n,k]$-linear code with distance d exists iff a binary $[n+1,k]$-linear code with distance $d+1$ exists.*

Proof. Suppose C is a binary $[n,k]$-linear code with distance d. C' is the extended code obtained from C. Since $wt(X')$ is even for every codeword X'

of $\mathcal{C}'$, by Lemma 5.1, $d(X', Y')$ is even for all $X', Y' \in \mathcal{C}'$. Therefore $d(\mathcal{C}')$ is even. Also $d \leq d(\mathcal{C}') \leq d+1$. By assumption, d is odd. Hence $d(\mathcal{C}') = d+1$. Thus, $\mathcal{C}'$ is a binary $[n+1, k]$-linear code with distance $d+1$.

Suppose $\mathcal{C}'$ is an $[n+1, k]$-linear code with distance $d+1$. Since $d(\mathcal{C}') = d+1$, there exist two codewords X and Y such that $d(X, Y) = d+1$. Choose a position in which X and Y differ and delete this from all the codewords. The result is an $[n, k]$-linear code $\mathcal{C}$ with distance d. □

A generator matrix of $\mathcal{C}'$ is obtained by adding an overall parity check to the rows of a generator matrix of $\mathcal{C}$. Thus, a generator matrix of the extended code $\mathbb{H}_4'$ is

$$\left[\begin{array}{ccccccc|c} 1 & 0 & 0 & 0 & 1 & 1 & 1 & 0 \\ 0 & 1 & 0 & 0 & 1 & 1 & 0 & 1 \\ 0 & 0 & 1 & 0 & 1 & 0 & 1 & 1 \\ 0 & 0 & 0 & 1 & 0 & 1 & 1 & 1 \end{array}\right].$$

5.9 Syndrome Decoding

Let $\mathcal{C}$ be an $[n, k]$-linear code over $GF(q) = F$. The standard array decoding scheme requires storage of q^n vectors of F^n and also comparisons of a received vector with the coset leaders. The number of such comparisons is at most q^{n-k}, the number of distinct cosets in the standard array. Hence any method that makes a sizeable reduction in storage and the number of comparisons is to be welcomed. One such method is the syndrome-decoding scheme.

Definition 5.8. *The* syndrome *of a vector $Y \in F^n$ with respect to a linear $[n, k]$-code over F with parity-check matrix H is the vector HY^t.*

As H is an $(n-k)$ by n matrix, the syndrome of Y is a column vector of length $n-k$. We denote the syndrome of Y by $S(Y)$. For instance, the syndrome of $Y = 1110001$ with respect to the Hamming code $\mathbb{H}_4$ is

$$\begin{bmatrix} 1 & 1 & 1 & 0 & 1 & 0 & 0 \\ 1 & 1 & 0 & 1 & 0 & 1 & 0 \\ 1 & 0 & 1 & 1 & 0 & 0 & 1 \end{bmatrix} \begin{bmatrix} 1 \\ 1 \\ 1 \\ 0 \\ 0 \\ 0 \\ 1 \end{bmatrix} = \begin{bmatrix} 1 \\ 0 \\ 1 \end{bmatrix}.$$

Theorem 5.7. *Two vectors of F^n belong to the same coset in the standard array decomposition of a linear code $\mathcal{C}$ iff they have the same syndrome.*

Proof. Let u and v belong to the same coset $a + \mathcal{C}$, $a \in F^n$, of $\mathcal{C}$. Then, $u = a + X$ and $v = a + Y$, where X, Y are in $\mathcal{C}$. Then, $S(u) = S(a + X) =$

$H(a+X)^t = Ha^t + HX^t = S(a)$ (Recall that as $X \in \mathcal{C}$, $S(X) = HX^t = 0$). Similarly $S(v) = S(a)$. Thus, $S(u) = S(v)$.

Conversely, let $S(u) = S(v)$. Then, $Hu^t = Hv^t$ and therefore $H(u-v)^t = 0$. This means that $u - v \in \mathcal{C}$, and hence the cosets $u + \mathcal{C}$ and $v + \mathcal{C}$ are equal. □

Theorem 5.7 shows that the syndromes of all the vectors of F^n are determined by the syndromes of the coset leaders of the standard array of $\mathcal{C}$. In case $\mathcal{C}$ is an $[n, k]$-binary linear code, there are 2^{n-k} cosets and therefore the number of distinct syndromes is 2^{n-k}. Hence in contrast to standard-array decoding, it is enough to store 2^{n-k} vectors (instead of 2^n vectors) in the syndrome decoding. For instance, if $\mathcal{C}$ is a $[100, 30]$-binary linear code, it is enough to store the 2^{70} syndromes instead of the 2^{100} vectors in $\mathbb{Z}_2^{100}$, a huge saving indeed.

5.10 Error Detection

Consider the binary code $\mathcal{C} = \{0000, 1100, 1111\}$. If we change any single coordinate of a codeword, it does not give another codeword. However, the same thing is not true if we change two coordinates. For instance, if we replace both 1s by 0 or both 0s by 1 in 1100, we end up in another codeword. Hence we say that $\mathcal{C}$ can detect one error but not 2 or more.

Definition 5.9. *Let $\mathcal{C}$ be any code, not necessarily linear. Then, $\mathcal{C}$ is t-error detecting, if any word of $\mathcal{C}$ incurs k errors, $1 \le k \le t$, then the resulting word does not belong to $\mathcal{C}$.*

Theorem 5.8. *A code $\mathcal{C}$ of length n over $\mathbb{F}$ of minimum distance d can detect at most $d-1$ errors.*

Proof. Suppose $X \in \mathcal{C}$, and $Y \in \mathbb{F}^n, Y \neq X$, such that $d(X, Y) \le t < d$. Then, Y cannot belong to $\mathcal{C}$. Hence $\mathcal{C}$ can detect up to $d-1$ errors. □

5.11 Sphere Packing Bound or Hamming Bound

We look at the following problems: Given a t-error correcting linear code of length n over the finite field $F = GF(q)$, where q is a prime power p^r (p being a prime), what is the number of elements of $F^n = \{(x_1, x_2, \dots, x_n) : x_i \in F, 1 \le i \le n\}$ that a (closed) sphere S of radius t with center at a codeword u can contain? Now any vector in S is at distance k from u, where $0 \le k \le t$. Let n_k be the number of vectors in S which are at distance k from u. Clearly,

$n_0 = 1$. Any vector in S which is at distance 1 from u can be obtained by replacing any one of the n coordinates of u by a new element of $GF(q)$, and this can be done in $(q-1)$ ways. Each single coordinate can be chosen in n ways, as the vector is of length n. Hence there are $\binom{n}{1}(q-1)$ vectors in S which are at a distance 1 from u. Next to find the point in S which are at distance 2 from u, we change two of the coordinates of u. But then there are $\binom{n}{2}$ ways of choosing two coordinates out of the n coordinates of u. So there are $\binom{n}{2}(q-1)^2$ vectors in S which are at a distance 2 from u. The same argument extends to the number of points at any distance s, $1 \le s \le t$. As the code is t-error correcting, we have the following result.

Theorem 5.9. *The number of points that a (closed) sphere of radius t in the space F_q^n of n-tuples over the Galois field $\mathbb{F}_q$ of q elements is*

$$\binom{n}{0} + \binom{n}{1}(q-1) + \binom{n}{2}(q-1)^2 + \cdots + \binom{n}{t}(q-1)^t.$$

Corollary 5.3. *(Sphere-packing bound or Hamming bound.) An $(n, M, 2t+1)$-linear code C over $GF(q)$ satisfies the condition*

$$M\left\{\binom{n}{0}(q-1) + \binom{n}{2}(q-1)^2 + \cdots + \binom{n}{t}(q-1)^t\right\} \le q^n \ldots (*)$$

Proof. As the minimum distance of C is $d = 2t+1$, the closed spheres of radius $t = (d-1)/2$ about the codewords are pairwise disjoint. The total number of points of $\mathbb{F}_q^n$ that belong to these spheres is, by virtue of Theorem 5.9, the expression on the LHS of (*). Certainly, this number cannot exceed the total number of points of $\mathbb{F}_q^n$ which is q^n. This proves the result. □

Corollary 5.4. *If the $(n, M, 2t+1)$-code is binary, then the sphere packing bound is given by*

$$M\left(\binom{n}{0} + \binom{n}{1} + \cdots + \binom{n}{t}\right) \le 2^n \ldots (**)$$

Proof. Take $q = 2$ in Corollary 5.3. □

Consider again the binary case (that is, when $q = 2$), and let $n = 6$. Then, $|\mathbb{Z}_2^6| = 2^6 = 64$. We ask: Can an $(n, M, d) = (6, 9, 3)$, code exist in $\mathbb{Z}_2^6$ so that the closed unit spheres with centers at the 9 codewords contain all the 64 vectors of $\mathbb{Z}_2^6$? The answer is "no" since the nine spheres with centers at each of the nine vectors can contain at the most $M[\binom{n}{0} + \binom{n}{1}] = 9[\binom{6}{0} + \binom{6}{1}] = 63$ vectors and hence not all of the 64 vectors of $\mathbb{Z}_2^6$. This leads us to the concept of a perfect code.

Definition 5.10. *An (n, M, d)-code over the finite field $\mathbb{F}_q$ is called perfect if the spheres of radius $\lfloor (d-1)/2 \rfloor$ with the centers at the M codewords cover all the q^n vectors of length n over $\mathbb{F}_q$.*

In other words, an (n, M, d)-code is perfect iff equality is attained in (*) (of Corollary 5.3).

Example of a perfect code Let H be the m by $2^m - 1$ matrix, in which the columns of H are the $2^m - 1$ non-zero binary vectors of length m given in some order. Let $\mathcal{H}_m$ be the *binary* linear code with H as its parity-check matrix. As H contains I_m, the identity matrix of order m as a submatrix, $rank(H) = m$. As the number of columns of H is $2^m - 1$, and $\mathcal{H}_m$ is the null space of H,

$$\dim(\mathcal{H}_m) = 2^m - 1 - m. \text{ (see [20])}.$$

Taking $t = 1$, and $M = 2^{2^m-1-m}$ in (**) above, we get $M(\binom{n}{0} + \binom{n}{1}) = 2^{2^m-1-m}(1+n) = 2^{2^m-1-m}(1+(2^m-1)) = 2^{2^m-1} = 2^n =$ the RHS of (**). We thus conclude:

Theorem 5.10. *The binary Hamming code $\mathcal{H}_m$ is a single error correcting perfect code.*

Our earlier results show that no binary $(6, 9, 3)$ perfect code exists. (See also Exercise 9 at the end of this section). We now look at the general Hamming code $\mathcal{C}$ of length m over the field $GF(q)$. The column vectors of any parity-check matrix of $\mathcal{C}$ consists of non-zero vectors of length m over $GF(q)$. Now the space spanned by any such vector v over $GF(q)$ is the same as the space spanned by αv for any non-zero element α of $GF(q)$, that is, $\langle v \rangle = \langle \alpha v \rangle$ for $0 \neq \alpha \in GF(q)$. Hence we choose only one of these $q - 1$ vectors as a column vector of H. As there are $q^m - 1$ non-zero vectors of length m over $GF(q)$, we have $(q^m - 1)/(q - 1)$ distinct vectors of length m, no two of which spanning the same subspace over $GF(q)$. Hence the number of columns of the parity-check matrix of this generalized Hamming code is $(q^m - 1)/(q - 1)$.

Note: When $q = 2$, the number of columns of the Hamming code is $(2^m - 1)/(2 - 1) = 2^m - 1$, as seen earlier.

Another Example We now construct the Hamming code $\mathcal{H}_m$ with $m = 3$ over the field $GF(3)$. The elements of $GF(3)$ are $0, 1, 2$ and addition and multiplication are taken modulo 3. The number of non-zero vectors of length 3 over $GF(3)$ is $3^3 - 1 = 26$. But then for any vector of x of length 3 over $GF(3)$, $2x$ is also such a vector. Hence the number of distinct column vectors of $\mathbb{H} = \mathbb{H}_3$ in which no two column vectors span the same space is $26/2 = 13$. Hence H is a 3×13 matrix of rank 3 and it is therefore of dimension $13 - 3 = 10$. This shows that $\mathcal{H}_3$ is a $(13, 3^{10}, 3)$-ternary perfect code. In fact, the condition (*) is

$$3^{10}\left(\binom{13}{0} + \binom{13}{1}(3-1)\right) = 3^{10}(1 + 13 \times 2) = 3^{10} \cdot 3^3 = 3^{13},$$

so that equality is present in (*). Finally, we give the parity-check matrix H of the code as constructed above:

$$\begin{bmatrix} 0 & 0 & 0 & 0 & 1 & 1 & 1 & 1 & 1 & 1 & 1 & 1 & 1 \\ 0 & 1 & 1 & 1 & 0 & 0 & 0 & 1 & 1 & 1 & 2 & 2 & 2 \\ 1 & 0 & 1 & 2 & 0 & 1 & 2 & 0 & 1 & 2 & 0 & 1 & 2 \end{bmatrix}$$

We observe that the first non-zero entry in each column is 1. This makes H unique except for the order of the columns. We may as well replace any column vector v by $2v$ (where $0 \neq 2 \in GF(3)$).

5.12 Exercises

1. Show by means of an example that the syndrome of a vector depends on the choice of the parity-check matrix.

2.(a) Find all the codewords of the binary code with generator matrix $\begin{bmatrix} 1 & 0 & 1 & 1 & 1 \\ 1 & 0 & 0 & 1 & 1 \end{bmatrix}$.

(b) Find a parity-check matrix of the code.

(c) Write down the parity-check equations.

(d) Determine the minimum weight of the code.

3. Decode the received vector 1100011 in $\mathbb{H}_4$ using (i) the standard array decoding, and (ii) syndrome decoding.

4. How many vectors of $\mathbb{Z}_2^7$ are there in $S(u, 3)$, where $u \in \mathbb{Z}_2^7$?

5. How many vectors of F^n are there in $S(u, 3)$, where $u \in F^n$, and $|F| = q$?

6. Show that a t-error-correcting binary perfect $[n, k]$-linear code satisfies the relation

$$\binom{n}{0} + \binom{n}{1} + \cdots + \binom{n}{t} = 2^{n-k}.$$

More generally, show that a t-error-correcting perfect $[n, k]$-linear code over $GF(q)$ satisfies the relation

$$\binom{n}{0} + \binom{n}{1} + \cdots + \binom{n}{t} = q^{n-k}.$$

7. Show that the function $d(X, Y)$ defined in Section 5.4 is indeed a metric.

8. Show that it is impossible to find nine binary vectors of length 6 such that the distance between any two of them is at least 3.

9. Let C be a linear code of length n over a field F, and let C be of minimum weight 5. Can there be two vectors of $\mathbb{F}_2^n$, each of weight 2, belonging to the same coset of C in $\mathbb{F}^n$?

10. Determine the number of errors that the linear code C over $\mathbb{F}^5$ with generators 01234, 12044, 13223 can detect.

11. Show by means of an example that the generator matrix of a linear code in $\mathbb{Z}_3^7$ (that is a code of length 7 over the field $\mathbb{Z}_3$) need not be unique.

12. Let $\mathcal{C} \subseteq \mathbb{Z}_3^3$ be given by:
 $\mathcal{C} = \{(x_1, x_2, x_3) \in \mathbb{Z}_3^3 : x_1 + x_2 + 2x_3 = 0\}$.

 A. List all the codewords of $\mathcal{C}$.

 B. Give a generator matrix of $\mathcal{C}$.

 C. Give another generator matrix of $\mathcal{C}$ (different from what you gave in (ii)).

 D. Calculate the minimum distance $d(\mathcal{C})$ of $\mathcal{C}$.

 E. How many errors can $\mathcal{C}$ correct?

 F. How many errors can $\mathcal{C}$ detect?

 G. Give a parity-check matrix for $\mathcal{C}$ using (c) above.

13. Show by means of an example that the coset leader of a coset of a linear code need not be unique.

We close this chapter with a brief discussion on a special class of linear codes, namely, cyclic codes.

5.13 Cyclic Codes

Definition 5.11. *A linear code $\mathcal{C}$ over a field F is called cyclic if every cyclic shift of a codeword in $\mathcal{C}$ also belongs to $\mathcal{C}$.*

To be precise, $(a_0, a_1, a_2, \ldots, a_{n-1}) \in \mathcal{C} \Rightarrow (a_{n-1}, a_0, a_1 \ldots, a_{n-2}) \in \mathcal{C}$. This definition of course implies that $(a_{n-2}, a_{n-1}, a_0, \cdots, a_{n-3}) \in \mathcal{C}$, etc. For example, if $n = 4$, and $(a_0, a_1, a_2, a_3) \in \mathcal{C}$, then $(a_3, a_0, a_1, a_2), (a_2, a_3, a_0, a_1), (a_1, a_2, a_3, a_0)$ are all in $\mathcal{C}$.

We now identify the codeword $(a_0, a_1, a_2, \ldots, a_{n-1})$ of $\mathcal{C}$ with the polynomial (See Section 3.20)

$$g(x) = a_0 + a_1 x + a_2 x^2 + \cdots + a_{n-1} x^{n-1} \in F[x]/(x^n - 1). \tag{5.7}$$

It is more precise to say that we identify the codeword $(a_0, a_1, \ldots, a_{n-1})$ with the residue class

$$a_0 + a_1 x \cdots + a_{n-1} x^{n-1} + (x^n - 1).$$

But then as mentioned in Remark 3.1, we can identify the residue class with the corresponding polynomial and do arithmetic modulo $(x^n - 1)$. The identification given by Equation 5.7 makes $xg(x) = x(a_0 + a_1 x + \cdots + a_{n-1} x^{n-1}) = a_0 x + a_1 x^2 + a_2 x^3 + \cdots + a_{n-1} x^n = a_{n-1} + a_0 x + a_1 x^2 + a_2 x^3 + \cdots + a_{n-2} x^{n-1}$

to correspond to the word $(a_{n-1}, a_0, a_1, \cdots + a_{n-2})$, $x^2g(x) = a_{n-2} + a_{n-1}x + a_0x^2 + \ldots, a_{n-3}x^{n-1}$ to correspond to the word $(a_{n-2}, a_{n-1}, a_0, \ldots, a_{n-3})$ and so on. Moreover, if $g(x) \in \mathcal{C}$, $kg(x)$ also belongs to $\mathcal{C}$ for every $k \in F$. Hence if $f(x) \in F[x]$, $f(x)g(x) \in \mathcal{C}$ (under the identification). In other words, $\mathcal{C}$ is an ideal in the commutative ring $F[x]/(x^n - 1)$. Recall that (Theorem 3.13), any ideal in $F[x]$ is principal and it is generated by the unique monic polynomial of least degree present in $\mathcal{C}$.

Example 5.2

Let V be the space of all binary 3-tuples of $\mathbb{R}$. Then, the cyclic codes in V are the following:

Code	Codewords in $\mathcal{C}$	Corresponding polynomials in $\mathbb{R}_3$
$\mathcal{C}_1$	(0,0,0)	0
$\mathcal{C}_2$	(0,0,0)	0
	(1,1,1)	$1 + x + x^2$
$\mathcal{C}_3$	(0,0,0)	0
	(1,1,0)	$1 + x$
	(0,1,1)	$x + x^2$
	(1,0,1)	$1 + x^2$
$\mathcal{C}_4$	All of V	All of $\mathbb{R}_3$

Note: Not every linear code is cyclic. For instance, the binary code $\mathcal{C}' = \{(0,0,0), (1,0,1)\}$ is the code generated by $(1,0,1)$ but it is not cyclic (as $(1,1,0) \notin \mathcal{C}'$).
To summarize, we have the following result.

Theorem 5.11. *Let $\mathcal{C}$ be a cyclic code of length n over a field F, and let $R_n = F[x]/(x^n - 1)$. Then, the following are true:*

i. There exists a unique monic polynomial $g(x)$ of least degree $k \leq n - 1$ in $\mathcal{C}$.

ii. $\mathcal{C} = \langle g(x) \rangle$, the ideal generated by $g(x) \in R_n$.

iii. $g(x)/(x^n - 1)$.

(For a proof of (iii), see Theorem 3.13.) We observe that the constant term in $g(x) \neq 0$. For assume that $g(x) = a_0 + a_1x + \cdots + a_{n-k}x^{n-k}$, and $a_0 = 0$. Then, $x^{n-1}g(x) \in \mathcal{C} \Rightarrow x^{n-1}(a_1x + \cdots + a_{n-k}x^{n-k}) \in \mathcal{C}$ ($\mathcal{C}$ being an ideal in R_n) $\Rightarrow a_1 + a_2x + \cdots + a_kx^{n-k-1} \in \mathcal{C}$, contradicting (i).

We now determine the generator polynomial of a cyclic code.

Theorem 5.12. *Let $\mathcal{C} = \langle g(x) \rangle$ be a cyclic code of length n, equivalently an ideal in $R_n = F[x]/(x^n - 1)$. Let $g(x) = a_0 + a_1 x + \cdots + a_{n-k}x^{n-k}$, $a_{n-k} \neq 0$. Then, a generator matrix of $\mathcal{C}$ is*

$$G : \begin{bmatrix} a_0 & a_1 & a_2 & . & a_{n-k} & 0 & 0 & 0 & . & 0 \\ 0 & a_0 & a_1 & a_2 & . & a_{n-k} & 0 & 0 & . & 0 \\ 0 & . & . & . & . & . & . & . & . & . \\ . & . & . & . & . & . & . & . & . & . \\ 0 & 0 & 0 & 0 & a_0 & a_1 & . & . & a_{n-k} & 0 \\ 0 & 0 & 0 & 0 & 0 & a_0 & a_1 & . & . & a_{n-k} \end{bmatrix}.$$

Consequently $\dim \mathcal{C} = k$.

Proof. $(a_0, a_1, a_2, \ldots, a_{n-1}) \in \mathcal{C}$ is equivalent to: $g(x) = a_0 + a_1 x + \cdots + a_{n-1}x^{n-1} \in$ (ideal) $\mathcal{C} \subseteq\in R_n$. Hence $g(x), xg(x), \ldots, x^{k-1}g(x)$ all belong to $\mathcal{C}$.
Claim: The above polynomials are linearly independent over F. Indeed, $c_0 g(x) + c_1 x g(x) + \cdots + c_{k-1}x^{k-1}g(x) = 0$, where $c_i \in F$, for each i, $0 \leq i \leq k-1$,
$\Rightarrow (c_0 + c_1 x + \cdots + c_{k-1}x^{k-1})g(x) = 0$
$\Rightarrow (c_0 + c_1 x + \cdots + c_{k-1}x^{k-1})(a_0 + a_1 x + \cdots + a_{n-1}x^{n-1}) = 0$ (the zero polynomial)
$\Rightarrow c_0 a_0 = 0, c_0 a_1 + c_1 a_0 = 0, c_0 a_2 + c_1 a_1 + c_0 a_0 = 0$, etc.
Recall $a_0 \neq 0 \Rightarrow c_0 = 0 \Rightarrow c_1 = 0 \Rightarrow c_2 = 0$, etc. Hence $c_i = 0$ for each i, $0 \leq i \leq k-1$.
Again, the k polynomials span $\mathcal{C}$. To see this, assume that $s(x) \in \mathcal{C}$. Then, $s(x)$ is a multiple of $g(x)$.
$\Rightarrow \exists$ a polynomial $c(x)$ of degree $\leq (n-1) - (n-k) = k-1$, say, $c(x) = c_0 + c_1 x + \cdots + c_{k-1}x^{k-1}$, such that $s(x) = c(x)g(x) = c_0 g(x) + c_1 x g(x) + \cdots + c_{k-1}x^{k-1}g(x) \Rightarrow s(x)$ belongs to the row space of the matrix G. This proves the result.

□

Consider $\mathcal{C} = \langle g(x) \rangle$, where $g(x) = 1 + x^2 + x^3$. Then, $\dim\mathcal{C} = 7 - 3 = 4$. A generator matrix G of $\mathcal{C}$ is obtained by writing the vectors for the polynomials $g(x), xg(x), x^2 g(x)$ and $x^3 g(x)$. Thus, as $g(x) = 1 + 0.x + 1.x^2 + 1.x^3 + 0.x^4 + 0.x^5 + 0.x^6$,

$$G = \begin{bmatrix} 1 & 0 & 1 & 1 & 0 & 0 & 0 \\ 0 & 1 & 0 & 1 & 1 & 0 & 0 \\ 0 & 0 & 1 & 0 & 1 & 1 & 0 \\ 0 & 0 & 0 & 1 & 0 & 1 & 1 \end{bmatrix}.$$

Hence C is of dimension $4 =$ number of rows of G. (Note: The last four columns form a lower triangular non-singular matrix. In the binary case, it is in fact unit lower triangular).

Example 5.3

Binary cyclic codes of length 7.

If $C = \langle g(x) \rangle$ is a binary cyclic code of length 7, $g(x) \in \mathbb{Z}_2[x]$ and $g(x)/(x^7 - 1)$. Hence, to determine all binary cyclic codes of length 7, we first factorize $x^7 - 1$ into irreducible factors over $\mathbb{Z}_2$. In fact we have:

$$x^7 - 1 = (1 + x)(1 + x + x^3)(1 + x^2 + x^3).$$

As there are three irreducible factors on the right, there exist $2^3 = 8$ binary cyclic codes of length 7.

5.14 Dual Codes

Let C be a linear code of length n and dimension k over $\mathbb{R}$. Let $C^\perp$ denote the set of vectors in F^n which are orthogonal to all the codewords of C. If v and w are orthogonal to u, then any linear combination of v and w over F is also orthogonal to u. Hence $C^\perp$ is also a linear code. It is the null space of a generator matrix of C and hence its dimension is $n - k$.

Suppose C is a cyclic code. The natural question is: Is the code $C^\perp$ also cyclic? The answer is "yes" as shown below.

Let $g(x)$ be the generator polynomial of C. By Theorem 5.11 $g(x)/(x^n - 1)$. Let $x^n - 1 = g(x)h(x)$. However, $h(x)$ need not be the generator polynomial of $C^\perp$. For example, consider again the polynomial $g(x) = 1 + x^2 + x^3$ of Example 5.3 which corresponds to the vector $(1, 0, 1, 1, 0, 0, 0) \in \mathbb{Z}_2$. Now $h(x) = (x^7 - 1)/(1 + x^2 + x^3) = (1 + x)(1 + x + x^3) = 1 + x^2 + x^3 + x^4$ and this corresponds to the vector $(1, 0, 1, 1, 1, 0, 0)$. The inner product of these two vectors is 1 and not zero. It turns out that the generator polynomial of $C^\perp$ is the reciprocal polynomial of $h(x)$.

Definition 5.12. *Let $h(x) = c_0 + c_1x + \cdots + c_{n-1}x^{n-1}$. then the reciprocal polynomial $h_R(x)$ of $h(x)$ is defined by: $h_R(x) = c_{n-1} + c_{n-2}x + \cdots + c_0x^{n-1} = x^{n-1}(h(1/x))$.*

Lemma 5.2. *Let $u(x) = a_0 + a_1x + \cdots + a_{n-1}x^{n-1}$ and $v(x) = b_0 + b_1x + \cdots + b_{n-1}x^{n-1}$ be two polynomials in $R_n = F[x]/(x^n - 1)$. Then, $u(x)v(x) = 0$ in R_n iff the vector $u_1 = (a_0, a_1, \ldots, a_{n-1})$ is orthogonal to $v_1 = (b_{n-1}, b_{n-2}, \ldots, b_0)$ and to all the cyclic shifts of v_1.*

Proof.

$$\begin{aligned}
u(x)v(x) &= 0 \\
&\iff (a_0b_{n-1} + a_1b_{n-2} + \cdots + a_{n-1}b_0)x^{n-1} + (a_1b_{n-1} + a_2b_{n-2} + \ldots \\
&\quad + a_0b_0)x^n + \cdots + (a_{n-1}b_{n-1} + a_0b_{n-2} + \ldots a_{n-2}b_0)x^{n-2} = 0 \\
&\iff \text{(on rearranging)}(a_0b_{n-1} + a_1b_{n-2} + \cdots + a_{n-1}b_0)x^{n-1} + \\
&\quad (a_0b_0 + a_1b_{n-1} + a_2b_{n-2} + \cdots + a_{n-1}b_1)x^{n-2} + \cdots + (a_0b_{n-2} \\
&\quad + a_1b_{n-3} + \cdots + a_{n-2}b_0 + a_{n-1}b_{n-1})x^{n-2} = 0 \\
&\iff \text{each coefficient of } x^i, 0 \le i \le n-1, \text{ is zero.} \\
&\iff u_1 = (a_0, a_1, \ldots, a_{n-1}) \text{is orthogonal to} \, v_1 = (b_0, b_1, \ldots, b_{n-1}) \\
&\quad \text{and to all the cyclic shifts of } v_1.
\end{aligned}$$

□

Corollary 5.5. *(with the same notation as in Lemma 5.2)* $u(x)v(x) = 0$ *in* R_n *iff* $(b_0, b_1, \ldots, b_{n-1})$ *is orthogonal to* $v_1 = (a_{n-1}, a_{n-2}, \ldots, a_0)$ *and all its cyclic shifts of* v_1.

Proof. $a(x)b(x) = 0 \iff b(x)a(x) = 0$(in R_n). Now apply Lemma 5.2 □

Theorem 5.13. *Let* $\mathcal{C}$ *be a cyclic code of length* n *with generator polynomial* $g(x) = a_0 + a_1x + a_2x^2 + \cdots + a_{n-k}x^{n-k}$ *so that* $\mathcal{C}$ *is of dimension* k. *Suppose* $x^n - 1 = g(x)h(x)$, *and let* $h_R(x)$ *be the reciprocal polynomial of* $h(x)$. *If* G *and* H *are respectively the* k *by* n *and* $n-k$ *by* n *matrices:*

$$G = \begin{bmatrix} a_0 & a_1 & a_2 & . & a_{n-k} & 0 & 0 & 0 & . & 0 \\ 0 & a_0 & a_1 & a_2 & . & a_{n-k} & 0 & 0 & . & 0 \\ 0 & . & . & . & . & . & . & . & . & . \\ . & . & . & . & . & . & . & . & . & . \\ 0 & 0 & 0 & 0 & a_0 & a_1 & . & . & a_{n-k} & 0 \\ 0 & 0 & 0 & 0 & 0 & a_0 & a_1 & . & . & a_{n-k} \end{bmatrix}, \quad \text{and}$$

$$H = \begin{bmatrix} h_k & h_{k-1} & h_{k-2} & . & h_0 & 0 & 0 & 0 & . & 0 \\ 0 & h_k & h_{k-1} & . & h_1 & h_0 & 0 & 0 & . & 0 \\ 0 & 0 & . & . & . & . & . & . & . & . \\ . & . & . & . & . & . & . & . & . & . \\ . & . & . & . & . & . & . & . & . & . \\ 0 & 0 & 0 & 0 & 0 & h_k & . & & h_1 & h_0 \end{bmatrix},$$

then G *is a generator matrix of* $\mathcal{C}$, *and* H *is a generator matrix of* $\mathcal{C}^\perp$. *Consequently,* $\mathcal{C}^\perp$ *is also a cyclic code.*

Proof. That G is a generator matrix of $\mathcal{C}$ has already been seen in Theorem 5.12. By hypothesis, $g(x)h(x) = x^n - 1$ and hence $g(x)h(x) = 0$ in R_n. This means, by virtue of Lemma 5.2, every row of G is orthogonal to every row of H. Moreover, rank of $G = k = \dim(\mathcal{C})$, and rank of $H = n - k$. Hence H is a generator matrix of $\mathcal{C}^{\perp}$ and $\dim(\mathcal{C}^{\perp}) = (n - k)$. Further, as the rows of H are cyclic shifts of its first row, $\mathcal{C}^{\perp}$ is also cyclic. □

Note that of the matrices G and H, the transpose of one is the parity-check matrix of the linear code defined by the other.

5.15 Exercises

1. Let G be the generator matrix of a binary linear code $\mathcal{C}$ of length n and dimension k. Let G' be the matrix obtained by adding one more parity check at the end of each row vector of G (that is add 0 or 1 according to whether the row vector is of even or odd weight). Let $\mathcal{C}^{'}$ be the code generated by $G^{'}$. Show that $G^{'}$ is an even-weight code (that is, every word is of even weight). Determine $H^{'}$, a parity-check matrix of $\mathcal{C}^{'}$.

2. Let $\mathcal{C}_3$ be the code of Example 5.2. Show that $\mathcal{C}_3$ is cyclic. Determine $\dim(\mathcal{C}_3)$ and $\mathcal{C}_3^{\perp}$.

3. Determine all the binary cyclic codes of length 8.

4. Determine the binary cyclic code of length n with generator polynomial $1 + x$.

5. Factorice $x^7 - 1$ over $\mathbb{Z}_2$. Hence determine two cyclic codes each of dimension 3. Show that these two codes are equivalent.

6. Let $\mathcal{C}$ be a binary code of odd length n and $d(\mathcal{C}) \geq 3$. Prove that $\mathcal{C}$ has no codewords of weight $n - 2$.

Chapter 6

Cryptography

Three may keep a secret, if two are dead.

Benjamin Franklin
Statesman and Scientist

6.1 Introduction

To make a message secure, the sender usually sends the message in a disguised form. The intended receiver removes the disguise and then reads off the original message. The original message of the sender is the *plaintext*, and the disguised message is the *ciphertext*. The plaintext and the ciphertext are usually written in the same alphabet. The plaintext and the ciphertext are divided, for the sake of computational convenience, into units of a fixed length. The process of converting a plaintext to a ciphertext is known as *enciphering* or *encryption*, and the reverse process is known as *deciphering* or *decryption*. A message unit may consist of a single letter or any ordered k-tuple, $k \geq 2$. Each such unit is converted into a number in a suitable arithmetic and the transformations are then carried out on this set of numbers. An enciphering transformation f converts a plaintext message unit P (given by its corresponding number) into a number that represents the corresponding ciphertext message unit C while its inverse transformation, namely, the deciphering transformation just does the opposite by taking C to P. We assume that there is a 1–1 correspondence between the set of all plaintext units $\mathcal{P}$ and the set of all ciphertext units $\mathcal{C}$. Hence each plaintext unit gives rise to a unique ciphertext unit and vice versa. This can be represented symbolically by

$$\mathcal{P} \xrightarrow{f} \xi \xrightarrow{f^{-1}} \mathcal{C}.$$

Such a setup is known as a *cryptosystem*.

6.2 Some Classical Cryptosystems

6.2.1 Caesar cryptosystem

One of the earliest of the cryptosystems is the Caesar cryptosystem attributed to the Greek emperor Julius Caesar of the first century B.C. In this cryptosystem, the alphabet is the set of English characters A, B, C, $\ldots$,

TABLE 6.1: Numerical equivalents of English characters

A	B	C	D	E	F	G	H	I	J	K	L	M
0	1	2	3	4	5	6	7	8	9	10	11	12
N	O	P	Q	R	S	T	U	V	W	X	Y	Z
13	14	15	16	17	18	19	20	21	22	23	24	25

X, Y, Z labelled 0, 1, 2, ..., 23, 24, 25, respectively, so that 0 corresponds to A, 1 corresponds to B and so on, and finally 25 corresponds to Z. In this system, each message unit is of length 1 and hence consists of a single character. The encryption (transformation) $f : \mathcal{P} \longrightarrow \mathcal{C}$ is given by

$$f(a) = a + 3 \pmod{26}, \tag{6.1}$$

while the decryption (transformation) $f^{-1} : \mathcal{C} \longrightarrow \mathcal{P}$ is given by

$$f^{-1}(b) = b - 3 \pmod{26}. \tag{6.2}$$

Table 6.1 gives the 1–1 correspondence between the characters A to Z and the numbers 0 to 25.

For example, the word "OKAY" corresponds to the number sequence "(14) (10) (0) (24)" and this gets transformed, by Equation 6.1 to "(17) (13) (3) (1)" and so the corresponding ciphertext is "RNDB." The deciphering transformation applied to "RNDB" then gives back the message "OKAY."

6.2.2 Affine cryptosystem

Suppose we want to encrypt the message "I LIKE IT." In addition to the English characters, we have in the message two spaces in between words. So we add "space" to our alphabet by assigning to it the number 26. We now do arithmetic modulo 27 instead of 26. Suppose, in addition, each such message unit is an ordered pair (sometimes called a digraph). Then, each unit corresponds to a unique number in the interval $[0,\ 27^2 - 1]$. Now, in the message, "I LIKE IT," the number of characters including the two spaces is 9, an odd number. As our message units are ordered pairs, we add an extra blank space at the end of the message. This makes the number of characters as 10 and hence the message can be divided into 5 units $U_1, U_2, \ldots, U_5$:

$$|I-|\ |LI|\ |KE|\ |-I|\ |T-|$$

where $U_1 = I-$ etc. (here, $-$ stands for space). Now U_1 corresponds to the number (see Table 6.1) $8 \cdot 27^1 + 26 = 242$. Assume that the enciphering transformation that acts now on ordered pairs is given by

$$C \equiv aP + b \pmod{27^2} \tag{6.3}$$

where a and b are in the ring Z_{27}, $a \neq 0$ and $(a, 27) = 1$. In Equation 6.3, P and C denote a pair of corresponding plaintext and ciphertext units.

The Extended Euclidean Algorithm (see chapter 3 of [25]) ensures that as $(a, 27) = 1$, $(a, 27^2) = 1$ and hence a has a unique inverse a^{-1} (mod 729) so that

$$aa^{-1} \equiv 1 \pmod{27^2}.$$

This enables us to solve for P in terms of C from the Congruence (6.3). Indeed, we have

$$P \equiv a^{-1}(C - b) \pmod{27^2}. \tag{6.4}$$

As a specific example, let us take $a = 4$ and $b = 2$. Then,

$$C \equiv 4P + 2 \pmod{27^2}.$$

Further as $(4, 27) = 1$, 4 has a unique inverse $\pmod{27^2}$; in fact $4^{-1} = 547$ (mod 729) as $4 \cdot 547 \equiv 1 \pmod{27^2}$. (Indeed, if $4x \equiv 1 \pmod{729}, 4(-x) \equiv -1 \equiv 728 \Rightarrow -x \equiv 182 \Rightarrow x \equiv -182 \equiv 547 \pmod{729}$.) This when substituted in the congruence (6.4) gives

$$P \equiv 547\,(C - 2) \pmod{27^2}.$$

Getting back to $P =$ "I (space)" $= 242$ in I LIKE IT, we get

$$\begin{aligned} C &\equiv 4 \cdot 242 + 2 \pmod{27^2} \\ &\equiv 241 \pmod{27^2}. \end{aligned}$$

Now $241 = 8 \cdot 27 + 25$, and therefore it corresponds to the ordered pair IZ in the ciphertext. (Here, I corresponds to 8 and Z corresponds to 25.) Similarly, "LI," "KE" and "(space)I" and "T(space)" correspond to SH, NS, YH and ZZ, respectively. Thus, the ciphertext that corresponds to the plaintext "I LIKE IT" is "IZSHNSYHZZ." To get back the plaintext, we apply the inverse transformation (6.4). As the numerical equivalent of "IZ" is 241, relation (6.4) gives $P \equiv 547\,(241 - 2) \equiv 243 \pmod{27^2}$ and this, as seen earlier, corresponds to "I (space)." Similarly, the other pairs can be deciphered in order.
An equation of the form $C = aP + b$ is known as an affine transformation. Hence such cryptosystems are called *affine cryptosystems*.

In the Ceasar cryptosystem, given by the transformation $f(a) \equiv a + 3$ (mod 26), 3 is known as the *key* of the transformation. In the affine transformation, given by Equation 6.3, there are two keys, namely, a and b.

6.2.3 Private key cryptosystems

In the Caesar cryptosystem and the affine cryptosystem, the keys are known to the sender and the receiver in advance. That is to say that whatever information the sender has with regard to his encryption, it is shared by the receiver. For this reason, these cryptosystems are called private key cryptosystems.

6.2.4 Hacking an affine cryptosystem

Suppose an intruder I (that is a person other than the sender A and the receiver B) who has no knowledge of the private keys wants to hack the message, that is, decipher the message stealthily. We may suppose that the intruder knows the type of cryptosystem used by A and B including the unit length of the system, though not the keys. Such an information may get leaked out over a passage of time or may be obtained even by spying. How does I go about hacking? He does it by a method known as *frequency analysis*.

Assume for a moment that the message units are of length 1. Look at a long string of the ciphertext and find out the most-repeated character, the next most-repeated character and so on. Suppose, for the sake of precision, they are U, V, X, ... Now in the English language, the most common characters of the alphabet of 27 letters consisting of the English characters A to Z and "space" are known to be "space" and E in order. Then, "space" and E of the plaintext correspond to U and V of the ciphertext, respectively. If the cryptosystem used is the affine system given by the equation (see Table 6.1).

$$\begin{aligned} C &= aP + b \pmod{27}, \\ \text{we have} \quad 20 &= a \cdot 26 + b \pmod{27}, \\ \text{and} \quad 21 &= a \cdot 4 + b \pmod{27}. \end{aligned}$$

Subtraction yields

$$22a \equiv -1 \pmod{27} \tag{6.5}$$

As $(22, 27) = 1$, (6.5) has a unique solution, namely, $a = 11$. This gives $b = 21 - 4a = 21 - 44 = -23 = 4 \pmod{27}$. The cipher has thus been hacked. □

Suppose now the cryptosystem is based on an affine transformation $C = aP + b$ with unit length 2. If the same alphabet consisting of 27 characters of this section (namely, A to Z and space) is used, each unit corresponds to a unique non-negative integer less that 27^2. Suppose the frequency analysis of the ciphertext reveals that the most commonly occurring ordered pairs are "CA" and "DX" in their decreasing orders of their frequencies. The decryption transformation is of the form

$$P \equiv a'C + b' \pmod{27^2} \tag{6.6}$$

Here, a and b are the enciphering keys and a', b' are the deciphering keys. Now it is known that in the English language, the most frequently occurring order pairs, in their decreasing orders of their frequencies, are "E(space)" and "S(space)." Symbolically,

$$\begin{aligned} \text{"E(space)"} &\longrightarrow \text{CA}, \quad \text{and} \\ \text{"S(space)"} &\longrightarrow \text{DX}. \end{aligned}$$

Writing these in terms of their numerical equivalents, we get

$$\begin{aligned}(4 \times 27) + 26 = 134 &\longrightarrow (2 \times 27) + 0 = 54, \quad \text{and} \\ (18 \times 27) + 26 = 512 &\longrightarrow (3 \times 27) + 23 = 104.\end{aligned} \tag{6.7}$$

These, when substituted in Equation 6.5, give the congruences:

$$\begin{aligned}134 &\equiv 54a' + b' \pmod{729}, \quad \text{and} \\ 512 &\equiv 104a' + b' \pmod{729}.\end{aligned}$$

Note that $729 = 27^2$. Subtraction gives

$$50a' \equiv 378 \pmod{729}. \tag{6.8}$$

As $(50, 729) = 1$, this congruence has a unique solution by the Extended Euclidean Algorithm. In fact, $a' = 270$ as Equation 6.8 implies that $50a' \equiv 3^3 \times 14 \pmod{3^6} \Rightarrow a' = 3^3c$, where $50c \equiv 14 (\text{mod } 3^3) \Rightarrow c = 10$. This gives that $b' \equiv 134 - 54a' \pmod{27^2} \Rightarrow b' \equiv 134 (\text{mod } 27^2)$, as $27^2 | 54a' \Rightarrow b' = 4 \times 27 + 26 \Rightarrow b' =$ ESpace. Thus, the deciphering keys a' and b' have been determined and the cryptosystem has been hacked.

In our case, the $gcd(50, 729)$ happened to be 1 and hence we had no problem in determining the deciphering keys. If not, we have to try all the possible solutions for a' and take the plaintext that is meaningful. Instead, we can also continue with our frequency analysis and compare the next most repeated ordered pairs in the plaintext and ciphertext and get a third congruence and try for a solution in conjunction with one or both of the earlier congruences. If these also fail, we may have to adopt ad hoc techniques to determine a' and b'.

6.3 Encryption Using Matrices

Assume once again that the message units are ordered pairs in the same alphabet of size 27 of Section 6.2. We can use 2 by 2 matrices over the ring Z_{27} to set up a private key cryptosystem in this case. In fact if A is any 2 by 2 matrix with entries from Z_{27}, and (X, Y) is any plaintext unit, we encipher it as $B = A \begin{bmatrix} X \\ Y \end{bmatrix}$, where B is again a 2 by 1 matrix and therefore a ciphertext unit of length 2. If $B = \begin{bmatrix} X' \\ Y' \end{bmatrix}$, we have the equations

$$\begin{aligned}\begin{bmatrix} X' \\ Y' \end{bmatrix} &= A \begin{bmatrix} X \\ Y \end{bmatrix}, \\ \text{and} \quad \begin{bmatrix} X \\ Y \end{bmatrix} &= A^{-1} \begin{bmatrix} X' \\ Y' \end{bmatrix}.\end{aligned} \tag{6.9}$$

The first equation of (6.9) gives the encryption, while the second gives the decryption. Notice that A^{-1} must be taken in Z_{27}. For A^{-1} to exist, we must have $\gcd(\det A,\ 27) = 1$. If this were not the case, we may have to try once again ad hoc methods.

As an example, take $A = \begin{bmatrix} 2 & 1 \\ 4 & 3 \end{bmatrix}$. Then, $\det A = 2$, and $\gcd(\det A,\ 27) = \gcd(2,\ 27) = 1$. Hence 2^{-1} exists: in fact, $2^{-1} = 14 \in Z_{27}$. This gives (Recall $A^{-1} = (1/\det A)(adj\, A))$, where $(adj\, A)$ = adjugate of the matrix $A = (B_{ij})$. Here, $B_{ij} = (-1)^{i+j} A_{ji}$ where A_{ji} is the cofactor of a_{ji} in $A = (a_{ij})$ (see Chapter 3). Hence

$$A^{-1} = 14 \begin{bmatrix} 3 & -1 \\ -4 & 2 \end{bmatrix} = \begin{bmatrix} 42 & -14 \\ -56 & 28 \end{bmatrix} = \begin{bmatrix} 15 & 13 \\ 25 & 1 \end{bmatrix} \quad \text{over} \quad Z_{27}. \tag{6.10}$$

(The reader can verify that $AA^{-1} = A^{-1}A = I_2 \pmod{27}$.)
Suppose, for instance, we want to encipher "HEAD" using the above matrix transformation. We proceed as follows: "HE" corresponds to the vector $\begin{bmatrix} 7 \\ 4 \end{bmatrix}$, and "AD" to the vector $\begin{bmatrix} 0 \\ 3 \end{bmatrix}$. Hence the enciphering transformation gives the corresponding ciphertext as

$$\begin{aligned} A \begin{bmatrix} 7 & 0 \\ 4 & 3 \end{bmatrix} &= A \begin{bmatrix} 7 \\ 4 \end{bmatrix}, \quad A \begin{bmatrix} 0 \\ 3 \end{bmatrix} \pmod{27} \\ &= \begin{bmatrix} 2 & 1 \\ 4 & 3 \end{bmatrix} \begin{bmatrix} 7 \\ 4 \end{bmatrix}, \quad \begin{bmatrix} 2 & 1 \\ 4 & 3 \end{bmatrix} \begin{bmatrix} 0 \\ 3 \end{bmatrix} \pmod{27} \\ &= \begin{bmatrix} 18 & 3 \\ 40 & 9 \end{bmatrix} \pmod{27} \\ &= \begin{bmatrix} 18 & 3 \\ 13 & 9 \end{bmatrix} = \begin{bmatrix} \text{S} \\ \text{N} \end{bmatrix} \begin{bmatrix} \text{D} \\ \text{J} \end{bmatrix} \end{aligned}$$

Thus, the ciphertext of "HEAD" is "SNDJ." We can decipher "SNDJ" in exactly the same manner by taking A^{-1} in Z_{27}. This gives the plaintext

$$A^{-1} \begin{bmatrix} 18 \\ 13 \end{bmatrix}, \quad A^{-1} \begin{bmatrix} 3 \\ 9 \end{bmatrix}, \quad \text{where} \quad A^{-1} = \begin{bmatrix} 15 & 13 \\ 25 & 1 \end{bmatrix}, \quad \text{as given by (6.10)}$$

Therefore the plaintext is

$$\begin{bmatrix} 450 \\ 463 \end{bmatrix}, \quad \begin{bmatrix} 162 \\ 84 \end{bmatrix} \pmod{27} = \begin{bmatrix} 7 \\ 4 \end{bmatrix}, \quad \begin{bmatrix} 0 \\ 3 \end{bmatrix}$$

and this corresponds to "HEAD."

6.4 Exercises

1. Find the inverse of $A = \begin{bmatrix} 17 & 5 \\ 8 & 7 \end{bmatrix}$ in Z_{27}.

2. Find the inverse of $A = \begin{bmatrix} 12 & 3 \\ 5 & 17 \end{bmatrix}$ in Z_{29}.

3. Encipher the word "MATH" using the matrix A of Exercise 1 above as the enciphering matrix in the alphabet A to Z of size 26. Check your result by deciphering your ciphertext.

4. Solve the simultaneous congruences

$$\begin{aligned} x - y &= 4 \pmod{26} \\ 7x - 4y &= 10 \pmod{26}. \end{aligned}$$

5. Encipher the word "STRIKES" using the affine transformation $C \equiv 4P + 7 \pmod{27^2}$ acting on units of length 2 over an alphabet of size 27 consisting of A to Z and the exclamation mark ! with 0 to 25 and 26 as the corresponding numerals.

6. Suppose that we know that our adversary is using a 2 by 2 enciphering matrix with a 29-letter alphabet, where A to Z have numerical equivalents 0 to 25, (space)=26, ?=27 and !=28. We receive the message

   AMGQTZAFJVMHQV

   Suppose we know by some means that the last four letters of the plaintext are our adversary's signature "MIKE." Determine the full plaintext.

6.5 Other Private Key Cryptosystems

We now describe two other private key cryptosystems.

6.5.1 Vigenere cipher

In this cipher, the plaintext is in the English alphabet. The key consists of an ordered set of d letters for some fixed positive integer d. The plaintext is divided into message units of length d. The ciphertext is obtained by adding the key to each message unit using modulo 26 addition.

For example, let $d = 3$ and the key be XYZ. If the message is "ABANDON," the ciphertext is obtained by taking the numerical equivalence of the plaintext, namely,

$$[0][1][0]\ [13]\ [3]\ [14]\ [13],$$

and the addition modulo 26 of the numerical equivalence of "XYZ," namely, [23] [24] [25] of the key. This yields

$$\begin{aligned}&[23]\ [25]\ [25]\ [36]\ [27]\ [39]\ [36]\ (\text{mod } 26)\\&=[23]\ [25]\ [25]\ [10]\ [1]\ [13]\ [10]\\&=\text{X Z Z K B N K}\end{aligned}$$

as the ciphertext.

6.5.2 The one-time pad

This was introduced by Frank Miller in 1882 and reinvented by Gilbert S. Vernam in 1917. The alphabet Σ for the plaintext is the set of 26 English characters. If the message M is of length N, the key K is generated as a pseudo-random sequence of characters of Σ also of the same length N. The ciphertext is then obtained by the equation

$$C \equiv M + K \pmod{26}$$

Notwithstanding the fact that the key K is as long as the message M, the system has its own drawbacks.

i. There are only standard methods of generating pseudo-random sequences from out of Σ, and their number is not large.

ii. The long private key K must be communicated to the receiver in advance.

Despite these drawbacks, this cryptosystem was said to be used in some highest levels of communication such as the Washington-Moscow hotline.

There are several other private key cryptosystems. The interested reader can have a look at these in public domains.

6.6 Public Key Cryptography

All cryptosystems described so far are private key cryptosystems. "This means that someone who has enough information to encipher messages has enough information to decipher messages as well." As a result, in private key cryptography, any two persons in a group who want to communicate messages in a secret way must have exchanged keys in a safe way (for instance, through a trusted courier).

In 1976, the face of cryptography got altered radically with the invention of *public key cryptography* by Diffie and Hellman [36]. In this cryptosystem, the encryption can be done by anyone. But the decryption can be done only by the intended recipient who alone is in possession of the secret key.

At the heart of this cryptography is the concept of a "one-way function." Roughly speaking, a one-way function is a 1–1 function f which is such that whenever k is given, it is possible to compute $f(k)$ "rapidly" while it is "extremely difficult" to compute the inverse of f in a "reasonable" amount of time. There is no way of asserting that such and such a function is a one-way function since the computations depend on the technology of the day—the hardware and the software. So what passes for a one-way function today may fail to be a one-way function a few years later.

As an example of a one-way function, consider two large primes p and q each having at least 500 digits. Then, it is "easy" to compute their product $n = pq$. However, given n, there is no efficient factoring algorithm as on date that would give p and q in a reasonable amount of time. The same problem of forming the product pq with p and q having 100 digits had passed for a one-way function in the 1980s but is no longer so today.

6.6.1 Working of public key cryptosystems

A public key cryptosystem works in the following way: Each person A in a group has a public key P_A and a secret key S_A. The public keys are made public as in a telephone register with P_A given against the name A. A computes his own secret key S_A and keeps it within himself. The security of the system rests on the fact that no person of the group other than A or an intruder would be able to find out S_A. The keys P_A and S_A are chosen to be inverses of each other in that for any message M,

$$(P_A \circ S_A)\, M = M = (S_A \circ P_A)\, M.$$

6.6.1.1 Transmission of messages

Suppose A wants to send a message M to B in a secured fashion. The public key of B is S_B which is known to everyone. A sends "$P_B \cdot M$" to B. Now, to decipher the message, B applies S_B to it and gets $S_B(P_B M) = (S_B \cdot P_B)M = M$. Note that none other than B can decipher the message sent by A since B alone is in possession of S_B. □

6.6.1.2 Digital signature

Suppose A wants to send some instruction to a bank (for instance, transfer an amount to Mr. C out of his account). If the intended message to the bank is M, A applies his secret key S_A to M and sends $S_A M$ to the bank. He also gives his name for identification. The bank applies A's public key P_A to it and gets the message $P_A(S_A M) = (P_A \circ S_A)M = M$. This procedure also authenticates A's digital signature. This is in fact the method adopted in credit cards. □

We now describe two public key cryptosystems. The first is RSA, after their inventors, Rivest, Shamir and Adleman. In fact, Diffie and Hellman, though they invented public key cryptography in 1976, did not give the procedure to implement it. Only Rivest, Shamir and Adleman did it in 1978, two years later.

6.6.2 RSA public key cryptosystem

Suppose there is a group of people who want to communicate among themselves secretly. In such a situation, RSA is the most commonly used public key cryptosystem. The length of the message units is fixed in advance as also the alphabet in which the cryptosystem is operated. If, for instance, the alphabet consists of the English characters and the unit length is k, then any message unit is represented by a number less than 26^k.

6.6.2.1 Description of RSA

We now describe RSA.

i. Each person A (traditionally called Alice) chooses two large distinct primes p and q and computes their product $n = pq$, where p and q are so chosen that $n > N$, where N is a very large positive integer.

ii. Each A chooses a small positive integer e, $1 < e < \phi(n)$, such that $(e,\ \phi(n)) = 1$, where the Euler function $\phi(n) = \phi(pq) = \phi(p)\phi(q) = (p-1)(q-1)$. ($e$ is odd as $\phi(n)$ is even).

iii. As $(e,\ \phi(n)) = 1$, by Extended Euclidean Algorithm [25], e has a multiplicative inverse d modulo $\phi(n)$, that is,

$$ed \equiv 1 \pmod{\phi(n)}.$$

iv. A (Alice) gives the ordered pair $(n,\ e)$ as her public key and keeps d as her private (secret) key.

v. Encryption $P(M)$ of the message unit M is done by

$$P(M) \equiv M^e \pmod{n}, \tag{6.11}$$

while decryption $S(M')$ of the cipher text unit M' is given by

$$S(M') \equiv {M'}^d \pmod{n}. \tag{6.12}$$

Thus, both P and S (of A) act on the ring Z_n. Before we establish the correctness of RSA, we observe that d (which is computed using the Extended Euclidean Algorithm) can be computed in $O(\log^3 n)$ time. Further powers of M^e and ${M'}^d$ modulo n in Equations 6.11 and 6.12 can also be computed in $O(\log^3 n)$ time [25]. Thus, all computations in RSA can be done in polynomial time.

Theorem 6.1 (Correctness of RSA). *Equations 6.11 and 6.12 are indeed inverse transformations.*

Proof. We have

$$S\left(P(M)\right) = \left(S(M^e)\right) \equiv M^{ed} \pmod{n}.$$

Hence it suffices to show that

$$M^{ed} \equiv M \pmod{n}.$$

Now, by the definition of d,

$$ed \equiv 1 \pmod{\phi(n)}.$$

But $\phi(n) = \phi(pq) = \phi(p)\phi(q) = (p-1)(q-1)$,
and therefore, $ed = 1 + k(p-1)(q-1)$ for some integer k. Hence

$$\begin{aligned} M^{ed} &= M^{1+k(p-1)(q-1)} \\ &= M \cdot M^{k(p-1)(q-1)}. \end{aligned}$$

By Fermat's Little Theorem (FLT), if $(M, p) = 1$,

$$M^{p-1} \equiv 1 \pmod{p}$$

and therefore,

$$M^{ed} = M \cdot M^{k(p-1)(q-1)} \equiv M \cdot (M^{(p-1)})^{k(q-1)} \equiv M \cdot (1)^{k(q-1)} \equiv M \pmod{p}.$$

If, however, $(M, p) \neq 1$, then (as p is a prime) $(M, p) = p$, and trivially (as p is a divisor of M)

$$M^{ed} \equiv M \pmod{p}.$$

Hence, in both the cases,

$$M^{ed} \equiv M \pmod{p}. \tag{6.13}$$

For a similar reason,

$$M^{ed} \equiv M \pmod{q}. \tag{6.14}$$

As p and q are distinct primes, the congruences (6.13) and (6.14) imply that

$$M^{ed} \equiv M \pmod{pq},$$

$$\text{so that } M^{ed} \equiv M \pmod{n}.$$

□

The above description shows that if Bob wants to send the message M to Alice, he will send it as $M^e \pmod{n}$ using the public key of Alice. To decipher the message, Alice will raise this number to the power d and get $M^{ed} \equiv M \pmod{n}$, the original message of Bob.

The security of RSA rests on the supposition that none other than Alice can determine the private key d of Alice. A person can compute d if he/she knows $\phi(n) = (p-1)(q-1) = n - (p+q) + 1$, that is to say, if he/she knows the sum $p + q$. For this, he should know the factors p and q of n. Thus, in essence, the security of RSA is based on the assumption that factoring a large number n that is a product of two distinct primes is "difficult." However, to quote Koblitz [30], "no one can say with certainty that breaking RSA requires factoring n. In fact, there is even some indirect evidence that breaking RSA cryptosystem might not be quite as hard as factoring n. RSA is the public key cryptosystem that has had by far the most commercial success. But, increasingly, it is being challenged by elliptic curve cryptography."

6.6.3 The ElGamal public key cryptosystem

We have seen that RSA is based on the premise that factoring a very large integer which is a product of two "large" primes p and q is "difficult" compared to forming their product pq. In other words, given p and q, finding their product is a one-way function. ElGamal public key cryptosystem uses a different one-way function, namely, a function that computes the power of an element of a large finite group G. In other words, given G, $g \in G$, $g \neq e$, and a positive integer a, ElGamal cryptosystem is based on the assumption that computation of $g^a = b \in G$ is "easy" while given $b \in G$ and $g \in G$, it is "difficult" to find the exponent a.

Definition 6.1. *Let G be a finite group and $b \in G$. If $y \in G$, then the discrete logarithm of y with respect to base b is any non-negative integer x less than $o(G)$, the order of G, such that $b^x = y$, and we write $\log_b y = x$.*

As per the definition, $\log_b y$ may or may not exist. However, if we take $G = F_q^*$, the group of non-zero elements of a finite field F_q of q elements and g, a generator of the cyclic group F_q^* (see [25]), then for any $y \in F_q^*$, the discrete logarithm $\log_g y$ exists.

Example 6.1

5 is a generator of F_{17}^*. In F_{17}^*, the discrete logarithm of 12 with respect to base 5 is 9. In symbols: $\log_5 12 = 9$. In fact, in F_{17}^*,

$$\langle 5 \rangle = \{5^1 = 5,\ 5^2 = 8, 5^3 = 6, 5^4 = 13, 5^5 = 14, 5^6 = 2, 5^7 = 10,\ 5^8 = -1,$$

$$5^9 = 12, 5^{10} = 9, 5^{11} = 11, 5^{12} = 4, 5^{13} = 3, 5^{14} = 15, 5^{15} = 7,\ 5^{16} = 1$$

This logarithm is called "discrete logarithm," as it is taken in a finite group.

6.6.4 Description of ElGamal system

The ElGamal system works in the following way: All the users in the system agree to work in an already chosen large finite field F_q. A generator g of F_q^* is fixed once and for all. Each message unit is then converted into a number of F_q. For instance, if the alphabet is the set of English characters and if each message unit is of length 3, then the message unit BCD will have the numerical equivalent $26^2 \cdot 1 + 26^1 \cdot 2 + 3 \pmod{q}$. It is clear that in order that these numerical equivalents of the message units are all distinct, q should be quite large. In our case, $q \geq 26^3$. Now each user A in the system randomly chooses an integer $a = a_A$, $0 < a < q-1$, and keeps it as his or her secret key. A declares $g^a \in F_q$ as his public key.

If B wants to send the message unit M to A, he chooses a random positive integer k, $k < q - 1$, and sends the ordered pair

$$(g^k,\ Mg^{ak}) \tag{6.15}$$

to A. Since B knows k, and since g^a is the public key of A, B can compute g^{ak}. How will A decipher B's message? She will first raise the first number of the pair given in Equation 6.15 to the power a and compute it in F_q^*. She will then divide the second number Mg^{ak} of the pair by g^{ak} and get M. A can do this as she has a knowledge of a. An intruder who gets to know the pair $(g^k,\ Mg^{ak})$ cannot find $a = \log_{g^k}(g^{ak}) \in F_q^*$, since the security of the system rests on the premise that finding discrete logarithm is “difficult,” that is, given h and h^a in F_q, there is no efficient algorithm to determine a.

There are other public key cryptosystems as well. The interested reader can refer to [30].

6.7 Primality Testing

6.7.1 Non-trivial square roots (mod n)

We have seen that the most commonly applied public key cryptosystem, namely, the RSA, is built up on very large prime numbers (numbers having, say, 500 digits and more). So there arises the natural question: Given a large positive integer, how do we know that it is a prime or not? A ‘primality test’ is a test that tells if a given number is a prime or not.

Let n be a prime, and a, a positive integer with $a^2 \equiv 1 \pmod{n}$. Then, a is called a square root mod n. This means that n divides $(a-1)(a+1)$, and so, $n|(a-1)$ or $n|(a+1)$; in other words, $a \equiv \pm 1 \pmod{n}$. Conversely, if $a \equiv \pm 1 \pmod{n}$, then $a^2 \equiv 1 \pmod{n}$. Hence a prime number has only the trivial square root 1 and -1 modulo n. However, the converse may not be true, that is, there exist composite numbers m having only trivial square roots modulo m; for instance, $m = 10$ is such a number. On the other hand, 11 is a non-trivial square root of the composite number 20 since $11 \not\equiv \pm 1 \pmod{20}$ while $11^2 \equiv 1 \pmod{20}$.

Consider the modular exponentiation algorithm which determines $a^c \pmod n$ in $O(\log^2 n \log c)$ time [25]. At any intermediate stage of the algorithm, the output i is squared, taken modulo n and then multiplied to a or 1 as the case may be. If the square $i^2 \pmod n$ of the output i is 1 modulo n, then already we have determined a non-trivial square root modulo n, namely, i. Therefore, we can immediately conclude that n is not a prime and therefore a composite number. This is one of the major steps in the Miller-Rabin Primality Testing Algorithm to be described below.

6.7.2 Prime Number Theorem

For a positive real number x, let $\pi(x)$ denote the number of primes less than or equal to x. The Prime Number Theorem states that $\pi(x)$ is asymptotic to $x/\log x$; in symbols, $\pi(x) \approx x/\log x$. Here, the logarithm is with respect to base e. Consequently, $\pi(n) \approx n/\log n$, or, equivalently, $\pi(n)/n \approx 1/\log n$. In other words, in order to find a 100-digit prime, one has to examine roughly $\log_e 10^{100} \approx 230$ randomly chosen 100-digit numbers for primality (Recall that any 100-digit number k satisfies the inequalities $10^{99} \leq k < 10^{100}$) (this figure may drop down by half if we omit even numbers), (cf. [37]).

6.7.3 Pseudo-primality testing

Fermat's Little Theorem (FLT) states that if n is prime, then for each a, $1 \leq a \leq n-1$,

$$a^{n-1} \equiv 1 \pmod n. \tag{6.16}$$

Note that for any given a, $a^{n-1} \pmod n$ can be computed in polynomial time using the repeated squaring method [25]. However, the converse of FLT is not true. This is because of the presence of Carmichael numbers. A Carmichael number is a composite number n satisfying (6.16) for each a prime to n. They are sparse but are infinitely many. The first few Carmichael numbers are 561, 1105, 1729 (the smallest 4-digit Ramanujan number).

Since we are interested in checking if a given large number n is prime or not, n is certainly odd and hence $(2, n) = 1$. Consequently, if $2^{n-1} \not\equiv 1 \pmod n$, we can conclude with certainty, in view of FLT, that n is composite. However, if $2^{n-1} \equiv 1 \pmod n$, n may be a prime or not. If n is not a prime but $2^{n-1} \equiv 1 \pmod n$, then n is called a pseudo-prime with respect to base b.

Definition 6.2. *n is called a pseudo-prime to base a, where $(a, n) = 1$, if*

i. *n is composite, and*

ii. $a^{n-1} \equiv 1 \pmod n$.

In this case, n is also called a base a pseudo-prime.

6.7.3.1 Base-2 Pseudo-prime test

Given an odd positive integer n, check if $2^{n-1} \not\equiv 1 \pmod{n}$. If yes, n is composite. If no, $2^{n-1} \equiv 1 \pmod{n}$ and n may be a prime.

But then there is a chance that n is not a prime. How often does this happen? For $n < 10000$, there are only 22 pseudo-primes to base 2. They are 341, 561, 645, 1105, ... Using better estimates due to Carl Pomerance (See [27]), we can conclude that the chance of a randomly chosen 50-digit (resp. 100-digit) number satisfies (6.16) but fails to be a prime is $< 10^{-6}$ (resp. $< 10^{-13}$).

More generally, if $(a, n) = 1$, $1 < a < n$, the pseudo-prime test with reference to base a checks if $a^{n-1} \not\equiv 1 \pmod{n}$. If true, a is composite; if not, a may be a prime.

6.7.4 Miller-Rabin Algorithm

The Miller-Rabin Algorithm is a *randomized algorithm* for primality testing. A randomized algorithm is an algorithm which during the execution of the algorithm makes at least one random choice. It is in contrast with the *deterministic algorithms* in which every step has a unique subsequent step. A *randomized polynomial* time algorithm is a randomized algorithm which runs in polynomial time (RP for short). We now give the precise definition of a randomized algorithm.

Definition 6.3 (RP algorithm). *Consider a decision problem, that is, a problem whose answer is either "yes" or "no." (For example, given an integer $n \geq 2$, we would like to know if n is prime or composite). A randomized polynomial time algorithm is a randomized algorithm which runs in polynomial time in the worst case such that for any input I, the probability that the answer is "yes" for the input I is $\geq 1/2$ and the probability that the answer is "no" is 0. That is, the algorithm errs in case of output "yes" and makes no error in case of ouput "no."*

A randomized algorithm for primality test can be devised subject to the following conditions:

1. The existence of a set S which contains a large number of "certificates" for the proof of primality of a given number p.

2. The set S can be sampled efficiently.

6.7.5 Horner's method to evaluate a polynomial

The Miller-Rabin method uses the Horner's method to evaluate a polynomial at a given point. It is a classical algorithm based on parenthesizing. To evaluate a polynomial of degree n, *Horner's method* uses only n additions and n multiplications unlike the *n*aive method which uses n additions but $2n - 1$ multiplications. This method is also called *synthetic division*. Let us illustrate this method by an example.

TABLE 6.2: Horner's method

```
t=p[n];
for(i = n − 1;i ≥ 0;−−i)
{   t=t*c; t=t+p[i];
}
printf("%d", t)
```

Example 6.2 Horner's method

Let us evaluate $p(x) = 3x^4 - 5x^3 + 6x^2 + 2x + 7$ at $x = 2$. First we factorize the polynomial $p(x)$ from right, that is, we write: $p(x) = (((3x - 5)x + 6)x + 2)x + 7$.

More generally, we write a polynomial $p(x) = \sum_{i=0}^{n} p_i x^i$ of degree n as

$$p(x) = (((...(p_n x + p_{n-1})x + p_{n-2})x + p_{n-3})x + \cdots)x + p_0$$

Note that there are $(n-1)$ opening parentheses "(" and $(n-1)$ closing parentheses ")" . Let us write a pseudo-code for Horner's method:

Input: A polynomial $p(x) = \sum_{i=0}^{n} p_i x^i$ of degree n and a constant c.

Output: The value of $p(c)$.

The algorithm is given in Table 6.2.

Clearly, in the loop of the algorithm of Table 6.2, there are n additions and n multiplications. The Miller-Rabin algorithm is based on the following result in number theory.

Theorem 6.2. *If there is a non-trivial square root of 1 with respect to modulo n, then n is a composite number. In notation, if the equation*

$$x^2 \equiv 1 \pmod{n}$$

has a solution other than $+1$ and -1, then n is a composite number.

For example, the equation $x^2 \equiv 1 \pmod 8$ has a solution $x = 3$ (other than ± 1) because $3^2 \equiv 1 \pmod 8$. Here, $n = 8$, a composite number. In the algorithm we use the method of converting an integer n into its binary representation: Just divide successively n by 2, keeping the remainders, until we arrive at the quotient 0. To get the binary representation of n, we now simply write the remainders obtained during the division process, in the reverse manner. For example, if $n = 25$, we get the successive remainders: 1,0,0,1,1. Writing these in reverse order, we get: $(11001)_2$ (the subscript says that the number is in base 2) as the binary representation of 25.

More generally, the binary representation of a non-negative integer b is written as

$$b = (b_p b_{p-1} b_{p-2} \dots b_2 b_1 b_0)_2$$

b_0 is the least significant digit and b_p is the most significant digit. We observe by Horner's method $b = b_p 2^p + b_{p-1} 2^{p-1} + \cdots + b_2 2^2 + b_1 2 + b_0$ which is also equal to (by Horner's method)

$$(((...(b_p 2 + b_{p-1})2 + b_{p-2})2 + b_{p-3})2 \cdots + b_1)2 + b_0.$$

We need one more algorithm to state Miller-Rabin algorithm: It is called the **modular-exponentiation algorithm**:

6.7.6 Modular exponentiation algorithm based on repeated squaring

Input: Three integers a, b, n where a and b are non-negative integers and n is a positive integers.

Output: $a^b \bmod n$.

Algorithm: The *naive strategy* of multiplying a by itself uses $b - 1$ multiplications which is an exponential algorithm because the number of bits to represent b in binary is $\lfloor \log_2 b \rfloor + 1$ and $b - 1 = 2^{log_2 b} - 1$ which is an exponential in $\log_2 b$.

Instead, the algorithm we study is a *polynomial* algorithm based on "repeated squaring." The algorithm is given in Table 6.3. It is based on the equation which uses the Horner's method:

$$a^b = a^{(((...(b_p 2 + b_{p-1})2 + b_{p-2})2 + b_{p-3})2 \cdots + b_1)2 + b_0}$$

where $b = (b_p b_{p-1} b_{p-2} \dots b_2 b_1 b_0)$ is the binary representation of b. The repeated multiplication by 2 in the power of the above equation for a^b translates into repeated squaring.

Complexity of the Modular exponentiation: The loop is executed exactly $p + 1$ times and $p = \lfloor \log_2 b \rfloor + 1$, which is polynomial in $\log_2 b$.

TABLE 6.3: Modular exponentiation

```
Write b = (b_p b_{p-1} b_{p-2} ... b_2 b_1 b_0)_2 in binary representation.
//initialization of v. At the end of algorithm v = a^b.
v=1;//empty product is defined as 1 in the same spirit as 0!=1 or 2^0 = 1.
for (i = p;i ≥ 0; --i)
{ v = v * v mod n;//square v
   if (b_i == 1)
         v = v * a mod n //v = a^{(b_p b_{p-1} ... b_i)_2}. b_i = 0 has no effect on v.
}
return v;
```

TABLE 6.4: Certificate of compositeness

```
int certificate(int a, int n)
{Express n − 1 = (b_p b_{p−1} b_{p−2} ... b_2 b_1 b_0)_2 in binary notation
v=1;
for (i = p; i ≥ 0;−−i)
    t=v; v = v * v mod n;//save in v square v
    if ((v == 1)&&(t! = 1)&&(t! = n − 1))// non-trivial square root of 1?
        return 1;//1=true
    if (b_i == 1) v = v * a mod n
}
if (v! = 1) return 1;//1=true because a^{n−1} mod n ≠ 1 by Fermat's theorem
return 0://0=false
}
```

TABLE 6.5: Miller-Rabin primality testing algorithm

```
int miller_rabin(int n, int m)
{ for (i = 1;i <= m;+ + i)
    { a=(rand()% (n − 1)+1;
    if (certificate(a,n)) return 1;//function call true. surely composite
return 0;//false. Not composite with probability >= 1 − 1/2^m
    }
}
```

We are now ready to present the Miller-Rabin algorithm. The algorithm uses a function called "certificate(a,n)." The function certificate(a,n) returns the value "true" if and only if a can be used to certify that n is a composite number, that is, we can use a to prove that n is composite. Table 6.4 gives the function "certificate(a,n)," where n is assumed to be an *odd* integer. For example, a divisor $a \neq 1, n$ of an integer n is a certificate of compositeness of n, because one can *indeed* verify if a divides n by the usual division. The Miller-Rabin algorithm is the same as the modular exponentiation, except that it checks if a non-trivial square root of 1 is obtained *after every squaring* and if so it stops with output "true," by Theorem 6.2. In the algorithm, we use Fermat's little theorem in the contrapositive form: If there is an integer $a < n$ such that $a^{n-1} \bmod n \neq 1$ then n is a composite number. Note that $n - 1 \equiv -1 \pmod{n}$.

We can now write the Miller-Rabin algorithm (See Table 6.5). The algorithm tests if n is a composite number. It generates randomly m integers a between 1 and $n - 1$ and tests if each random integer is a certificate for the compositeness of n. This algorithm is based on the following fact: By repeating a randomized algorithm several times, the error probability can be reduced arbitrarily. Of course, repetition increases the running time. Further the algorithm uses the following result:

Theorem 6.3. *For an odd composite number n, the number of certificates for the proof of compositeness of n is $\geq \frac{n-1}{2}$.*

6.8 The Agrawal-Kayal-Saxena (AKS) Primality Testing Algorithm

6.8.1 Introduction

The Miller-Rabin primality testing algorithm is a probabilistic algorithm that uses Fermat's Little Theorem. Another probabilistic algorithm is due to Solovay and Strassen. It uses the fact that if n is an odd prime, then $a^{(n-1)/2} \equiv (a/n) \pmod{n}$, where (a/n) stands for the Legendre symbol. The Miller-Rabin primality test is known to be the fastest randomized primality testing algorithm, to within constant factors. However, the question of determining a polynomial time algorithm [25] to test if a given number is prime or not remained unsolved until July 2002. In August 2002, Agrawal, Kayal and Saxena of the Indian Institute of Technology, Kanpur, India made a sensational revelation that they have found a polynomial time algorithm for primality testing which works in $\tilde{O}\left(\log^{10.5} n\right)$ time (where $\tilde{O}\left(f(n)\right)$ stands for $O\left(f(n) \cdot \text{polynomial in } \log f(n)\right)$). (Subsequently, this has been reduced to $\tilde{O}\left(\log^{7.5} n\right)$.) It is based on a generalization of Fermat's Little Theorem to polynomial rings over finite fields. Notably, the correctness proof of their algorithm requires only simple tools of algebra. In the following section, we present the details of the AKS algorithm.

6.8.2 The basis of AKS algorithm

The AKS algorithm is based on the following identity for prime numbers which is a generalization of Fermat's Little Theorem.

Lemma 6.1. *Let $a \in \mathbb{Z}$, $n \in \mathbb{N}$, $n \geq 2$, and $(a, n) = 1$. Then, n is prime if and only if*

$$(X + a)^n \equiv X^n + a \pmod{n}. \tag{6.17}$$

Proof. We have

$$(X + a)^n = X^n + \sum_{n=1}^{n-1} \binom{n}{i} X^{n-1} a^i + a^n.$$

If n is prime, each term $\binom{n}{i}$, $1 \leq i \leq n-1$, is divisible by n. Further, as $(a, n) = 1$, by Fermat's Little Theorem, $a^n \equiv a \pmod{n}$. This establishes (6.17) as we have $(X + a)^n \equiv X^n + a^n \pmod{n} \equiv X^n + a \pmod{n}$.

If n is composite, n has a prime factor $q < n$. Let $q^k || n$ (that is, $q^k | n$ but $q^{k+1} \nmid n$). Now consider the term $\binom{n}{q} X^{n-q} a^q$ in the expansion of $(X + a)^n$. We have

$$\binom{n}{q} = \frac{n(n-1)\cdots(n-q+1)}{1\cdot 2\cdots q}.$$

Then, $q^k \nmid \binom{n}{q}$. For if $q^k \left| \binom{n}{q} \right.$, (as $q^k || n$), $(n-1)\cdots(n-q+1)$ must be divisible by q, a contradiction (As q is a prime and $q|(n-1)(n-2)\ldots(n-q+1)$, q must divide at least one of the factors, which is impossible since each term of $(n-1)(n-2)\ldots(n-q+1)$ is of the form $n-(q-k), 1 \leq k \leq q-1$). Hence q^k, and therefore n does not divide the term $\binom{n}{q} X^{n-q} a^q$. This shows that $(X+a)^n - (X^n + a)$ is not identically zero over Z_n (note that if an integer divides a polynomial with integer coefficients, it must divide each coefficient of the polynomial). □

The above identity suggests a simple test for primality: Given input n, choose an a and test whether the congruence (6.17) is satisfied, However, this takes time $\Omega(n)$ because we need to evaluate n coefficients in the LHS of congruence 6.17 in the worst case. A simple way to reduce the number of coefficients is to evaluate both sides of Equation 6.17 modulo a polynomial of the form $X^r - 1$ for an appropriately chosen small r. In other words, test if the following equation is satisfied:

$$(X + a)^n = X^n + a \pmod{X^r - 1,\, n} \tag{6.18}$$

From Lemma 6.1, it is immediate that all primes n satisfy Equation 6.18 for all values of a and r. The problem now is that some composites n may also satisfy Equation 6.18 for a few values of a and r (and indeed they do). However, we can almost restore the characterization: we show that for an appropriately chosen r if Equation 6.18 is satisfied for several a's, then n must be a prime power. It turns out that the number of such a's and the appropriate r are both bounded by a polynomial in $\log n$, and this yields a deterministic polynomial time algorithm for testing primality.

6.8.3 Notation and preliminaries

F_p denotes the finite field with p elements, where p is prime. Recall that if p is prime and $h(x)$ is a polynomial of degree d irreducible over F_p, then $F_p[X]/(h(X))$ is a finite field of order p^d. We will use the notation, $f(X) = g(X) \pmod{h(X), n}$ to represent the equation $f(X) = g(X)$ in the ring $Z_n[X]/(h(X))$, that is, if the coefficients of $f(X)$, $g(X)$ and $h(X)$ are reduced modulo n, then $h(X)$ divides $f(X) - g(X)$.

As mentioned in Section 6.8.1, for any function $f(n)$ of n, $\tilde{O}(f(n))$ stands for $O\Big(f(n) \cdot \text{polynomial in } \log\big(f(n)\big)\Big)$. For example,

$$\begin{aligned}
\tilde{O}(\log^k n) &= O\Big(\log^k n \cdot \text{poly}\ \big(\log\big(\log^k n\big)\big)\Big) \\
&= O\Big(\log^k n \cdot \text{poly}\ \big(\log\big(\log n\big)^k\big)\Big) \\
&= O\Big(\log^k n \cdot \text{poly}\ \big(k \log\big(\log n\big)\big)\Big) \\
&= O\Big(\log^k n \cdot \text{poly}\ \big(\log\log n\big)\Big) \\
&= O\Big(\log^{k+\epsilon} n\Big) \quad \text{for any } \epsilon > 0.
\end{aligned}$$

All logarithms in this section are with respect to base 2.

Given $r \in \mathbb{N}$, $a \in \mathbb{Z}$ with $(a, r) = 1$, the order of a modulo r is the smallest number k such that $a^k \equiv 1 \pmod{r}$ (note that such a k exists by Euler's Theorem [6]). It is denoted $O_r(a)$. $\phi(r)$ is Euler's totient function. Since by Euler's theorem, $a^{\phi(r)} \equiv 1 \pmod{r}$, and since $a^{O_r(a)} \equiv 1 \pmod{r}$, by the definition of $O_r(a)$, we have $O_r(a) \mid \phi(r)$.

We need the following result of Nair (see [33]) on the least common multiple (lcm) of the first m natural numbers.

Lemma 6.2. *Let $LCM(m)$ denote the lcm of the first m natural numbers. Then, for $m \geq 9$, $\text{LCM}(m) \geq 2^m$.*

6.8.4 The AKS algorithm

Input, integer $n > 1$

1 If ($n = a^b$ for $a \in \mathbb{N}$ and $b > 1$), output COMPOSITE
2 Find the smallest r such that $O_r(n) > 4 \log^2 n$
3 If $1 < \gcd(a, n) < n$ for some $a \leq r$, output COMPOSITE
4 If $n \leq r$, output PRIME
5 For $a = 1$ to $\lfloor 2\sqrt{\phi(r)} \cdot \log n \rfloor$, do if $(X + a)^n \neq X^n + a \pmod{X^r - 1, n}$, output COMPOSITE
6 output PRIME

Theorem 6.4. *The AKS algorithm returns PRIME iff n is prime.*

We now prove Theorem 6.4 through a sequence of lemmas.

Lemma 6.3. *If n is prime, then the AKS algorithm returns PRIME.*

Proof. If n is prime, we have to show that AKS will not return COMPOSITE in steps 1, 3 and 5. Certainly, the algorithm will not return COMPOSITE in step 1 (as no prime n is expressable as $a^b, b > 1$). Also, if n is prime, there exists no a such that $1 < \gcd(a, n) < n$, so that the algorithm will not return COMPOSITE in step 3. By Lemma 6.1, the 'For loop' in step 5 cannot return

COMPOSITE. Hence the algorithm will identify n as PRIME either in step 4 or in step 6. □

We now consider the steps when the algorithm returns PRIME, namely, steps 4 and 6. Suppose the algorithm returns PRIME in step 4. Then, n must be prime. If n were composite, $n = n_1 n_2$, where $1 < n_1, n_2 < n$. Then, as $n \leq r$, if we take $a = n_1$, we have $a \leq r$. So in step 3, we would have had $1 < (a, n) = a < n$, $a \leq r$. Hence the algorithm would have output COMPOSITE in step 3 itself. Thus, we are left out with only one case, namely, the case if the algorithm returns PRIME in step 6. *For the purpose of subsequent analysis, we assume this to be the case.*

The algorithm has two main steps (namely, 2 and 5). Step 2 finds an appropriate r and step 5 verifies Equation 6.18 for a number of a's. We first bound the magnitude of r.

Lemma 6.4. *There exists an* $r \leq 16 \log^5 n + 1$ *such that* $O_r(n) > 4 \log^2 n$.

Proof. Let $r_1, \ldots, r_t$ be all the numbers such that $O_{r_i}(n) \leq 4 \log^2 n$ for each i and therefore r_i divides $\alpha_i = (n^{O_{r_i}(n)} - 1)$ for each i (recall that if $O_{r_i}(n) = k_i$, then $n^{k_i} - 1$ is divisible by r_i). Now for each i, α_i divides the product

$$P = \prod_{i=1}^{\lfloor 4 \log^2 n \rfloor} (n^i - 1)$$

We now use the fact that $\prod_{i=1}^{t} (n^i - 1) < n^{t^2}$, the proof of which follows readily by induction on t. Hence,

$$P < n^{16 \log^4 n} = (2^{\log n})^{16 \log^4 n} = 2^{16 \log^5 n}.$$

As r_i divides α_i and α_i divides P for each i, $1 \leq i \leq t$, the lcm of the r_i's also divides P. Hence (lcm of the r_i's)$< 2^{16 \log^5 n}$. However, by Lemma 6.2,

$$\text{lcm } \left\{1, 2, \ldots, \lceil 16 \log^5 n \rceil\right\} \geq 2^{\lceil 16\ 2^{\log^5 n} \rceil}.$$

Hence there must exist a number r in $\left\{1, 2, \ldots, \lceil 16 \log^5 n \rceil\right\}$, that is, $r \leq 16 \log^5 n + 1$ such that $O_r(n) > 4 \log^2 n$. □

Let p be a prime divisor of n. We must have $p > r$. For, if $p \leq r$, (then as $p < n$), n would have been declared COMPOSITE in step 3, while if $p = n \leq r$, n would have been declared PRIME in step 4. This forces that $(n, r) = 1$. Otherwise, there exists a prime divisor p of n and r, and hence $p \leq r$, a contradiction as seen above. Hence (p, r) is also equal to 1. *We fix* p *and* r *for the remainder of this section.* Also, let $l = \lfloor 2\sqrt{\phi(r)} \log n \rfloor$.

Step 5 of the algorithm verifies l equations. Since the algorithm does not output COMPOSITE in this step (recall that we are now examining step 6), we have

$$(X+a)^n = X^n + a \pmod{X^r - 1,\, n}$$

for every a, $1 \le a \le l$. This implies that

$$(X+a)^n = X^n + a \pmod{X^r - 1,\, p} \tag{6.19}$$

for every a, $1 \le a \le l$. By Lemma 6.1, we have

$$(X+a)^p = X^p + a \pmod{X^r - 1,\, p} \tag{6.20}$$

for $1 \le a \le l$. Comparing Equation 6.19 with Equation 6.20, we notice that n behaves like prime p. We give a name to this property:

Definition 6.4. *For polynomial $f(X)$ and number $m \in \mathbb{N}$, m is said to be introspective for $f(X)$ if*

$$[f(X)]^m = f(X^m) \pmod{X^r - 1,\, p}$$

It is clear from Equations 6.19 and 6.20 that both n and p are introspective for $X + a$, $1 \le a \le l$. Our next lemma shows that introspective numbers are closed under multiplication.

Lemma 6.5. *If m and m' are introspective numbers for $f(X)$, then so is mm'.*

Proof. Since m is introspective for $f(X)$, we have

$$f(X^m) = (f(X))^m \pmod{X^r - 1,\, p}$$

$$\text{and hence} \quad [f(X^m)]^{m'} = f(X)^{mm'} \pmod{X^r - 1,\, p}. \tag{6.21}$$

Also, since m' is introspective for $f(X)$, we have

$$f\left(X^{m'}\right) = f(X)^{m'} \pmod{X^r - 1,\, p}.$$

Replacing X by X^m in the last equation, we get

$$f\left(X^{mm'}\right) = f\left(X^m\right)^{m'} \pmod{X^{mr} - 1,\, p}.$$

$$\text{and hence} \quad f\left(X^{mm'}\right) \equiv f\left(X^m\right)^{m'} \pmod{X^p - 1,\, p} \tag{6.22}$$

(since $X^r - 1$ divides $X^{mr} - 1$). Consequently from Equations 6.21 and 6.22

$$(f(X))^{mm'} = f\left(X^{mm'}\right) \pmod{X^r - 1,\, p}.$$

Thus, mm' is introspective for $f(X)$. □

Next we show that for a given number m, the set of polynomials for which m is introspective is closed under multiplication.

Lemma 6.6. *If m is introspective for both $f(X)$ and $g(X)$, then it is also introspective for the product $f(X)g(X)$.*

Proof. The proof follows from the equation:

$$[f(X) \cdot g(X)]^m = [f(X)]^m[g(X)]^m = f(X^m)g(X^m) \pmod{X^r - 1, p}.$$

□

Equations 6.19 and 6.20 together imply that both n and p are introspective for $(X + a)$. Hence by Lemmas 6.5 and 6.6, every number in the set $I = \{n^i p^j : i, j \geq 0\}$ is introspective for every polynomial in the set $P = \left\{\prod_{a=1}^{l}(X + a)^{e_a} : e_a \geq 0\right\}$. We now define two groups based on the sets I and P that will play a crucial role in the proof.

The first group consists of the set G of all residues of numbers in I modulo r. Since both n and p are prime to r, so is any number in I. Hence $G \subset Z_r^*$, the multiplicative group of residues modulo r that are relatively prime to r. It is easy to check that G is a group. The only thing that requires verification is that $n^i p^j$ has a multiplicative inverse in G. Since $n^{O_r(n)} \equiv 1 \pmod{r}$, there exists i', $0 \leq i' < O_r(n)$ such that $n^i = n^{i'}$. Hence inverse of n^i $(= n^{i'})$ is $n^{(O_r(n)-i')}$. A similar argument applies for p as $p^{O_r(n)} = 1$. Let $|G| =$ the order of the group $G = t$ (say). As G is generated by n and p modulo r and since $O_r(n) > 4\log^2 n$, $t > 4\log^2 n$. (Recall that all our logarithms are w.r.t. base 2.)

To define the second group, we need some basic facts about cyclotomic polynomials over finite fields. Let $Q_r(X)$ be the r-th cyclotomic polynomial over the field F_p [28]. Then, $Q_r(X)$ divides $X^r - 1$ and factors into irreducible factors of the same degree $d = O_r(p)$. Let $h(X)$ be one such irreducible factor of degree d. Then, $F = F_p[X]/(h(X))$ is a field. The second group that we want to consider is the group generated by $X + 1, X + 2, \ldots, X + l$ in the multiplicative group F^* of non-zero elements of the field F. Hence it consists of simply the residues of polynomials in P modulo $h(X)$ and p. Denote this group by $\mathcal{G}$.

We claim that the order of $\mathcal{G}$ is exponential in either $t = |G|$ or l.

Lemma 6.7. $|\mathcal{G}| \geq \min\{2^t - 1, 2^l\}$.

Proof. First note that $h(X) \mid Q_r(X)$ and $Q_r(X) \mid (X^r - 1)$. Hence X may be taken as a primitive r-th root of unity in $F = F_p[X]/(h(X))$.

We claim:

(*) if $f(X)$ and $g(X)$ are polynomials of degree less than t and if $f(X) \neq g(X)$ in P, then their images in F (got by reducing the coefficients modulo p and then taking modulo $h(X)$) are distinct.

To see this, assume that $f(X) = g(X)$ in the field F (that is, the images of $f(X)$ and $g(X)$ in the field F are the same). Let $m \in I$. Recall that every number of I is introspective with respect to every polynomial in P. Hence m is introspective with respect to both $f(X)$ and $g(X)$. This means that

$$\begin{aligned} & f(X)^m = f(X^m) \pmod{X^r - 1,\, p}, \\ \text{and} \quad & g(X)^m = g(X^m) \pmod{X^r - 1,\, p}. \\ \text{Consequently,} \quad & f(X^m) = g(X^m) \pmod{X^r - 1,\, p}, \end{aligned}$$

and since $h(X) \mid (X^r - 1)$,

$$f(X^m) = g(X^m) \pmod{h(X),\, p}.$$

In other words $f(X^m) = g(X^m)$ in F, and therefore X^m is a root of the polynomial $Q(Y) = f(Y) - g(Y)$ for each $m \in G$. As X is a primitive r-th root of unity and for each $m \in G$, $(m, r) = 1$, X^m is also a primitive r-th root of unity. Since each $m \in G$ is reduced modulo r, the powers X^m, $m \in G$, are all distinct. Thus, there are at least $t = |G|$ primitive r-th roots of unity X^m, $m \in G$, and each one of them is a root of $Q(Y)$. But since f and g are of degree less than t, $Q(Y)$ has degree less than t. This contradiction shows that $f(X) \neq g(X)$ in F. This establishes (*).

We next observe that the numbers 1, 2, ..., t are all distinct in F_p. This is because

$$\begin{aligned} l &= \lfloor 2\sqrt{\phi(r)} \log n \rfloor < 2\sqrt{r} \log n \\ &< 2\sqrt{r}\frac{\sqrt{t}}{2} \quad (\text{as } t > 4\log^2 n) \\ &< r \quad (\text{as } t < r; \text{ recall that } G \text{ is a subgroup of } Z_r^*) \\ &< p \quad (\text{by assumption on } p). \end{aligned}$$

Hence $\{1, 2, \ldots, l\} \subset \{1, 2, \ldots, p-1\}$. This shows that the elements $X + 1$, $X + 2$, ..., $X + l$ are all distinct in $F_p[X]$ and therefore in $F_p[X]/h(X) = F$. If $t \leq l$, then all the possible products of the polynomials in the set $\{X + 1, X + 2, \ldots X + t\}$ except the one containing all the t of them are all distinct and each of them is of degree less than t. Their number is $2^t - 1$ and all of them belong to P. By (*), their images in F are all distinct, that is, $|\mathcal{G}| \geq 2^t - 1$. If $t > l$, then there exist at least 2^l such polynomials (namely, the product of all subsets of $\{X + 1, \ldots, X + l\}$). These products are all of degree at most l and hence of degree $< t$. Hence in this case, $|\mathcal{G}| \geq 2^l$. Thus, $|\mathcal{G}| \geq \min\{2^t - 1,\, 2^l\}$. □

Finally we show that if n is not a prime power, then $|\mathcal{G}|$ is bounded above by a function of t.

Lemma 6.8. *If n is not a prime power, $|\mathcal{G}| \leq (1/2)n^{2\sqrt{t}}$.*

Proof. Set $\hat{I} = \{n^i \cdot p^j : 0 \le i, j \le \lfloor t \rfloor\}$. If n is not a prime power (recall that $p|n$), the number of terms in $\hat{I} = (\lfloor t \rfloor + 1)^2 > t$. When reduced mod r, the elements of $\hat{I}$ give elements of G. But $|G| = t$. Hence there exist at least two distinct numbers in $\hat{I}$ which become equal when reduced modulo r. Let them be m_1, m_2 with $m_1 > m_2$. So we have (since r divides $(m_1 - m_2)$),

$$X^{m_1} = X^{m_2} \pmod{X^r - 1} \tag{6.23}$$

Let $f(X) \in P$. Then,

$$\begin{aligned}
[f(X)]^{m_1} &= f(X^{m_1}) \quad (\text{mod } X^r - 1,\ p) \\
&= f(X^{m_2}) \quad (\text{mod } X^r - 1,\ p) \quad \text{by (6.23)} \\
&= [f(X)]^{m_2} \quad (\text{mod } X^r - 1,\ p) \\
&= [f(X)]^{m_2} \quad (\text{mod } h(X),\ p)(\text{since} \\
&\qquad\qquad h(X) \mid (X^r - 1)).
\end{aligned}$$

$$\text{This implies that} \quad [f(X)]^{m_1} = [f(X)]^{m_2} \quad \text{in the field } F. \tag{6.24}$$

Now $f(X)$ when reduced modulo $(h(X), p)$ yields an element of $\mathcal{G}$. Thus, every polynomial of $\mathcal{G}$ is a root of the polynomial

$$Q_1(Y) = Y^{m_1} - Y^{m_2} \quad \text{over } F.$$

Thus, there are at least $|\mathcal{G}|$ distinct roots in F. Naturally, $|\mathcal{G}| \le$ degree of $Q_1(Y)$. Now the degree of $Q_1(Y)$

$$= m_1 \quad (\text{as } m_1 > m_2) \tag{6.25}$$

$$\le (n\,p)^{\lfloor \sqrt{t} \rfloor}, \text{the greatest number in } \hat{I} \tag{6.26}$$

$$< \left(\frac{n^2}{2}\right)^{\sqrt{t}} \left(p < \frac{n}{2}, \quad \text{since} \quad p \mid n, \quad \text{and} \quad p \ne n\right) \tag{6.27}$$

$$= \frac{n^{2\sqrt{t}}}{2^{\sqrt{t}}} < \frac{n^{2\sqrt{t}}}{2}\ . \tag{6.28}$$

Hence $|\mathcal{G}| \le m_1 < (n^{2\sqrt{t}}/2)$. □

Lemma 6.7 gives a lower bound for $|\mathcal{G}|$, while Lemma 6.8 gives an upper bound for $|\mathcal{G}|$. These bounds enable us to prove the correctness of the algorithm. Armed with these estimates on the size of $\mathcal{G}$, we are now ready to prove the correctness of the algorithm.

Lemma 6.9. *If the AKS algorithm returns PRIME, then n is prime.*

Proof. Suppose that the algorithm returns PRIME. Lemma 6.8 implies that for $t = |G|$ and $l = \lfloor 2\sqrt{\phi(r)} \log n \rfloor$,

$$
\begin{aligned}
|\mathcal{G}| &\geq \min\{2^t - 1, 2^l\} \\
&\geq \min\left\{2^t - 1, 2^{\log^2 \sqrt{\phi(r)}}\right\} \\
&\geq \min\left\{2^t - 1, \frac{1}{2} n^{2\sqrt{\phi(r)}}\right\} \\
&\geq \min\left\{2^t - 1, \frac{1}{2} n^{2\sqrt{t}}\right\} \quad \text{(since } t \text{ divides } \phi(r)\text{)} \\
&\geq \min\left\{2^{2\sqrt{t}\log n}, \frac{1}{2} n^{2\sqrt{t}}\right\} \quad \text{(since } t > (2\log n)^2\text{)} \\
&\geq \frac{1}{2} n^{2\sqrt{t}}.
\end{aligned}
$$

By Lemma 6.8, $|\mathcal{G}| < (1/2)n^{2\sqrt{t}}$ if n is not a power of p. Therefore, $n = p^k$ for some $k > 0$. If $k > 1$, then the algorithm would have returned COMPOSITE in step 1. Therefore, $n = p$, a prime. This completes the proof of Lemma 6.9 and hence of Theorem 6.4. □

It is straightforward to calculate the time complexity of the algorithm. In these calculations, we use the fact that addition, multiplication, and division operations between two m-bit numbers can be performed in time $O\ (m)$. Similarly, these operations on two degree d polynomials with coefficients at most m bits in size can be done in time $O\ (d \cdot m)$ steps.

Theorem 6.5. *The asymptotic time complexity of the algorithm is* $\widetilde{O}(\log^{10.5} n)$.

Proof. The first step of the algorithm takes asymptotic time $\widetilde{O}(\log^3 n)$.

In step 2, we find an r with $o_r(n) > 4\log^2 n$. This can be done by trying out successive values of r and testing if $n^k \neq 1 (\text{mod } r)$ for every $k \leq 4\log^2 n$. For a particular r, this will involve at most $O(\log^2 n)$ multiplications modulo r and so will take time $\widetilde{O}(\log^2 n \log r)$. By Lemma 6.4, we know that only $O(\log^5 n)$ different r's need to be tried. Thus, the total time complexity of step 2 is $\widetilde{O}(\log^7 n)$.

The third step involves computing gcd of r numbers. Each gcd computation takes time $O(\log n)$, and therefore, the time complexity of this step is $O(r \log n) = O(\log^6 n)$. The time complexity of step 4 is just $O(\log n)$.

In step 5, we need to verify $\lfloor 2\sqrt{\phi(r)} \log n \rfloor$ equations. Each equation requires $O(\log n)$ multiplications of degree r polynomials with coefficients of size $O(\log n)$. So each equation can be verified in time $\widetilde{O}(r \log^2 n)$ steps. Thus, the time complexity of step 5 is $\widetilde{O}(r\sqrt{\phi(r)} \log^3 n) = \widetilde{O}(r^{\frac{3}{2}} \log^3 n) = \widetilde{O}(\log^{10.5} n)$. This time complexity dominates all the rest and is therefore the time complexity of the algorithm. □

As mentioned earlier, certain improvements on the time complexity of the algorithm are available (see, for instance, [26,32]). Our presentation is based on the lecture given by Professor Manindra Agrawal at the Institute of Mathematical Sciences, Chennai on February 4, 2003 and reproduced in the Mathematics Newsletter of Vol. 13 (2003) published by the Ramanujan Mathematical Society.

We now present some illustrative examples. (The computations have been done in C++. Readers who require the details of the computation may write to any of the two authors):

Examples using AKS algorithm

1. Check whether the following numbers are prime or composite.

   i. 25
   Solution: Here, $n = 25$. According to step 1 of the algorithm, $n = a^b$ for $a = 5$ and $b = 2$. Hence, the given number is composite.

   ii. 34
   Solution: Here, $n = 34$.

   Step 1: Since $n \neq a^b\ \forall\ a, b \in \mathbb{N}$ and $b > 1$. Hence, step 1 does not give anything decisive.

   Step 2: Smallest value of r such that $O_r(n) > 4\log^2 n$ is 113.

   Step 3: For $a = 2 \leq r$, $\gcd(a, n) = 2$, and hence $1 < \gcd(a, n) < n$. It concludes that 34 is composite.

   iii. 23
   Solution: Here, $n = 23$.

   Step 1: Since $n \neq a^b\ \forall\ a, b \in \mathbb{N}$ and $b > 1$. Hence, step 1 does not give anything decisive.

   Step 2: Smallest value of r such that $O_r(n) > 4\log^2 n$ is 89.

   Step 3:
   $gcd(a, 23) = 1$ for all $a \leq 22$ and
   $gcd(a, 23)$ is either 1 or 23 for all $a \geq 23$. Hence, step 3 does not give anything decisive.

   Step 4: $n = 23$ and $r = 89$, so $n \leq r$. It concludes that 23 is prime.

   iv. 31
   Solution: Here, $n = 31$.

   Step 1: Since $n \neq a^b\ \forall\ a, b \in \mathbb{N}$ and $b > 1$. Hence, step 1 does not give anything decisive.

   Step 2: Smallest value of r such that $O_r(n) > 4\log^2 n$ is 107.

Step 3:
$gcd(a,31) = 1$ for all $a \leq 30$ and $\gcd(a,31)$ is either 1 or 31 for all $a \geq 31$. Hence, step 3 does not give anything decisive.

Step 4 : $n = 31$ and $r = 107$, so $n \leq r$. It concludes that 31 is prime.

v. 271
Solution: Here, $n = 271$.

Step 1: Since $n \neq a^b \ \forall \ a, b \in \mathbb{N}$ and $b > 1$. Hence, step 1 does not give anything decisive.

Step 2: Smallest value of r such that $O_r(n) > 4\log^2 n$ is 269.

Step 3: $\gcd(a, 271) = 1$ for all $a \leq 269$. Hence, step 3 does not give anything decisive.

Step 4: $n = 271$ and $r = 269$, so $n > r$. Hence, step 4 does not give anything decisive.

Step 5: $\phi(r) = 268$, and $\lfloor 2\sqrt{\phi_r} \cdot \log n \rfloor = 264$.
For all a, where $1 \leq a \leq 264$,
$(X + a)^{271} = X^{271} + a(\text{mod } X^{269} - 1, 271)$.

Step 6: The algorithm outputs that 271 is prime.

Appendix A: Answers to Chapter 1—Graph Algorithms I

Exercises 1.13

Exercise 2: We suppose the graph has been already represented by n linked lists $L[i]$ of nodes for $i = 1, 2, \ldots, n$ where we assume the declaration struct node {int v; struct node *succ;}; L[i] is the pointer to the list of all successors of the node i in the graph. Note that a node consists of two fields: v of type integer representing a vertex and succ is a pointer to a node in linked list. Here, n is the number of vertices. When we delete the vertex k, we not only delete the vertex k, but also the arcs going into the vertex k and leaving out of k. The removed vertex k will point to an artificial node whose "v" field is -1, which cannot be a vertex. We now write a fragment of a program in C. The reader is asked to write a complete program in C based on this fragment and execute on some examples.

```
/* remove the nodes in the list L[k]*/
struct node *t,*p;/* t,p pointers to node*/
t=L[k];
{while(t!=NULL){p=t; t=t->succ; free(p)}
t=(struct node *) malloc(sizeof *t);/*create a node pointed by t*/
t->v=-1; t->succ=NULL;L[k]=t;
/*scan all other lists to remove the node containing the vertex k*/
int i;
for(i = 1;i <= n;i++)
if (i!=k)
{t=L[i];while(t!=NULL)if (t->v==k) {p=t;t=t->succ; free(p)};else
{t=t->succ};
```

Complexity: $O(n + m)$

Exercise 4: Note that when we remove an arc (directed edge) (i, j), the vertices i and j remain in the graph. Removing the arc (i, j) means that we remove the node containing the field j from the liked list $L[i]$. We now have a fragment of a program in C. The reader is asked to write a complete program and execute on some examples.

```
struct node *t,*p;/* t,p are temporary pointers to node*/
t=L[i];
while(t!=NULL)
```

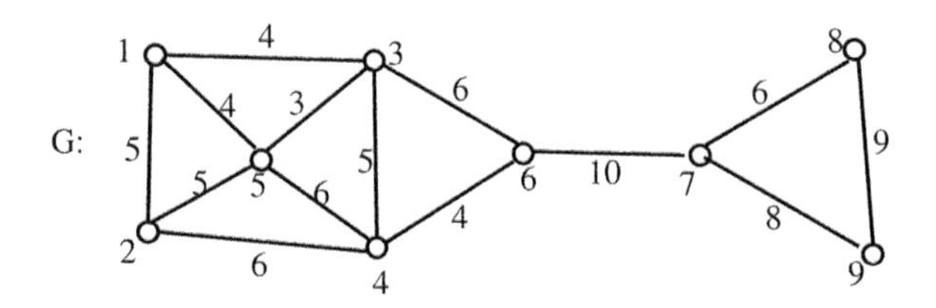

FIGURE A.1: A weighted simple graph.

```
if (t− > v==j){ p=t;t=t− >succ; free(p)};else {t=t− >succ};
```

Complexity: $O(n)$

Exercise 6: True. Suppose a spanning tree T of G does not contain a bridge $e = xy$ of G. Since, T is a spanning connected graph it contains the vertices x and y. Since the only elementary path between x and y in G is the edge xy and this edge is not in T. This means that in the tree T, the vertices x and y are not connected by any path, which is impossible.

Exercise 8: The reader is asked to draw the tree obtained by Table A.1.

TABLE A.1: Execution of Prim's algorithm on graph of Figure A.1

Iteration Number	$S \neq X$?	s	t	$S \leftarrow S \cup \{t\}$	$T \leftarrow T \cup \{st\}$
0 (Initial)	–	–	–	$\{1\}$	$\emptyset$
1	yes	1	3	$\{1,3\}$	$\{13\}$
2	yes	3	5	$\{1,3,5\}$	$\{13,35\}$
3	yes	1	2	$\{1,3,5,2\}$	$\{13,35,12\}$
4	yes	3	4	$\{1,3,5,2,4\}$	$\{13,35,12,34\}$
5	yes	4	6	$\{1,3,5,2,4,6\}$	$\{13,35,12,34,46\}$
6	yes	6	7	$\{1,3,5,2,4,6,7\}$	$\{13,35,12,34,46,67\}$
7	yes	7	8	$\{1,3,5,2,4,6,7,8\}$	$\{13,35,12,34,46,67\}$
8	yes	7	9	$\{1,3,5,2,4,6,7,8,9\}$	$\{13,35,12,34,46,67,79\}$
Exit the loop	no				

TABLE A.2: Execution of Dijkstra's algorithm on the graph of Figure A.2

Iteration Number	$S \neq X$	y	$S \leftarrow S \cup \{y\}$	D [13]	D [14]	D [15]	D [16]
0(Initial)	–	–	$\{1\}$	5	∞	∞	5
1	yes	2	$\{1,2\}$	5	12	∞	5
2	yes	5	$\{1,2,5\}$	5	11	15	5
3	yes	3	$\{1,2,5,3\}$	5	11	14	5
4	yes	4	$\{1,2,5,3,4\}$	5	11	14	5
Exit loop	no						

Exercise 10: The execution is given in Tables A.2 and A.3. $X = \{1, 2, 3, 4, 5\}$. The reader is asked to draw the arborescence using the dad array.

Exercise 12: Initialization:

$$M_0 = \begin{array}{c} \\ 1 \\ 2 \\ 3 \\ 4 \\ 5 \end{array} \begin{array}{c} \begin{array}{ccccc} 1 & 2 & 3 & 4 & 5 \end{array} \\ \begin{pmatrix} 0 & 5 & \infty & \infty & 5 \\ \infty & 0 & 7 & \infty & \infty \\ \infty & \infty & 0 & 3 & \infty \\ \infty & \infty & \infty & 0 & \infty \\ 6 & \infty & 6 & 10 & 0 \end{pmatrix} \end{array}. \quad M_1 = \begin{array}{c} \\ 1 \\ 2 \\ 3 \\ 4 \\ 5 \end{array} \begin{array}{c} \begin{array}{ccccc} 1 & 2 & 3 & 4 & 5 \end{array} \\ \begin{pmatrix} 0 & 5 & \infty & \infty & 5 \\ \infty & 0 & 7 & \infty & \infty \\ \infty & \infty & 0 & 3 & \infty \\ \infty & \infty & \infty & 0 & \infty \\ 6 & 11_1 & 6 & 10 & 0 \end{pmatrix} \end{array}.$$

$$M_2 = \begin{array}{c} \\ 1 \\ 2 \\ 3 \\ 4 \\ 5 \end{array} \begin{array}{c} \begin{array}{ccccc} 1 & 2 & 3 & 4 & 5 \end{array} \\ \begin{pmatrix} 0 & 5 & 12_2 & \infty & 5 \\ \infty & 0 & 7 & \infty & \infty \\ \infty & \infty & 0 & 3 & \infty \\ \infty & \infty & \infty & 0 & \infty \\ 6 & 11_1 & 6 & 10 & 0 \end{pmatrix} \end{array}. \quad M_3 = \begin{array}{c} \\ 1 \\ 2 \\ 3 \\ 4 \\ 5 \end{array} \begin{array}{c} \begin{array}{ccccc} 1 & 2 & 3 & 4 & 5 \end{array} \\ \begin{pmatrix} 0 & 5 & 12_2 & 15_3 & 5 \\ \infty & 0 & 7 & 10_3 & \infty \\ \infty & \infty & 0 & 3 & \infty \\ \infty & \infty & \infty & 0 & \infty \\ 6 & 11_1 & 6 & 9_3 & 0 \end{pmatrix} \end{array}.$$

$$M_4 = M_3$$

because $d^+(4) = 0$.

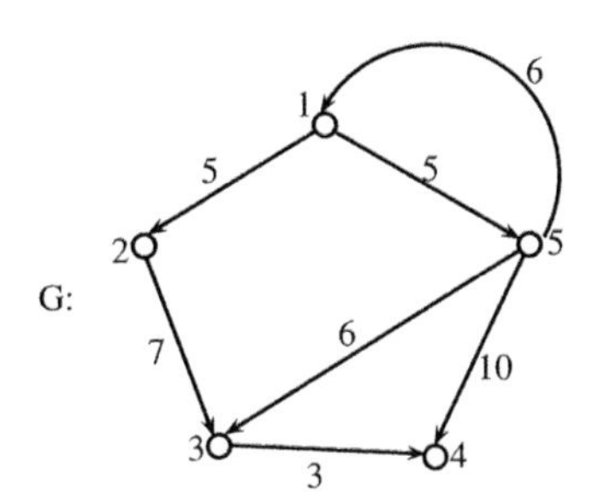

FIGURE A.2: A weighted directed graph.

TABLE A.3: Evaluation of dad array in Dijkstra's algorithm of Figure A.2

Dad [13]	Dad [14]	Dad [15]	Dad [16]
1	1	1	1
1	2	1	1
1	5	5	1
1	5	3	1
1	5	3	1

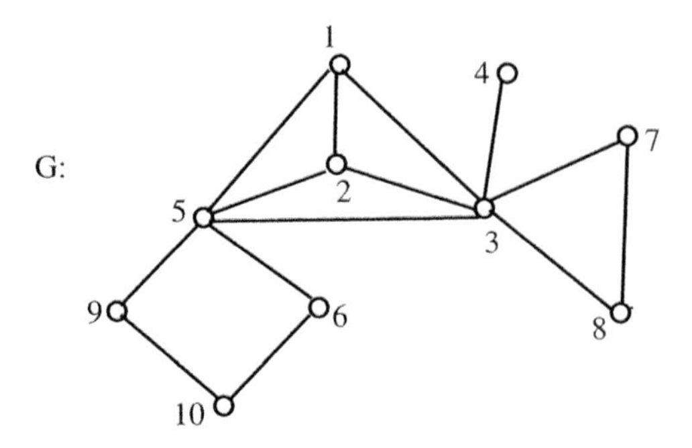

FIGURE A.3: A simple connected graph.

$$M_5 = \begin{array}{c} \\ 1 \\ 2 \\ 3 \\ 4 \\ 5 \end{array} \begin{array}{c} \begin{array}{ccccc} 1 & 2 & 3 & 4 & 5 \end{array} \\ \begin{pmatrix} 0 & 5 & 11_5 & 14_5 & 5 \\ \infty & 0 & 7 & 10_3 & \infty \\ \infty & \infty & 0 & 3 & \infty \\ \infty & \infty & \infty & 0 & \infty \\ 6 & 11_1 & 6 & 9_3 & 0 \end{pmatrix} \end{array}.$$

$$\text{INTER} = \begin{array}{c} \\ 1 \\ 2 \\ 3 \\ 4 \\ 5 \end{array} \begin{array}{c} \begin{array}{ccccc} 1 & 2 & 3 & 4 & 5 \end{array} \\ \begin{pmatrix} 0 & 0 & 5 & 5 & 0 \\ 0 & 0 & 0 & 3 & 0 \\ 0 & 0 & 0 & 0 & 0 \\ 0 & 0 & 0 & 0 & 0 \\ 0 & 1 & 0 & 3 & 0 \end{pmatrix} \end{array}.$$

Call to interpath(1,4) prints the intermediate vertices between 1 and 4 in a shortest path from 1 to 4. The output is 5, 3.

Exercise 14: The following Table A.4 gives us the dfsn of different vertices of the graph of Figure A.4.

TABLE A.4: Dfsn table of graph of Figure A.4

	1	2	3	4	5	6	7	8	9	10
dfsn	1	2	3	4	5	6	9	10	8	7

The following Table A.5 gives the LOW function of different vertices.

TABLE A.5: LOW function table of the graph of Figure A.4

	1	2	3	4	5	6	7	8	9	10
low	1	1	1	4	1	5	3	3	5	5

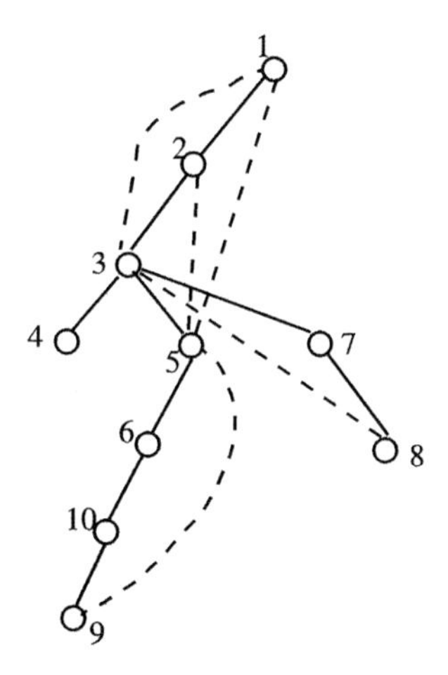

FIGURE A.4: A dfs drawing of the graph of Figure A.3.

The evolution of stack of edges (the rows from left to right):
We push the edges, 12,23,31,34. Stack = (12,23,31,34). Now at the vertex 4, low[4] $\geq$ $dfsn[3]$. Hence, we pop the edge 34 from the stack, which is the first biconnected component. Now the stack grows to: stack= $(12, 23, 31, 35, 51, 52, 56, (6, 10), (10, 9), 95)$. At the vertex 6, low[6] $\geq dfsn[5]$. Hence, we pop the edges from the stack till and including edge 56, that is, 95,(10,9),(6,10),56 which is the second biconnected component emitted. The new stack after popping: stack = (12,23,31,35,51,52). The stack grows as: stack = (12,23,31,35,51,52,37,78,83). At vertex 7, we have low[7] $\geq dfsn[3]$. Hence, we pop till the edge 37, that is, 83,78,37 which is the third biconnected component. The stack shrinks to stack = (12,23,31,35,51,52). At vertex 2, we have low[2] $\geq dfsn[1]$. We pop the edges till the edge 12, that is, 12,23,31,35,51,52 which is the fourth and last biconnected component. Note that the stack is empty.

Exercise 16: The reader is first asked to draw the graph with direction of the arcs as given in the exercise. As usual, we process the vertices in increasing order and we suppose the vertices are listed in increasing order in each list $L[i]$ for $1 = 1, 2, \ldots, 10$. The following Table A.6 gives the dfsn of different vertices.

TABLE A.6: dfsn table of graph of Figure 1.38

	1	2	3	4	5	6	7	8	9	10
dfsn	1	2	3	4	5	6	9	10	8	7

The dfs drawing of the graph is given in Figure A.5.
The following Table A.7 gives the function LOWLINK.

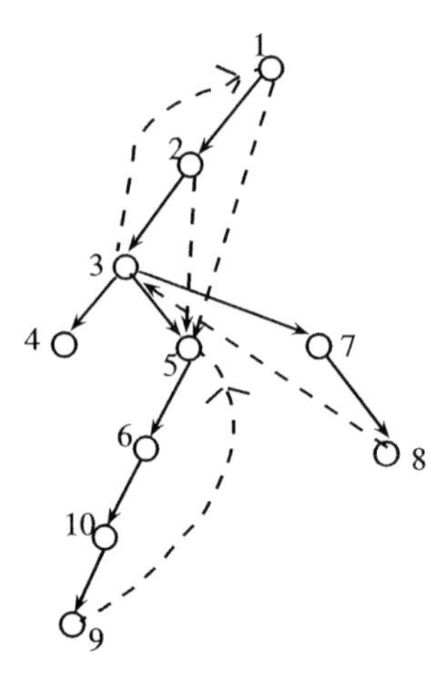

FIGURE A.5: dfs: strong components algorithm.

TABLE A.7: LOWLINK function table of graph of Figure 1.38

	1	2	3	4	5	6	7	8	9	10
LOWLINK	1	1	1	4	5	5	3	3	5	5

We use a stack of vertices. The stack grows from left to right. Stack=$(1, 2, 3, 4)$. At vertex 4, we find LOWLINK$[4]$ $= dfsn[4]$. Hence, we pop the vertex 4 which forms a strongly connected component consisting of a single vertex. Then, we push the vertices $5, 6, 10, 9$ onto the stack. Hence the stack is, stack=$(1, 2, 3, 5, 6, 10, 9)$. While climbing the tree, we find at vertex 5, LOWLINK$[5]$ $= dfsn[5]$. Hence we pop the stack till the vertex 5, that is, we output: 9,10,6,5 which is the second strongly connected component. The stack now shrinks to: stack=$(1, 2, 3)$. The vertices 7 and 8 are pushed onto the stack. Stack=$(1, 2, 3, 7, 8)$. We climb the tree till the vertex 1. Since, LOWLINK$[1]$ $= dfsn[1]$, we pop the stack till the vertex 1, that is, we output: 8,7,3,2,1 which is the third and last strongly connected component.

Exercise 18: After assigning the given orientations, we observe that the resulting directed graph is without circuits. Hence the topological sort is possible. As usual, we process the vertices in increasing order and in each linked list $L[i]$, with $i = 1, 2, \dots, 10$, the vertices are listed in increasing order. We perform dfs of the graph by drawing only the tree arcs. We obtain two arborescences in Figure A.6 (the reader is asked to perform the dfs):

Note that the second arborescence consists of only one vertex 4. We now write the vertices of the forest of arborescence in postfix order: 9,10,6,5,8,7,3,2,1,4. We now take the mirror image, that is, write the postfix order obtained from right to left: 4,1,2,3,7,8,5,6,10,9. This is the topological

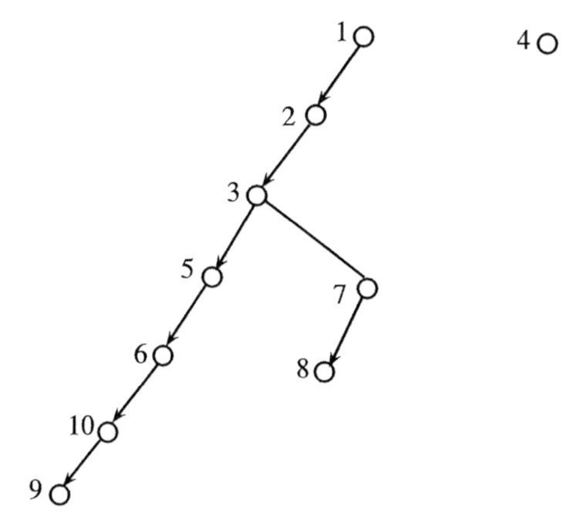

FIGURE A.6: dfs: topological sort.

order required. The reader is asked to draw the entire graph by aligning the vertices horizontally and verify that there are no arcs going from right to left.

Exercise 20: See the text.

Exercise 22: See the text.

Appendix B: Answers to Chapter 2—Graph Algorithms II

Exercises 2.5

Exercise 2: "Yes." Let the source vertex be 1. Perform a breadth-first search (bfs) from 1. This search partitions the vertex set $X = \{1, 2, \ldots, n\}$ as:

$$X = L_0(1) \cup L_1(1) \cdots \cup L_{e(1)}(1)$$

where $L_0(1) = \{1\}$, $L_i(1) = \{x | d(1, x) = i\}$, $e(1)$ is the eccentricity of the vertex 1. Here, $d(1, x)$ is the minimum length of a path from the vertex 1 to the vertex x. Let us recall that the length of a path is the number of edges in it. We use an array D of n reals.

Arrange the vertices of the graph in non-decreasing order of distances from the vertex 1. For example, write the source vertex 1 first, then the vertices of $L_1(1)$ (in any order), then the vertices of $L_2(1)$ (in any order), etc., till the vertices of $L_{e(1)}(1)$. $D[1] = 0$; //initialization for each vertex $x \neq 1$ in non-decreasing order from the vertex 1 to $D[x] = \min\{D[y] + c(y, x) | y \in L_{i-1}(1)\}$ $//c(y, x)$ is the cost associated with the edge yx.

Exercise 4: Perform the bfs from each vertex i, with $1 \leq i \leq n$. The largest integer k with $L_k(i) \neq emptyset$ is the eccentricity of the vertex i. (See the solution of Exercice 2.) The minimum of the eccentricities is the radius of the graph, and the maximum of the eccentricities is the diameter of the graph. Since we perform n bfs, and the cost of a single bfs is $O(\max(m, n))$, the complexity of the algorithm is $O(n \max(m, n))$ where n and m are the number of vertices and edges, respectively, of the graph.

Exercise 6: A bipartite graph is geodetic if and only if it is a tree.

Exercise 8: Let us first draw the Petersen graph (see Figure B.1). The execution of the algorithm is illustrated in Table B.1. We find an augmenting path visually. The reader is asked to draw the edges of the matching during the execution by a color, say, red, the other edges black. Hence, an augmenting path is an elementary path with alternating colors black and red, beginning with a black edge and ending with another black edge. Note that the origin and the terminus of such a path are both unsaturated by red edges. Note that

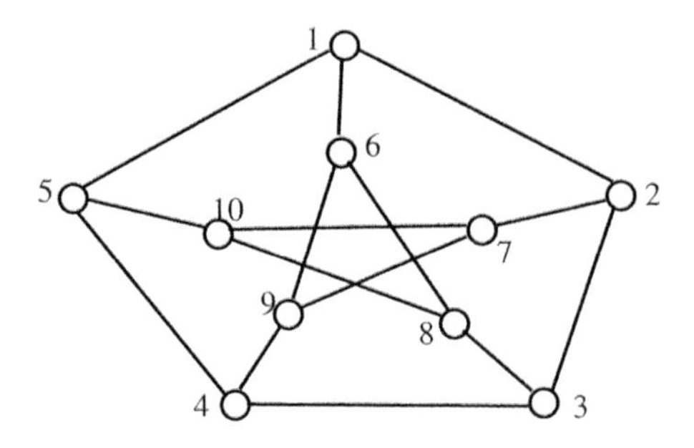

FIGURE B.1: the Petersen graph.

TABLE B.1: Execution of matching algorithm

Iteration number	**Is there an M augmenting path P?**	$M := (M \cup P) \setminus (M \cap P)$
0(initialization)	not applicable	$\{34\}$
1	"yes," $P = (2,3,4,5)$	$\{23,45\}$
2	"yes," $P = (7,2,3,4,5,1)$	$\{15,34,27\}$
3	"yes," $P = (9,7,2,1,5,10)$	$\{97,21,(5,10),34\}$
4	"yes," $P = (6,8)$	$\{97,21,(5,10),34,68\}$
	"no," quit the loop	

a black edge alone with its end vertices unsaturated is an augmenting path. The matching obtained at the 4th iteration is a perfect matching.

Exercise 10: M_1 and M_2 are two different perfect matchings of a graph G. The spanning subgraph whose edge set is $M_1 \delta M_2 = (M_1 \cup M_2) - (M_1 \cap M_2)$ has connected components either isolated vertices or an elementary cycle of *even* length whose edges are alternatively in M_1 and M_2. (The reader is asked to draw a graph with two different perfect matchings M_1 and M_2 and find the graph $(X, M_1 \Delta M_2)$).

Exercise 12: A 3-regular graph G has always an even number of vertices. (Because, the number of vertices of odd degree in a graph is always even.) Since the graph contains a Hamiltonian cycle C, and the number of vertices in C is even (note that C is a spanning connected subgraph of G with the degree of each vertex in C is 2), the *alternating* edges of C form two disjoint perfect matchings. Now by removing the edges of these two perfect matchings from the original graph G, we obtain another (third) disjoint perfect matching. Note that these three mutually disjoint perfect matchings exhaust all the edges of the graph.

Exercise 14: See an example in the text.

Exercise 16: See the reference K. R. Parthasarathy [11]

Appendix C: Answers to Chapter 3—Algebraic Structures I

Exercises 3.3.7

Exercise 2: The proof is by induction in n. The result is true for $n = 1$. Assume the result for n so that

$$M^n = \begin{pmatrix} \cos(n\alpha) & \sin(n\alpha) \\ -\sin(n\alpha) & \cos(n\alpha) \end{pmatrix}.$$

Then,

$$\begin{aligned} M^{n+1} = M^n \cdot M^1 &= \begin{pmatrix} \cos(n\alpha) & \sin(n\alpha) \\ -\sin(n\alpha) & \cos(n\alpha) \end{pmatrix} \begin{pmatrix} \cos(\alpha) & \sin(\alpha) \\ -\sin(\alpha) & \cos(\alpha) \end{pmatrix} \\ &= \begin{pmatrix} \cos(n\alpha)\cos\alpha - \sin(n\alpha)\sin\alpha & \cos(n\alpha)\sin(\alpha) + \sin(n\alpha)\cos\alpha \\ -\sin(n\alpha)\cos\alpha - \cos(n\alpha)\sin\alpha & -\sin(n\alpha)\sin\alpha + \cos(n\alpha)\cos\alpha \end{pmatrix} \\ &= \begin{pmatrix} \cos(n+1)\alpha & \sin(n+1)\alpha \\ -\sin(n+1)\alpha & \cos(n+1)\alpha \end{pmatrix}. \end{aligned}$$

Exercise 4: If

$$A = \begin{pmatrix} 1 & 3 \\ -2 & 2 \end{pmatrix}, \quad A^2 = \begin{pmatrix} 1 & 3 \\ -2 & 2 \end{pmatrix} \begin{pmatrix} 1 & 3 \\ -2 & 2 \end{pmatrix} = \begin{pmatrix} -5 & 9 \\ -6 & -2 \end{pmatrix},$$

and hence

$$A^2 - 3A + 8I = \begin{pmatrix} -5 & 9 \\ -6 & -2 \end{pmatrix} - 3\begin{pmatrix} 1 & 3 \\ -2 & 2 \end{pmatrix} + 8\begin{pmatrix} 1 & 0 \\ 0 & 1 \end{pmatrix} = \begin{pmatrix} 0 & 0 \\ 0 & 0 \end{pmatrix}.$$

Now $det(A) \neq 0$ and hence A^{-1} exists. Multiplying by A^{-1}, we get

$$A - 3I + 8A^{-1} = 0 \Rightarrow A^{-1} = \frac{-1}{8}(A - 3I) = \frac{-1}{8}\begin{pmatrix} -2 & 3 \\ -2 & -1 \end{pmatrix} = \begin{pmatrix} \frac{1}{4} & \frac{-3}{8} \\ \frac{1}{4} & \frac{1}{8} \end{pmatrix}.$$

Exercise 6:

(i) Take

$$A = \begin{pmatrix} a_{11} & a_{12} & \cdots & a_{1n} \\ \vdots & & & \\ a_{m1} & a_{m2} & \cdots & a_{mn} \end{pmatrix}, \quad B = \begin{pmatrix} b_{11} & b_{12} & \cdots & b_{1p} \\ \vdots & & & \\ b_{n1} & b_{n2} & \cdots & b_{np} \end{pmatrix}$$

so that product AB is defined. Now check that $(AB)^t = B^t A^t$.

(ii) Trivially $(AB)(B^{-1}A^{-1}) = A(BB^{-1})A^{-1} = AIA^{-1} = I$. Similarly, $(B^{-1})(A^{-1})(AB) = I$.

Exercise 8:

(i) Recall that if $A = (a_{ij}), A^* = (b_{ij})$, where $b_{rs} = \bar{a}_{sr}$. Here, $iA = (ia_{rs})$, (Recall $i = \sqrt{-1}$) and therefore $(iA)^* = (ia_{sr})^* = -(i\bar{a}_{sr}) = -i(\bar{a}_{sr}) = -iA^*$.

(ii) H is Hermitian $\Leftrightarrow H^* = H$. iH is skew-Hermitian $\Leftrightarrow (iH)^* = -iH \Leftrightarrow -iH^* = -iH \Leftrightarrow H^* = H$.

Exercise 10: Let A be a complex square matrix. Then, $A = ((A + A^*)/2) + ((A - A^*)/2)$. Here, $(A + A^*)/2$ is Hermitian since

$$\frac{(A + A^*)^*}{2} = \frac{A^* + (A^*)^*}{2} = \frac{A^* + A}{2}.$$

Further

$$\left(\frac{A - A^*}{2}\right)^* = \frac{A^* - (A^*)^*}{2} = \frac{A^* - A}{2} = -\left(\frac{A - A^*}{2}\right).$$

Uniqueness: Suppose $A = A_1 + A_2 = B_1 + B_2$, where A_1 and B_1 are Hermitian, while B_1 and B_2 are skew-Hermitian. Then, $A_1 - B_1 = B_2 - A_2$, and $A_1 - B_1$ is Hermitian while $B_2 - A_2$ is skew-Hermitian. But if a matrix is both Hermitian and skew-Hermitian, it must be the zero matrix (check!). Hence $A_1 = B_1$ and $A_2 = B_2$.

Exercises 3.16

Exercise 2: Closure:

$$\begin{pmatrix} a & 0 \\ b & 1 \end{pmatrix} \begin{pmatrix} a' & 0 \\ b' & 1 \end{pmatrix} = \begin{pmatrix} aa' & 0 \\ ba' + b' & 1 \end{pmatrix} \in G.$$

Note: As $a \neq 0, a' \neq 0, aa' \neq 0$). Associative law is true for matrix products. Identity: $\begin{pmatrix} 1 & 0 \\ 0 & 1 \end{pmatrix}$ is of the form $\begin{pmatrix} a & 0 \\ b & 1 \end{pmatrix}$ with $a = 1$ and $b = 0$.
Inverse: The inverse of

$$\begin{pmatrix} a & 0 \\ b & 1 \end{pmatrix} \quad \text{is} \quad \begin{pmatrix} \frac{1}{a} & 0 \\ \frac{-b}{a} & 1 \end{pmatrix}.$$

Exercise 4: Let $S = \{x_1, x_2, \ldots, x_n\}$ be a finite semigroup satisfying both cancellation laws. Let $x_i \in S$. Then, the set $\{x_i x_1, x_i x_2, \ldots, x_i x_n\}$ is a permutation of S. This is because $x_i x_j = x_i x_k \Rightarrow x_j = x_k$. Hence there exists a

p, $1 \leq p \leq n$, such that $x_i x_p = x_i$. We claim that x_p is the identity element of S. This is because if $x_r \in S$, $x_r = x_k x_i$ for some k, $1 \leq k \leq n$. (Note i is fixed). Hence $x_r x_p = (x_k x_i)x =$ (as S is associative) $x_k(x_i x_p) = x_k x_i = x$. Thus, $x_r x_p = x_r$, and similarly $x_p x_i = x_i$. Indeed, as $x_i x_p = x_i$, for any i, $x_i x_p x_j = x_i x_j \Rightarrow$ (cancelling x_i) $x_p x_j = x_j$ for each j. Hence x_p is the identity element of S. Again, for any i, $1 \leq i \leq n$, there exists j, $1 \leq j \leq n$ such that $x_i x_j = x_p$. Also, if $x_i x_j = x_p$, $x_j x_i x_j = x_j x_p = x_p x_j$ (as x_p is the identity element) $\Rightarrow$ (by right cancellation law) $x_j x_i = x_p$. Thus, x_j is the inverse of x_i. This proves that S is a group.

Exercise 6: Recall that the order of a non-singular square matrix A is the least positive integer k such that $A^k = I$

(i) Order is 4.

(ii) If

$$A = \begin{pmatrix} 1 & 1 \\ 0 & 1 \end{pmatrix}, \quad \text{then} \quad A^2 = \begin{pmatrix} 1 & 2 \\ 0 & 1 \end{pmatrix}, \quad A^3 = \begin{pmatrix} 1 & 3 \\ 0 & 1 \end{pmatrix}$$

and so on, and, in general, $A^k = \begin{pmatrix} 1 & k \\ 0 & 1 \end{pmatrix}$ and hence for no finite k, $A^k = I$. Thus, $\circ(A) = zero$.

(iii) Order $= 4$

(iv) If $A^k = I$, then $(detA)^k = 1$. But this is not the case. Hence $\circ(A) = 0$.

Exercise 8: If $\circ(a) = 2, a^2 = e$, and therefore $a = a^{-1}$. Conversely, if $a = a^{-1}$, and $a \neq e, \circ(a) = 2$. Hence if $a \neq a^{-1}, \circ(a) \neq 2$. Pairing off such elements (a, a^{-1}), we see that since the group is of even order and $\circ(e) = 1$, the group must contain an odd number of elements, and hence at least one element must be of order 2.

Exercise 10: Suppose to the contrary that a group G is the union of two of its proper subgroups, say, G_1 and G_2. Then, there exist $x \in G \setminus G_2$ (and hence $x \in G_1$) and $y \in G \setminus G_1$ (and hence $y \in G_2$). Then, look at xy. As $G = G_1 \cup G_2, xy \in G_1$ or G_2. If $xy = x_1 \in G_1$, then $y = x^{-1}x_1 \in G_1$, a contradiction. A similar argument applies if $xy \in G_2$.

Exercise 12: The set of all invertible 2×2 real matrices or the set of all diagonal 2×2 matrices with real entries and determinant not equal to zero.

Exercise 14:

(i) $(123)(456) = (12)(13)(45)(46)$, a product of an even number of transpositions and hence the permutation is even.

(ii) $(1516)(2)(3) = (15)(1)(16)(2)(3) = (15)(16)$ is even. [Note: The numbers that do not appear are all fixed]

(iii)

$$\begin{pmatrix} 1 & 2 & 3 & 4 & 5 & 6 & 7 & 8 & 9 \\ 2 & 5 & 4 & 3 & 1 & 7 & 6 & 9 & 8 \end{pmatrix} = (125)(34)(67)(89) = (12)(15)(34)(67)(89)$$

is an odd permutation, as it is a product of an odd number, namely 5, of transpositions.

Exercise 16: Let $a, b \in G$. Then, $a * b$ also belongs to G as $a * b = -1 \Rightarrow a + b + ab = -1 \Rightarrow (a + b)(b + 1) = 0$ that is, $a = -1$ or $b = -1$ which is not the case. So closure property is satisfied. A routine verification shows that $(a * b) * c = a * (b * c)$ for any three $a, b, c \in G$. Hence associative property is true. Further, $a * 0 = 0 * a = a$. Hence 0 acts as the identity in G. Finally, given $a \in G$, we have $a * ((-a)/(1 + a)) = a - (a/(1 + a)) - (a^2/(1 + a)) = 0$, and hence $(-a)/(1 + a)$ is the inverse of a. $\bar{G}$ under $*$ is commutative. Hence right inverse of a is automatically its left inverse.

Exercise 18: $\sigma = (i_1 i_2 \ldots i_r)$. Hence $\sigma^2 = (i_1 i_3 i_5 \ldots), \sigma^3 = (i_1 i_4 i_5 \ldots)$ and so $\sigma^r =$ identity. Hence $\circ(\sigma) = r$.

Exercise 20:

(i)

$$(\alpha)^{-1} = \begin{pmatrix} 1 & 4 & 3 & 2 \\ 1 & 2 & 3 & 4 \end{pmatrix} = \begin{pmatrix} 1 & 2 & 3 & 4 \\ 1 & 4 & 3 & 2 \end{pmatrix}$$

(ii)

$$\alpha^{-1}\beta\gamma = \begin{pmatrix} 1 & 2 & 3 & 4 \\ 1 & 4 & 3 & 2 \end{pmatrix} \begin{pmatrix} 1 & 2 & 3 & 4 \\ 2 & 1 & 4 & 3 \end{pmatrix} \begin{pmatrix} 1 & 2 & 3 & 4 \\ 3 & 1 & 2 & 4 \end{pmatrix} = \delta,$$

say then

$$\delta(1) = \alpha^{-1}\beta\gamma(1) = \alpha^{-1}\beta(3) = \alpha^{-1}(4) = 2,$$

(iii)

$$\beta\gamma^{-1} = \begin{pmatrix} 1 & 2 & 3 & 4 \\ 2 & 1 & 4 & 3 \end{pmatrix} \begin{pmatrix} 3 & 1 & 2 & 4 \\ 1 & 2 & 3 & 4 \end{pmatrix} = \begin{pmatrix} 1 & 2 & 3 & 4 \\ 1 & 4 & 2 & 3 \end{pmatrix}.$$

Exercise 22: For $m, n \in \mathbb{Z}, e^{im} \cdot e^{in} = e^{i(m+n)}, m + n \in \mathbb{Z}$. Further $e^{i \cdot 0} = 1$ is the identity element. $e^{in} \cdot e^{-in} = 1 \Rightarrow e^{-in}$ is the inverse of $e^{in}, -n \in \mathbb{Z}$. Associative law is trivially true. Hence the first part. Now consider the map $\phi : e^{in} \to n$ ϕ is 1–1. Since $\phi(e^{in}) = \phi(e^{im}) \Rightarrow n = m \Rightarrow e^{in} = e^{im}$. Trivially, ϕ is onto. ϕ is a homomorphism. Since for $m, n \in \mathbb{Z}$, $\phi(e^{in} \cdot e^{im}) = \phi(e^{i(n+m)}) = n + m = \phi(e^{in}) + \phi(e^{im})$. Hence ϕ is a 1–1, onto homomorphism from the multiplicative group $\{e^{in} : n \in \mathbb{Z}\}$ to the additive group $(\mathbb{Z}, +)$. This establishes the isomorphism. The given group is cyclic as $(\mathbb{Z}, +)$ is.

Exercise 24: The map $\phi : (\mathbb{Z}, +) \to (2\mathbb{Z}, +)$ from the additive group of integers to the additive group of even integers is a group isomorphism. (Check).

Exercise 26: Suppose there exists an isomorphism ϕ from $(\mathbb{R}^*, \cdot)$ onto $(\mathbb{C}^*, \cdot)$. Then, $\phi(1) = 1$ and hence $\phi(-1) = -1$ since $(-1)^2 = 1$ in $\mathbb{R}^*$ and $\mathbb{C}^*$. But then $i^2 = -1$ in $\mathbb{C}^*$ while there exists no element $a \in \mathbb{R}^*$ with $a^2 = -1$.

Exercise 28: Note: The center of S_3 consists of identity element only (Table C.1).

TABLE C.1: Numerical equivalents for English characters

	e	(12)	(23)	(31)	(123)	(132)
e	e	(12)	(23)	(31)	(123)	(132)
(12)	(12)	e	(123)	(132)	(23)	(13)
(213)	(23)	(132)	e	(123)	(13)	(12)
(31)	(31)	(123)	(132)	e	(12)	(23)
(123)	(123)	(13)	(12)	(23)	(132)	–
–						

Exercise 30: If G is Abelian, trivially $C(G)$ is G. If $C(G) = G$, then every element of G, that is, of $C(G)$ certainly commutes with all elements of G.

Exercise 32: $ab = ba \Leftrightarrow aba^{-1}b^{-1} = e$ and hence $[G, G] = \{e\}$.

Exercise 34: For $a \in A$ and $b \in B$, $aba^{-1}b^{-1} = (aba^{-1})b^{-1} \in B$ (as B is normal in G and so $aba^{-1} \in B$). Similarly $aba^{-1}b^{-1} = a(ba^{-1}b^{-1}) \in A$ as $a^{-1} \in A$ and as A is a normal subgroup of G, $ba^{-1}b^{-1} \in A$ and hence $aba^{-1}b^{-1} \in A$. Thus,

$$aba^{-1}b^{-1} \in A \cap B = \{e\}$$

$$\Rightarrow ab = ba \text{ for all } a, b \in G$$

$\Rightarrow G$ is an Abelian group.

Exercise 36: Let H be a subgroup of index 2 in G. Then, for all $h \in H$, $hH = H = Hh$, while if $g \in G \setminus H, G = H \cup Hg = H \cup gH$, where the unions are disjoint.

Exercise 38: Follows immediately from Exercise 36.

Exercises 3.19

Exercise 2: Routine verification.

Exercise 4: Result is true for $n = 2$ (from the axioms of a ring). Assume that a $a(b_1 + b_2 + \cdots + b_n) = ab_1 + ab_2 + \cdots ab_n$. Then, for elements $a, b_1, b_2, \ldots, b_n, b_{n+1}$ of A,
$a(b_1 + \cdots + b_n + b_{n+1}) = a(b_1 + \cdots + b_n) + (b_{n+1})$

$$\begin{aligned}
&= a(b_1 + \cdots + b_n) + ab_{n+1} \text{ (by the axioms for a ring)} \\
&= (ab_1 + \cdots + ab_n) + ab_{n+1} \text{ (by induction hypothesis)} \\
&= ab_1 + \cdots + ab_n + ab_{n+1} \text{ (by the associative property of } +).
\end{aligned}$$

Exercise 6: Suppose u is a unit and $ux = 0$. As u^{-1} exists, $u^{-1}(ux) = 0 \Rightarrow (u^{-1}u)x = 0 \Rightarrow 1_A \cdot x = 0 \Rightarrow x = 0$ (1_A is the unit element of A). Similarly, $xu = 0 \Rightarrow x = 0$.

Exercise 8: Look at the elements $1, 1 + 1, \ldots, 1 + 1 + \cdots + 1(p \text{ times})$. These elements must all be distinct. If not, there exist r, s with $0 < r < s \leq p$ such that $1 + 1 + \cdots + (r \text{ times}) = 1 + 1 + \cdots + 1(s \text{ times}) \Rightarrow 1 + 1 + \cdots + 1(s - r \text{ times}) = 0$. So there must exist a least positive integer $k(<p)$ such that $1 + 1 + \cdots + 1(k \text{ times}) = 0 \Rightarrow k|p$, a contradiction to the fact that p is a prime.

Appendix D: Answers to Chapter 4—Algebraic Structures II

Exercises 4.8

Exercise 2: If $f(x) = a_0 + a_1x + \cdots + a_nx^n \in \mathbb{R}[x], a_n \neq 0$, and $g(x) = b_0 + b_1x + \cdots + b_nx^n \in \mathbb{R}[x], b_n \neq 0$ are polynomials of degree n, then $b_nf(x) - a_ng(x)$ is a polynomial in $R[x]$ of degree $< n$. Hence the result.

Exercise 4: As in the Hint, if

$$e_{11} = \begin{pmatrix} 1 & 0 & 0 \\ 0 & 0 & 0 \\ 0 & 0 & 0 \end{pmatrix}, \ldots, e_{33} = \begin{pmatrix} 0 & 0 & 0 \\ 0 & 0 & 0 \\ 0 & 0 & 1 \end{pmatrix},$$

then if

$$A = \begin{pmatrix} a_{11} & a_{12} & a_{13} \\ a_{21} & a_{22} & a_{23} \\ a_{31} & a_{32} & a_{33} \end{pmatrix}$$

is any real matrix, then $A = \sum_{i,j=1}^{3} a_{ij}e_{ij}$. Further, if $A = 0$, all the 3^2 coefficients a_{ij} are zero. Hence the 3^2 matrices e_{ij} are linearly independent and span the space of all 3×3 real matrices over $\mathbb{R}$. Hence dimension of the vector space of all real 3×3 matrices over $\mathbb{R}$ is 3^2. Now generalize to $m \times n$ matrices.

Exercise 6: Let V be the vector space of all real polynomials in X. Suppose V is finite dimensional. Let n be the maximum of the degrees of the polynomials in a basis $\mathcal{B}$ of V. As every polynomial of V is a linear combination of the polynomials in $\mathcal{B}$, each polynomial in V is of degree n, at the most, a contradiction.

Exercise 8: Note $v_1 = 3v_3 - v_2$ and $v_4 = v_3 - v_2$. Hence the space $\langle v_1, v_2, v_3, v_4\rangle = \langle v_2, v_3\rangle$ (here $\langle u_1, u_2, \ldots, u_n\rangle$ stands for the space spanned by $\{u_1, u_2, \ldots u_n\}$). Now v_1 and v_2 are linearly independent over $\mathbb{R}$ since otherwise one would be a scalar multiple of the other, (here the scalars are the elements of $\mathbb{R}$). Hence $\dim\langle v_1, v_2, v_3, v_4\rangle = 2$.

Exercises 4.12

Exercise 2: The given system of equations is equivalent to

$$\begin{pmatrix} 4 & 4 & 3 & -5 \\ 1 & 1 & 2 & -3 \\ 2 & 2 & -1 & 0 \\ 1 & 1 & 2 & -2 \end{pmatrix} \begin{pmatrix} X_1 \\ X_2 \\ X_3 \\ 4X_4 \end{pmatrix} = \begin{pmatrix} 0 \\ 0 \\ 0 \\ 0 \end{pmatrix}.$$

Perform the row operations: $R_1 - (R_2 + R_3 + R_4)$. Followed by $R_3 - (R_2 + R_4)$. These give

$$\begin{pmatrix} 0 & 0 & 0 & 0 \\ 1 & 1 & 2 & -3 \\ 0 & 0 & -5 & 5 \\ 1 & 1 & 2 & -2 \end{pmatrix} \begin{pmatrix} X_1 \\ X_2 \\ X_3 \\ X_4 \end{pmatrix} = \begin{pmatrix} 0 \\ 0 \\ 0 \\ 0 \end{pmatrix}.$$

The matrix on the left is of rank 3 and hence there must be 4(number of columns)-3(rank)=1 linearly independent solution over $\mathbb{R}$.

The given equations are equivalent to $X_1 + X_2 + 2X_3 - 3X_4 = 0,\ -5X_3 + 5X_4 = 0,\ X_1 + X_2 + 2X_3 - 2X_4 = 0 \Rightarrow X_3 = X_4$.

Subtracting the 3rd equation from the 1st, we get $X_4 = 0$. Hence $X_3 = X_4 = 0$. So we can take $(1, -1, 0, 0)$ as a generator of the solution space.

Exercise 4(a): The given set of equations is equivalent to $AX = B$, where

$$A = \begin{pmatrix} 2 & 3 & -5 & 4 \\ 3 & 1 & -4 & 5 \\ 7 & 3 & -2 & 1 \\ 4 & 1 & -1 & 3 \end{pmatrix}, \quad X = \begin{pmatrix} X_1 \\ X_2 \\ X_3 \\ X_4 \end{pmatrix} \quad \text{and} \quad B = \begin{pmatrix} -8 \\ -8 \\ 56 \\ 20 \end{pmatrix}.$$

The Schur complement of A is

$$\begin{aligned} A_1 &= \begin{pmatrix} 1 & -4 & 5 \\ 3 & -2 & 1 \\ 1 & -1 & 3 \end{pmatrix} - \begin{pmatrix} \frac{3}{2} \\ \frac{7}{2} \\ \frac{4}{2} \end{pmatrix} \begin{pmatrix} 3 & -5 & 4 \end{pmatrix} \\ &= \begin{pmatrix} 1 & -4 & 5 \\ 3 & -2 & 1 \\ 1 & -1 & 3 \end{pmatrix} - \begin{pmatrix} \frac{9}{2} & \frac{-15}{2} & 6 \\ \frac{21}{2} & \frac{-35}{2} & 14 \\ 6 & -10 & 8 \end{pmatrix} = \begin{pmatrix} \frac{-7}{2} & \frac{7}{2} & -1 \\ \frac{-15}{2} & \frac{31}{2} & -13 \\ -5 & 9 & -5 \end{pmatrix}. \end{aligned}$$

The Schur complement of A_1 is A_2, where

$$\begin{aligned} A_2 &= \begin{pmatrix} \frac{31}{2} & -13 \\ 9 & -5 \end{pmatrix} - \begin{pmatrix} \frac{15}{7} \\ \frac{10}{7} \end{pmatrix} \begin{pmatrix} \frac{7}{2} & -1 \end{pmatrix} \\ &= \begin{pmatrix} 8 & \frac{-76}{7} \\ 4 & \frac{-25}{7} \end{pmatrix}. \end{aligned}$$

The Schur complement of A_2 is

$$\begin{aligned} A_3 &= \left(\tfrac{-25}{7}\right) - \left(\tfrac{4}{8}\right)\left(\tfrac{-76}{7}\right) \\ &= \left(\tfrac{13}{7}\right) \\ &= (1)\left(\tfrac{13}{7}\right) \\ &= L_3U_3 \end{aligned}$$

where $L_3 = (1)$ is unit lower-triangular, and $U_3 = \left(\frac{13}{7}\right)$ is upper-triangular.

This gives

$$\begin{aligned} A_2 &= \begin{pmatrix} 1 & 0 \\ \frac{4}{8} & L_3 \end{pmatrix} \begin{pmatrix} 8 & \frac{-76}{7} \\ 0 & U_3 \end{pmatrix} \\ &= \begin{pmatrix} 1 & 0 \\ \frac{1}{2} & 1 \end{pmatrix} \begin{pmatrix} 8 & \frac{-76}{7} \\ 0 & \frac{13}{7} \end{pmatrix} \\ &= L_2U_2. \end{aligned}$$

Therefore,

$$A_1 = \begin{pmatrix} 1 & 0 & 0 \\ \frac{15}{7} & 1 & 0 \\ \frac{10}{7} & \frac{1}{2} & 1 \end{pmatrix} \begin{pmatrix} \frac{-7}{2} & \frac{7}{2} & -1 \\ 0 & 8 & \frac{-76}{7} \\ 0 & 0 & \frac{13}{7} \end{pmatrix} = L_1U_1.$$

Consequently,

$$A = \begin{pmatrix} 1 & 0 & 0 & 0 \\ \frac{3}{2} & 1 & 0 & 0 \\ \frac{7}{2} & \frac{15}{7} & 1 & 0 \\ 2 & \frac{10}{7} & \frac{1}{2} & 1 \end{pmatrix} \begin{pmatrix} 2 & 3 & -5 & 4 \\ 0 & \frac{-7}{2} & \frac{7}{2} & -1 \\ 0 & 0 & 8 & \frac{-76}{7} \\ 0 & 0 & 0 & \frac{13}{7} \end{pmatrix} = LU.$$

Note that L is lower unit-triangular and U is upper-triangular.

Therefore, the given system of linear equations is equivalent to $LUX = B$. Set

$$UX = Y = \begin{pmatrix} Y_1 \\ Y_2 \\ Y_3 \\ Y_4 \end{pmatrix}.$$

We then have $LY = B$. We now solve for Y and then for X. Now $LY = B$ gives

$$\begin{pmatrix} 1 & 0 & 0 & 0 \\ \frac{3}{2} & 1 & 0 & 0 \\ \frac{7}{2} & \frac{15}{7} & 1 & 0 \\ 2 & \frac{10}{7} & \frac{1}{2} & 1 \end{pmatrix} \begin{pmatrix} Y_1 \\ Y_2 \\ Y_3 \\ Y_4 \end{pmatrix} = \begin{pmatrix} -8 \\ -8 \\ 56 \\ 20 \end{pmatrix}.$$

$$\Rightarrow Y_1 = -8, \frac{3}{2}Y_1 + Y_2 = -8, \frac{7}{2}Y_1 + \frac{15}{7}Y_2 + Y_3 = 56 \text{ and}$$
$$2Y_1 + \frac{10}{7}Y_2 + \frac{1}{2}Y_3 + Y_4 = 20$$
$$\Rightarrow Y_1 = -8, Y_2 = 4, Y_3 = \frac{528}{7}, \text{ and } Y_4 = \frac{-52}{7}.$$

We now determine X from the equation $UX = Y$.

$$\begin{aligned} UX = Y \Rightarrow 2X_1 + 3X_2 - 5X_3 + 4X_4 &= -8 \\ \frac{-7}{2}X_2 + \frac{7}{2}X_3 - X_4 &= 4 \\ 8X_3 - \frac{76}{7}X_4 &= \frac{528}{7} \\ \frac{13}{7}X_4 &= \frac{-52}{7} \end{aligned}$$

Solving backward, we get $X_4 = -4$, $8X_3 - (76/7)(-4) = (528/7) \Rightarrow 8X_3 = (528 - 304)/7 = 32 \Rightarrow X_3 = 4$. Hence $((-7)/2)X_2 + (7/2)X_3 - X_4 = 4$ gives $((-7)/2)X_2 + 14 + 4 = 4 \Rightarrow X_2 = 4$.

Finally, $2X_1 + 3X_2 - 5X_3 + 4X_4 = 8$ gives $X_1 = 8$. Thus,

$$X = \begin{pmatrix} X_1 \\ X_2 \\ X_3 \\ X_4 \end{pmatrix} = \begin{pmatrix} 8 \\ 4 \\ 4 \\ -4 \end{pmatrix}.$$

4(b): Similar to 4(a). Answer:

$$\begin{pmatrix} X_1 \\ X_2 \\ X_3 \end{pmatrix} = \begin{pmatrix} 2 \\ 3 \\ 7 \end{pmatrix}.$$

4(c): Since there are zeros in the principal diagonal of the matrix A, interchange the first two and the last two equations to avoid zeros in the main diagonal.

Exercises 4.15

Exercise 2: Let $F_1 = GF(2^5)$ and $F_2 = GF(2^3)$. If F_2 is a subfield of F_1, then F_2^*, the multiplicative group of the non-zero elements of F_2 should be a subgroup of F_1^*. But $|F_1^*| = 2^5 - 1 = 31$, and $|F_2^*| = 2^3 - 1 = 7$ and $7 \nmid 31$.

Exercise 4: $X^{3^2} - X = X(X^8 - 1) = X(X^4 - 1)(X^4 + 1) = X(X - 1)(X + 1)(X^2 + 1)(X^4 + 1) = X(X + 2)(X + 1)(X^2 + 1)(X^4 + 1)$. Note that $X^2 + 1$ is irreducible over $\mathbb{Z}_3$ since neither 1 nor 2 is a root $X^2 + 1$ in $\mathbb{Z}_3$. A similar statement applies to $X^4 + 1$ as well.

Appendix E: Answers to Chapter 5—Introduction to Coding Theory

Exercises 5.12

Exercise 2(a): Take

$$G = \begin{pmatrix} 1 & 0 & 1 & 1 & 1 \\ 1 & 0 & 0 & 1 & 1 \end{pmatrix} \text{ over } \mathbb{Z}_2.$$

G is of rank 2, and so H is of rank $5 - 2 = 3$ and each row of H is orthogonal to every row of G. So H can be taken either as

$$H_1 = \begin{pmatrix} 1 & 1 & 0 & 1 & 0 \\ 1 & 0 & 0 & 0 & 1 \\ 0 & 1 & 0 & 0 & 0 \end{pmatrix} \text{ or } H_2 = \begin{pmatrix} 1 & 1 & 0 & 0 & 1 \\ 1 & 0 & 0 & 0 & 1 \\ 0 & 1 & 0 & 1 & 1 \end{pmatrix}$$

Let

$$X = \begin{pmatrix} 1 \\ 0 \\ 1 \\ 1 \\ 1 \end{pmatrix}$$

Then, $S(x)$ with respect to H_1 is

$$H_1 X^T = \begin{pmatrix} 0 \\ 0 \\ 0 \end{pmatrix},$$

while $S(x)$ with respect to H_2 is

$$H_2 X^T = \begin{pmatrix} 0 \\ 0 \\ 1 \end{pmatrix}.$$

Solutions to 2(b), 2(c) and 2(d) are similar to the steps in Example 5.1.

Exercise 4: u has 7 coordinates. $s(u, 3)$ contains all binary vectors of length 7 which are at distances $0, 1, 2$ or 3 from u and hence their number is $1 + \binom{7}{1} + \binom{7}{2} + \binom{7}{3} = 1 + 7 + 21 + 35 = 64$. [Here, $\binom{7}{1}$ arises from vectors which are at a distance 1 from u etc.]

Exercise 6: Let C be such a code.Then, $|C| = 2^k$. Further, the spheres with centers at these 2^k codewords and of radii t cover $\mathbb{Z}_2^n$, where $|\mathbb{Z}_2^n| = 2^n$. Each codeword is of length n. Hence, as in the solution of Exercise 2, a sphere of radius t with center at a codeword will contain $1+\binom{n}{1}+\binom{n}{2}+\cdots+\binom{n}{t}$ vectors of $\mathbb{Z}_2^n$. Hence, as there are 2^k spheres,

$$2^k\left[\binom{n}{0}+\binom{n}{1}+\cdots+\binom{n}{t}\right]=2^n.$$

The general case is similar.

Exercise 8: Assume the contray, that is, there exists a set of 9 vectors in $\mathbb{Z}_2^6$ (the space of binary vectors of length 6) which are pairwise at a distance at least 3. Look at the last two coordinates of these 9 vectors. They should be from $\{00, 01, 10, 11\}$. As there are 9 vectors, at least one of these pairs must occur at least 3 times. Consider such a set of 3 vectors which have the same last two coordinates. These coordinates do not contribute to the distance between the corresponding three vectors. Hence if we drop these last two coordinates from these 3 vectors of $\mathbb{Z}_2^6$, we get 3 binary vectors of length 4 such that the distance between any two of them is at least 3. This is clearly impossible. This contradiction proves the result.

Exercise 10: Let $e_1 = 01234, e_2 = 12044$ and $e_3 = 13223$. Note that $e_3 = e_1 + e_2$ over $\mathbb{F}^5$. Hence $\text{Span}\langle e_1, e_2, e_3\rangle = \text{Span}\langle e_1, e_2\rangle$. Now write down the $(5 \times 5) - 1 = 24$ non-zero codewords of the code and find the minimum weight d. Then, the code can detect $d - 1$ errors (Refer to Theorem 5.8).

Exercise 12:

A. Since there is a single constraint $\dim(C) = 3 - 1 = 2$. Hence $|C| = 3^2 = 9$. To get all 9 words, give all possible 3^2 values to the pair (x_2, x_3). These are (0,0),(1,0),(2,0); (0,1),(1,1),(2,1); (0,2),(1,2),(2,2). These give, as $x_1 + x_2 + 2x_3 = 0$, the triads, (0,0,0),(2,1,0),(1,2,0),(1,0,1),(0,1,1), (2,2,1),(2,0,2),(1,1,2),(0,2,2).

B. As $\dim(C) = 2$, any two linearly independent codewords over $\mathbb{Z}_3$ would form a generator matrix for C. For instance,

$$G_1 = \begin{pmatrix} 1 & 0 & 1 \\ 1 & 1 & 2 \end{pmatrix}.$$

C.

$$G_2 = \begin{pmatrix} 1 & 0 & 1 \\ 2 & 2 & 1 \end{pmatrix}.$$

D. d = minimum weight of a non-zero codeword = 2 (Look at the nine codewords in (A)).

E. C can correct $(d-1)/2\rfloor = 0$ error.

F. C can detect $d-1 = 2-1 = 1$ error.

G. Rank of $G_2 = 2 = k$. Hence rank of the pairity-check matrix H is $3-2 = 1$. Every row of H is orthogonal to every row of G over $\mathbb{Z}_3$. Hence we can take $H = [112]$.

Exercises 5.15

Exercise 2: $\mathcal{C}_3$ is cyclic since the three non-zero codewords are the cyclic shifts of any one of them. For example, $(1,1,0) \longrightarrow (0,1,1) \longrightarrow (1,0,1)$. Since one of them is a linear combination (over $\mathbb{Z}_2$) of the other 2, as a generator matrix for $\mathbb{C}_3$, we can take any two of these codewords. Hence

$$G = \begin{pmatrix} 1 & 1 & 0 \\ 0 & 1 & 1 \end{pmatrix}$$

is a generator matrix of $\mathcal{C}_3$. As $\text{rank}(G) = 2$, its null space is of dimension 1. Hence the dual code of G, namely $G^{\perp} = [1\ 1\ 1]$, and so $G^{\perp}$ is also cyclic and $\dim(G^{\perp}) = 1$.

Exercise 4: The codeword $1+x$ corresponds to the n-vector $(1,1,0,\ldots,0)$. Hence the generator matrix of the binary code generated by $1+x$ is

$$\begin{pmatrix} 1 & 1 & 0 & 0 & \ldots & 0 \\ 0 & 1 & 1 & 0 & \ldots & 0 \\ 0 & 0 & 1 & 1 & \ldots & 0 \\ 0 & 0 & \ldots & 0 & 1 & 1 \end{pmatrix}$$

(Note: G is an $(n-1) \times n$ matrix of rank $n-1$. Hence the dimension of the cyclic code is $n-1$).

Exercise 6: Suppose $\mathcal{C}$ contains a word X of weight $(n-2)$. Then, $\mathcal{C}$ contains exactly two zeros and all the rest are 1s. Cyclically shift X and get the word $X_0 = (x_0, x_1, \ldots, x_{n-1})$, where $x_0 = 0 = x_i$, and $x_j = 1$ for every $j \neq 0, i$. Cyclically shift X_0 until x_i moves to the 0-th position. This can be effected in $n-i$ shifts. These $n-i$ shifts will not move the first zero of X_0 to the second zero of X_0. For this requires i shifts, and this means that $i = n-i \Rightarrow n$ is even, a contradiction to our assumption that n is odd. Let X_1 be the new word obtained after these $n-i$ shifts. Clearly, $d(X_0, X_1) = 2$, again a contradiction to our assumption that $d(\mathcal{C}) \geq 3$.

Appendix F: Answers to Chapter 6—Cryptography

Exercises 6.4

Exercise 2: $\det A = 12.7 - 5.3 = 204 - 15 = 189 = 15$ in $\mathbb{Z}_{29}$, and hence it is prime to 29. Now $15 \times 2 \equiv 30 \equiv 1 (\text{mod } 29)$. Hence $(\det A)^{-1} = 15^{-1} = 2$ in $\mathbb{Z}_{29}$. Therefore

$$A^{-1} = (\det A)^{-1}(adj A) = 2\begin{pmatrix} 17 & -3 \\ -5 & 12 \end{pmatrix} = \begin{pmatrix} 5 & 23 \\ 19 & 24 \end{pmatrix}$$

in $\mathbb{Z}_{29}$.

Exercise 4: The given equations are

$$A\begin{pmatrix} x \\ y \end{pmatrix} = \begin{pmatrix} 4 \\ 10 \end{pmatrix}, \quad \text{where} \quad A = \begin{pmatrix} 1 & -1 \\ 7 & -4 \end{pmatrix}.$$

Now $\det A = 3$ which is prime to 26. Hence 3^{-1} exists in $\mathbb{Z}_{26}$. In fact $3^{-1} = 9$ as $3 \cdot 9 \equiv 1 \pmod{26}$. Therefore,

$$A^{-1} = 9\begin{pmatrix} -4 & 1 \\ -7 & 1 \end{pmatrix} = \begin{pmatrix} -36 & 9 \\ -63 & 9 \end{pmatrix}$$

in $\mathbb{Z}_{26}$. One can check that $AA^{-1} = A^{-1}A = I_2$ in $\mathbb{Z}_{26}$.

Exercise 6: "MHQV" should be deciphered as "MIKE." Let the deciphering matrix be

$$\begin{pmatrix} a & b \\ c & d \end{pmatrix}.$$

Then,

$$\begin{pmatrix} a & b \\ c & d \end{pmatrix}\begin{pmatrix} M_n \\ H_n \end{pmatrix} = \begin{pmatrix} M_n \\ I_n \end{pmatrix}, \quad \text{and} \quad \begin{pmatrix} a & b \\ c & d \end{pmatrix}\begin{pmatrix} Q_n \\ V_n \end{pmatrix} = \begin{pmatrix} K_n \\ E_n \end{pmatrix}$$

where the suffix n in M_n stands for the numerical equivalence of M, namely, 12, etc. (See Table 6.1).Thus, we have

$$\begin{pmatrix} a & b \\ c & d \end{pmatrix}\begin{pmatrix} 12 \\ 7 \end{pmatrix} = \begin{pmatrix} 12 \\ 8 \end{pmatrix} \quad \text{and} \quad \begin{pmatrix} a & b \\ c & d \end{pmatrix}\begin{pmatrix} 16 \\ 21 \end{pmatrix} = \begin{pmatrix} 10 \\ 4 \end{pmatrix}.$$

These give a pair of simultaneous linear equations in a, b and c, d, namely,

$$12a + 7b = 12, 16a + 21b = 10 \text{ and}$$
$$12c + 7d = 8, 16c + 21d = 4.$$

Note that we have to solve these equations in $\mathbb{Z}_{29}$, and since 29 is a prime, $\mathbb{Z}_{29}$ is actually a field. We can add and subtract multiples of 29 to make our calculations become easier but must avoid multiplication by 29.

Now the first set of equations is equivalent to

$$\begin{aligned} 12a + 7b &= 12, \\ \text{and } 16a - 8b &= 10, \\ \Rightarrow 48a + 28b &= 48, \\ 48a - 24b &= 30. \end{aligned}$$

Subtraction gives $52b = 18 (\text{mod } 29) \Rightarrow -6b \equiv 18 \Rightarrow b \equiv -3 \equiv 26$ (all congruences are modulo 29). Now $12a + 7b = 12$ gives $12a - 21 \equiv 12 \Rightarrow 12a = 33 \equiv 4 \Rightarrow 3a \equiv 1 \Rightarrow a = 10$.

We now consider the second set of equations, namely, $12c + 7d = 8$ and $16c + 21d = 4$. Multiply the first equation by 3, and subtract from it the second equation. This gives $20c = 20 \Rightarrow c = 1$ and hence $d = 16$.
Hence the deciphering matrix is

$$A^{-1} = \begin{pmatrix} a & b \\ c & d \end{pmatrix} = \begin{pmatrix} 10 & 26 \\ 1 & 16 \end{pmatrix}.$$

To simplify over-working, we can take

$$A^{-1} = \begin{pmatrix} 10 & -3 \\ 1 & 16 \end{pmatrix}.$$

We now start deciphering pairs of letters in succession from left to right in the ciphertext. The first pair in the ciphertext is "*AM*," and its corresponding column vector is $\begin{pmatrix} 0 \\ 19 \end{pmatrix}$. Hence it represents the plaintext

$$\begin{aligned} \begin{pmatrix} 10 & -3 \\ 1 & 16 \end{pmatrix} \begin{pmatrix} 0 \\ 19 \end{pmatrix} &\equiv \begin{pmatrix} -57 \\ -160 \end{pmatrix} (\text{as } 19 \equiv -10 (\text{mod } 29)) \\ &\equiv \begin{pmatrix} 1 \\ 14 \end{pmatrix} (\text{as } -160 \equiv 14 (\text{mod } 29)) \\ &\backsimeq \text{"}BO\text{"}. \end{aligned}$$

Thus, "*AM*" represents to "*BO*" in the plaintext. In a similar way, it can be checked that the following is true:

Ciphertext	Corresponding Plaintext Pair
AM	BO
GQ	MB
TZ	!N
AF	OW
JV	?(space)
MH	MI
QV	KE

Hence the plaintext is: BOMB!NOW? MIKE.

Bibliography

1. Knuth D. E., *The Art of Computer Programming. Volume 1: Fundamental Algorithms*, 2nd edition, Addison-Wesley, Reading, MA, 1973; Volume 2, Seminumerical Algorithms, second edition, Addison-Wesley, Reading, MA, 1981; Volume 3: Sorting and Searching, second printing, Addison-Wesley, Reading, MA, 1975.

2. Sedgewick R., *Algorithms*, Addison-Wesley, Reading, MA, 1989; Algorithms in C, Addison-Wesley, Reading, MA, 1990; Algorithms in C++, Addison-Wesley, Reading, MA, 1998.

3. Aho A. V., Hopcroft J. E., Ullman J. D., *The Design and Analysis of Algorithms*, Addison-Wesley, Reading, MA, 1975; Data Structures and Algorithms, Reading, MA, 1983

4. Graham R. L., Knuth D. E., Pathashnik O., *Concrete Mathematics*, Addison-Wesley, Reading, MA, 1988.

5. Wirth N., *Algorithms + Data Structures = Programs*, Prentice-Hall, Inc., New Jersey, 1976; Systematic Programming, An Introduction, Prentice-Hall, Inc., New Jersey, 1973.

6. Balakrishnan R., Sriraman Sridharan, *Foundations of Discrete Mathematics with Algorithms and Programming*, CRC Press, Taylor & Francis, 2018.

7. M. Garey, D. S. Johnson, Computers and Intractability, *A Guide to the Theory of NP–Completeness*, W.H. Freeman Company, New York, 1979.

8. Balakrishnan R., Ranganathan K., *A Textbook of Graph Theory*, 2nd edition, Springer, 2012.

9. Berge C., *The Theory of Graphs and its Applications*, Wiley, New York, 1958; Graphs and Hypergraphs, North Holland, 1973.

10. Berge C., *Graphes*, Dunod, Paris, 1970.

11. Parthasarathy K. R., *Basic Graph Theory*, Tata-McGraw Hill Publishing Company, New Delhi, 1994.

12. Herstein I. N., *Topics in Algebra*, Wiley, 2005.

13. Clark A., *Elements of Abstract Algebra*, Dover Publications Inc., New York, 1984.

14. Fraleigh J. B., *A First Course in Abstract Algebra*, 7th edition, Wesley Pub. Co., 2003.

15. Bhattacharyya P. B., Jain S. K., Nagpaul S. R., *Basic Abstract Algebra*, 2nd edition, Cambridge University Press, 1997.

16. Lewis D. W., *Matrix Theory*, World Scientific Publishing Co: and Allied Publishers Ltd. (India), 1955.

17. Ryer H. J., *Combinatorial Mathematics*, **14**, Carus Mathematical Monograph, Wiley, New York, 1963.

18. Singh S., *Linear Algebra*, Vikas Publishing House, New Delhi, 2000.

19. www.ams.org/mcom/1962-16-079/S0025...1/S0025/S0025-5718-1962-0148256-1.pdf

20. Hill R., *A First Course in Coding Theory*, Clarendon Press, Oxford, 2004.

21. Ling S., *Coding Theory: A First Course*, Cambridge University Press, 2004.

22. Shannon C. E., A mathematical theory of communication, *Bell Syst. Tech. J.* 27(379–423), 623–656 (1948).

23. van Lint J. H., *Introduction to Coding Theory*, Springer GTM, 1999.

24. Pless V., *Introduction to the Theory of Error Correcting Codes*, 3rd edition, Cambridge University Press, 1998.

25. Agrawal M., Kayal N., Saxena N., PRIMES is in P (http:\www.cse-iitk.ac.in/users/manindra/primality.ps), August 2002.

26. Agrawal M., Lecture delivered on February 4, 2003 at the Institute of Mathematical Sciences, Chennai on: "PRIMES is in P". Reproduced in: Mathematics Newsletter (Published by Ramanujan Mathematical Society), 13, 11–19 (2003).

27. Pomerance C., On the distribution of pseudoprimes, *Math. Comput.*, 37(156), 587–593 (1981).

28. Lidl R., Niederreiter H., *Introduction to Finite Fields and Their Applications*, Cambridge University Press, 1986.

29. Motwani R., Raghavan P., *Randomized Algorithms*, Cambridge University Press, 1995.

30. Koblitz N., *A Course in Number Theory and Cryptography*, 2nd Edition, GTM, Springer, 1994.

31. Stinson D. R., *Cryptography: Theory and Practice*, Chapman & Hall, CRC, 2003.

32. Salembier R. G., Southerington P., An Implementation of the AKS Primality Test, CiteSeerX.

33. Nair M., On Chebyshev-type inequalities for primes, *Amer. Math. Monthly*, 89, 120–129 (1982).

34. Rotman J. J., *An Introduction to the Theory of Groups*, GTM, Springer, 1999.

35. Bose R. C., Parker E. T., Shrikhande S. S., Further results on the construction of mutually orthogonal latin squares and the falsity of Euler's conjecture, *Canadian Journal of Mathematics*, 12, 189–203 (1960).

36. Diffie W., Hellman M. E., New directions in cryptography, *IEEE Transactions on Information Theory*, II-22(6), 644–654 (1976).

37. Cormen T. H., Leiserson C. E., Rivest R. L., Stein C., *Introduction to Algorithms*, 3rd edition, MIT Press, 2009.

Index